What is science? How is scientific knowledge affected by the society that produces it? Does scientific knowledge directly correspond to reality? Can we draw a line between science and pseudo-science? Will it ever be possible for computers to undertake scientific investigation independently? Is there such a thing as feminist science?

In this book the author addresses questions such as these using a technique of 'cognitive play' which creates and explores new links between the ideas and results of contemporary history, philosophy, and sociology of science. New ideas and approaches are applied to a wide range of case studies, many of them from controversial and contested science.

The author adopts a stance of critical realism, according to which claims of empirical knowledge are constructed and reconciled with other beliefs, observations and actions, while being exposed to critical scrutiny. It is impossible to foresee the completion of knowledge according to this viewpoint, especially as apparent uncertainties can succumb to new forms of criticism: hence the book's title *Uncertain knowledge*. It is argued that, as we try to match our intellectual constructions to reality, we are simultaneously changing local aspects of reality into harmony with our concepts, as a result of knowledge-guided action. This inevitably complicates the task of a realist view of science, requiring the rethinking of many traditional conclusions about how to avoid error and illusion in science.

This book will be of interest to historians and sociologists of science, to anyone interested in science studies, and to educated general readers with an interest in the history, philosophy and social context of science.

UNCERTAIN KNOWLEDGE

UNCERTAIN KNOWLEDGE

An image of science for a changing world

R. G. A. DOLBY

Centre for the History and Cultural Studies of Science
Rutherford College
University of Kent at Canterbury

CAMBRIDGE
UNIVERSITY PRESS

Published by the Press Syndicate of the University of Cambridge
The Pitt Building, Trumpington Street, Cambridge CB2 1RP
40 West 20th Street, New York, NY 10011–4211, USA
10 Stamford Road, Oakleigh, Melbourne 3166, Australia

First published 1996

Printed in Great Britain at the University Press, Cambridge

A catalogue record for this book is available from the British Library

Library of Congress cataloguing in publication data

Dolby, R. G. A. (Riki G. A.)
Uncertain knowledge : an image of science for a changing world /
R. G. A. Dolby.
p.   cm.
Includes bibliographical references and index.
ISBN 0 521 56004 7
1. Science–Philosophy.   I. Title.
Q175.D688   1996
501–dc20   96-22588 CIP

ISBN 0 521 56004 7   Hardback

Contents

Preface

It is fun to think for oneself about grand intellectual issues. This book focuses upon some of the big questions about science. Little attention is paid to the personalities or politics of the science studies area on which it builds. It stands back from the literature of its field, avoiding reliance on arguments from disciplinary authority and ignoring the shorter swings of disciplinary fashion.

The big questions about science are those which try to assess what science has accomplished and what it might yet achieve. The traditional image of science as progressively building reliable knowledge of external reality on a foundation of secure fact, has had its underlying assumptions effectively criticised, especially by the present generation. There is a new image available. Science is merely a prominent form of socially legitimated belief in modern society. As society moves from depending on an aristocracy of learning to a free market in ideas, science moves from the consensual pronouncements of an institutionalised international orthodoxy to multi-faceted linguistic codifications of the culture of competing local forms of life.

The new image is not entirely persuasive. The practical ways in which science affects our everyday lives seems to require that science contains at least approximate truth about reality. In science studies, debates occur between the older philosophically oriented view and the newer view of science as a social construction (e.g. Kitcher, 1993, which follows a similar strategy to the present book). The older view asked the big questions but answered them unsatisfactorily, while the newer view has put much more limited questions in their place. I am reluctant to abandon the idea that science seeks knowledge of reality, and that its cultural importance comes from its limited achievement of that goal, but can such a view be sustained in the context of modern science studies? The book attempts to show that it can.

If enough unquestioned assumptions are made by those who seek to construct knowledge, their task is easy. Too easy. One strategy of this book is to expose the assumptions of thought, language, and social convention, so as to open up and explore alternative possibilities. If we open up our culture to competing alternatives, and if every possibility has equal status, there will be a proliferation of rival beliefs. Here, however, every effort is made to sustain the power of criticism. It is in this spirit that so much attention is given in this book is given to case studies of marginal and contested science. They frequently explore possibilities outside the limits of orthodox thought. The provisional uncriticised acceptance of a significant proportion of their findings would, however, turn one's image of science into a world of ideas in chaos.

Acknowledgements. I am grateful for the help I have received from those who have commented on the evolving manuscript. In particular, I thank Maurice Crosland, Riki Dolby, Ian Higginson, Adrian Johns, David Reason, Anne Seller, Crosbie Smith, and the anonymous referees of Cambridge University Press. I am also grateful to students who gave me feedback, particularly those in my course on 'Pseudo-Science', on whom I first tried out many of my ideas.

1

Introduction

1.1 The problem: understanding natural science as cognitive activity

Science is a victim of its own success. It has changed the world into one which judges more harshly the manner in which knowledge is created, maintained, and applied. Scientific knowledge becomes simultaneously easier to create and to criticise. We now live in a world which appears to have reduced the incalculable dangers of what once lay outside the horizons of our knowledge, but has made more worryingly obvious the risks inherent in our knowledge-guided actions. Although science is continually eliminating past errors while increasing the precision and range of its knowledge, it also undergoes infrequent revolutions which abandon earlier apparent certainties. The increasing power given to us by science to adapt the world to our desires is not accompanied by equal knowledge of the best way to implement such power. Indeed, science makes it easier to appreciate the problems that the power to change things has already produced. It is no wonder that people are increasingly ambivalent towards science. Perhaps it is a self-limiting system, the repeated operation of which cumulates problems it cannot handle.

And yet science still provides hope for the future. Within its limits, it is a powerful generator of new understanding, of resources for dealing with problems that threaten to overpower us. Indeed, we are now locked into a form of life that is only sustainable by the continuing growth of scientific knowledge. We cannot go back to any of our romantically reconstructed pre-scientific pasts without some intervening catastrophe to cut back our material expectations and our population. It is too late for anti-science to be the easy orthodoxy. Perhaps, however, we should not place too much faith in the *present* form of science. Science is not a fixed cultural activity. By changing society, it changes its own nature. Would not we be better off with a new

1

kind of science which gives us more of what we want with less problems? Can we change science for the better?

It would not be advisable to set about changing science without understanding it well. Several totalitarian regimes of the mid-twentieth century demonstrated the dangers when they sought to confine science within their own restricted ideological images.

Can we make science itself the object of our study in the hope of improving it, as early modern science sought to improve traditional arts and crafts by their learned study? Perhaps we could understand science without selective distortion, seeing where it works best and where it fails. If science were a comprehensive compendium of firm and final knowledge, then such questions could easily be answered with the aid of science itself. In reality, however, science is not complete, its knowledge claims are subject to revision and it cannot easily give a full account of itself.

A distinction can be made between the unselfconscious practice of science and the critical examination of its nature and processes. Our concern will be with the latter. One factor in the historical construction of images of the nature of science was dialogue between defenders of science, who constructed legitimating images, and anti-scientific opponents, who offered critical caricatures. Both sides worked within a shared philosophical language. Now, in addition, there are a number of disciplines that study science from a professional standpoint outside such conflicts of interest. I will refer to these new disciplines collectively as 'science studies'. An alternative name is 'cultural studies of science'. The disciplines which seek to extend the self-understanding of scientists include philosophy of science, professional history of science and sociology of science, along with ways of looking at science in its wider social, political, and economic context. We cannot afford to ignore the analyses offered by science studies, though we should not accept them uncritically. The insights offered are shaped by divergent frameworks of discourse. Some approaches emphasise the thought and action of the individual scientist, others look at science by foregrounding its collective social practice. Some concentrate on scientists as people, others examine verbal and written discourse. They do not readily come to a common viewpoint because they do not always appreciate that their selective perspectives lead to the study of rather different aspects of a wider cultural phenomenon. It is as if the schools of thought and the disciplinary perspectives blindly groping around an elephant were not merely studying distinct parts of the beast, but were also studying zones of radically different sizes.

The task of studying science is not made any easier by the fact that the object of the study is itself changing. For example, where, in the mid-twentieth century, it might have been seen as an autonomous intellectual activity, conducted within its own special institutions, it is now an essential component of much of economic life. The image of science is appealed to in everything from selling skin creams to the legitimation of new professions. We are in danger of losing our vision of science in the glare of its many reflections.

As science changes, its older images can get in the way of our understanding of its present nature. For example, one persisting but declining image still produces expanding ripples of distortion in cultural reactions to science. Positivism, a creation of the early nineteenth century, sought to restrict science to the facts and to reject any line of thought that failed to meet its anti-speculative standards. The more enthusiastic positivists sought to extend the approach beyond science to every pressing question of life. Positivism became a prominent ideology of natural science in the nineteenth century. In the twentieth century, logical positivism (especially the group known as the Vienna Circle) was even more hard-line in its proclamation of a rigorous and exclusive methodology. Thanks to Hitler, the message of logical positivism was spread around the Western intellectual world in the mid-century by refugees from central European universities. Positivist methodological proposals were never descriptive of natural science and finally proved unworkable as prescriptive principles. However, that judgement has not fully caught up with the influence of positivism. Even today, some present-day critics of science point to what they think are limitations of science when they are only defects of the positivist image.[1]

Since the 1970s, science studies has been undergoing major changes as a new generation moves to new intellectual enthusiasms, rapidly multiplying the range of sub-disciplinary perspectives on science.

1.2 The strategy employed

This book argues that there is no single satisfactory explanatory framework available for understanding natural science. That does not make science mysterious, for we can use the many frameworks as resources in our search for deeper understanding. The approach taken looks widely among the established approaches available without committing itself to any of them. It shows how to deal with the several sets of understanding of the elephant of science. The techniques it employs are rather less those of following chains of reasoning within a single intellectual discipline than of linking overlapping

---

[1] Among writers who attack science in the image of positivism are some of the modern feminist writers on science. See chapter 10.

disciplinary discourses through critical comparison, through metaphor, and through integration within higher-level frameworks. With such methods, synthetic insights can be obtained which go beyond the limits of the specific disciplinary forms on which the book builds.

Cognitive play with the idea of science

One way to go beyond the established limits of existing understanding is to explore the subject-matter playfully. A very effective strategy to learn to think in new ways about specific issues is through the fun of seeing cognitive incongruity in them. Humour is anchored in early forms of learning through play. One reason to laugh is to show those around us that we have identified an absurdity or incongruity, and one way to learn how to identify incongruity in a situation is by seeing those around us laughing. A fuller place could be given in natural science to cognitive play and to the humour of seeing incongruity. Cognitive play is particularly valuable if one wishes to be able to make one's own way through the network of scientific understanding, perhaps as a part of creative activity. Mathematics has an especially rich culture of amateur cognitive play, with a large range of mathematical puzzles and pastimes. There are cultures of intellectual play distantly related to natural science, such as science fiction. But science fiction is generally more concerned with the human consequences of science than with its creative processes.

The kind of cognitive play advocated here draws our attention away from the *result* and towards the *process* of what is done in science. To be able to play, we have to open up the completed processes of science, treat them as they were when first constructed, and see what variations are possible. We have the advantage of hindsight over those who first produced the science. For example, by being able to scrutinise the initial assumptions on which a scientific practice is grounded, we can explore the consequences of changes in those assumptions.

One form of cognitive play with science used in this book concerns the imaginative possibilities of constructing (or salvaging) alternative and perhaps preferable forms of science. Suppose we were to create a science of extra-sensory perception, or of Bermuda Triangle disappearances, or of UFOlogy. What would we need to do? Our imaginations in such whimsy can be stimulated by reading about people who have tried to create such a science. We can ask ourselves, 'Why have their efforts remained controversial? Is there some way to do it better?' Or if we are inclined to be sceptical that there are any genuine phenomena to investigate in such cases, what about trying to invent a new science of the biological basis of human social behaviour? Again, our imagination can be stimulated by looking at the actual story of (human) sociobiology.

The present study may be seen as an attempt to illuminate our understanding of science by providing resources for just this kind of cognitive play. The reader is encouraged to play actively with the ideas presented, treating them as a stimulus and challenge for his or her own thoughts.

The case studies – challenges to our understanding of science

Many case studies of contested or problematic science are discussed in this book. They illustrate and give substance to the general claims made and sometimes provide a test of the ideas offered. The appendix provides brief summaries of the case material. When a case is alluded to in the text, the presence of a summary will be indicated by an asterisk after the case's title.

Do cases of marginal or disputed science merely reveal the ignorance and incompetence of those who proclaim scientific alternatives? Every reader is likely to have made this judgement about at least some of the examples. However, by challenging unquestioned assumptions, the cases can throw light upon the production of scientific knowledge. Such cases can force us to re-examine the framework of our thought. It can be quite difficult to demonstrate any serious rational error in a radical alternative to scientific orthodoxy if it questions the very assumptions out of which a critique might be constructed. The criticism of a rival to orthodox science can be made even more difficult if it has its own account of scientific rationality. In its twentieth-century heyday, for example, the theory of Marxism* was packaged with a dialectical materialist conception of science.

The strategy employed to deal with the complex circularity of arguments about contested science in the context of rival theories of scientific rationality is to begin by taking these marginal cases seriously. If I study people who claim great authority in their contradictions of one another, I would normally anchor my thinking in the judgement of those whose opinions I most respect. In the case of the so-called pseudo-sciences, to do this would lock me into some specific framework of scientific orthodoxy, and severely limit what I could learn. Instead of doing that, the approach employed will be to look at each case as an open question, taking at face value the arguments put for each side. The 'neutral' analysis employed will seek to clarify what is going on at the various cognitive levels and how reflexivity questions are complicating the issue.

At some point, however, the role-play of neutrality must break down. Nobody could be so naive as to take seriously *all* the cases of marginal science that are covered in this book.

Uncertain knowledge

As its title implies, this book is a study of empirical scientific knowledge and the activities which generate and sustain prospective knowledge before it has reached a fully settled form. In the modern context, it is no longer so appropriate to see natural science as a coherent system of eternal and authoritative truths about reality. When a particular world view is complete in its own terms, its content tightly bound together in a consistent system by rigorous reasoning, it appears certain to its practitioners. Scientists have frequently claimed to be nearing completion of such a vision. But scientific knowledge is always being added to and reworked. New discoveries can take a generation or more before their significance is understood in a settled way. Sometimes, further discoveries change the story before the previous ones have been fully assimilated. Radical new discoveries require existing systematisations of knowledge to be reworked. Even apparently stable factual scientific conclusions are fallible and subject to rare revolutionary change.

Some may see the link of science with uncertainty as contentious. In science, rigorous arguments based on authoritative observation and experiment lead to long-lived results which work well when applied to practical matters. Although there are sometimes scientific revolutions, far more is kept than is ever abandoned. However, although science employs powerful methods of reasoning, it does so within frameworks of less formal assumption. If such frameworks change, the old certainties are lost and must be reconstructed. Science is, then, *uncertain knowledge*. It is no longer plausible to say that current scientific understanding represents reality directly, rather it is an approximate working representation of the reality we hope to know more fully in the future.

Science in the present study will be treated as a form of *working knowledge*. In the process of knowledge construction new ideas are embedded in the shared meanings of local forms of life. Only as the intellectual ferment of knowledge creation dies down and the initial features of the local context are suppressed, do more widely shared scientific meanings emerge. While knowledge remains in flux, its connection to context is at its most apparent.

1.3 The plan of the book

Part I develops a cynical form of critical realism. The content of our images of reality reflects the human culture in which they are constructed at least as much as it corresponds to external reality. We are far from representing reality in a definitive form. We criticise old ideas in our search for rigour and coherence. In addition, changes in our cultural circumstances modify what we regard as knowledge. Furthermore, the reality we represent by our concepts is not fixed. We change our immediate surroundings through our actions. We change reality to fit our concepts as much as we change our concepts. I argue for an account of science that builds upon these points.

Chapter 2 presents the core idea around which the book is structured. It is argued that when, as in science, knowledge is continually being transformed by the actions of those who sustain and perpetuate it, it is not merely distributed over space and time but also over a hierarchy of quasi-autonomous cognitive levels (cf., Bar-Tal & Kruglanski, 1988). Six levels are distinguished here: sensorimotor knowledge, personal knowledge, group knowledge, institutionally shared knowledge, common knowledge, and the knowledge of all possible knowing agents. As information flows back and forth between (all but the last of) these levels, the knowledge held at each is continually being reshaped. If particular forms of science studies reduce science to a subset of these levels then they miss the full story.

In chapter 3 it is argued that the task of understanding science involves making problems out of matters that would normally be treated as settled. It is a general feature of the identification and solution of problems that some kind of working distinction is made between what is to be taken for granted and what is open to revision in order to solve the problem. I refer to what is taken for granted as the 'framework', and show the diversity of factors involved. The present study treats the frameworks used in science as its own topic of study. When a problem is solved, its conclusions may be presented as 'fact'. In particular, science is full of facts about the world. Such facts are not simply about the world, for they also depend on the frameworks in which they were constructed. The relationship of fact to framework is discussed.

Chapter 4 looks at the issue of how we attribute rationality and irrationality to intellectual processes related to science. A distinction is made between the explicit rationality (that strives for the ideal of rigorous formal reasoning) and informal rationality (as carried out by individuals and social groups). Scientific rationality is a combination of explicit and informal rationality. It is argued that what we regard as rational depends on context, including framework assumptions. Attributions of rationality are simultaneously social instruments of legitimation and demonstrations of the absence of error. Similarly, attributions of irrationality are never merely descriptive. They are typically evaluations to support some wider purpose. These issues all appear to draw our investigation towards a relativist position on the rational basis of knowledge, and the chapter ends with an extended discussion of the problems of the relativism of knowledge in the context in which they are at their most extreme - in closed and totalitarian societies.

Chapter 5 attempts to redress the effect of the pull towards relativism in the previous chapter. The book takes the view that whatever its philosophical difficulties, scientific realism deserves sympathetic consideration. Even though the scientific enterprise is treated here as fallible and selective, it has had a highly successful record of accomplishment. As Ian Hacking (1983) suggested, realism is mostly to do with action. Chapter 5 takes up the idea that what we can be surest about in reality is how our actions affect our subsequent perceptions. The concept of 'immediate reality' is introduced to refer to this region of the functional link between actions and perception in which our ontological commitment (belief in what is real) is greatest. Immediate reality does not, however, offer a foundation of greatest certainty for empirical knowledge, for there is no such foundation. Empirical knowledge is always uncertain, always subject to revision. Material reality can most effectively impinge on the revision process if we learn from our perceptions in the context of actions guided by concepts. This process can, without any a priori limits, lead to increasingly adequate conceptualisations of reality.

In developing the consequences of the idea of immediate reality, chapter 5 goes on to show that there are special problems for knowledge of reality when we try to deal with an immediate reality that is being affected by human actions in ways which are not part of our conceptual representation. Some forms of immediate reality are misleadingly congenial for this reason. When we act upon things under the guidance of our ideas, we find it all too easy to discover that the world is as we believe. For example, quack medicines very often work and educational experiments run by enthusiasts mostly succeed. Other forms of immediate reality are less congenial - they are far more likely to expose the inadequacy of our understanding. For example, in situations involving competitive confrontation, whatever each side knows about the intentions and actions of the other affects its own actions and intentions. As reciprocal complications escalate, understanding can be left behind by the uncongenial reality.

Chapter 6 draws together the ideas of Part I in an account of the scientific process. This is offered as a philosophically defensible idealisation of the processes that actually go on in the practice of successful natural science.

An important part of the plan of this book is not merely to describe how science produces knowledge claims, but also to provide resources to evaluate the procedures involved. The evaluative task is clearest in cognitive discriminations like those between science and non-science and between good and bad science. Part II explores these issues.

Chapter 7 compares scientific, non-scientific, and near-scientific practices and accomplishments. Is science distinctive in some rational respect, is it different merely by being institutionalised in a different way, or is the contrast actively maintained by practitioners who find it to their advantage to construct boundaries around their practices?

Within science, the contrast between good and bad science raises further issues. The approach taken in Chapter 8 is to regard activities labelled as 'pseudo-science' as the most likely to show weaknesses in their knowledge-producing practices. However, beliefs and practices labelled as pseudo-science are generally valiantly defended by their supporters. Although examination of the case studies will begin by suspending judgements, when the time comes to make evaluations, I offer the view in chapter 9 that we can identify specific defective practices at all levels of cognitive activity. If we can agree about which practices weaken science, then we have a way of showing why a particular body of doctrine might deserve to be dismissed as pseudo-science, in spite of the protestations of its defenders. Chapter 9 attempts to extract prescriptive principles from the earlier more often neutral discussion.

Part III, applies the approach introduced in Parts I and II. Chapter 10 poses the question of what variations on the present practice of science we should regard as permissible. Can computers replace persons in science? Could there be a purely personal science? How would a properly feminist science diverge from present science? What kind of changes to the rules of scientific institutions would be permissible if we think that orthodox science fails to gain access to some aspects of reality (such as the domain of the spiritual)?

The book ends in chapter 11 with questions related to the longest time-scales. 'How does science change in the long term?' Does it progress towards the truth, or does it adapt to its changing environment by a process analogous to the evolution of species? I develop an extended evolutionary framework. Such a framework should pay attention to the environment as well as to the organisms within it. Human beings currently affect their survival chances more by what they do to the environment than by the differential selection of their genetic material. Our ability to change our environment has now reached the stage where, with genetic engineering, we have the ability to change our own genetic material. Scientific knowledge and the genetic material of life are becoming interconvertible.

The evolutionary account provides a general framework within which to place the more detailed studies of the rest of the book.

Part I

The nature of science

2

Levels of cognitive activity

2.1 Introduction

In this chapter I will present the argument central to the structure of the book, that the knowledge that is continually being constructed and reconstructed in science is usefully studied through the interaction of a hierarchy of levels of cognitive activity.

The post-Cartesian tradition of modern philosophy made knowledge a judgement of the individual mind. But knowledge is expressed in terms of the social instrument, language. There is a competing pull between analysing knowledge in terms of the individual and with respect to communities who share a language and a form of life. In societies as complex as ours, it is not always appropriate for social analyses to suppose that knowledge is constructed in large language-sharing communities Modern science studies seeks to understand science by drawing apart in space and time the local practices which constitute it. The present argument is that it is also helpful to distinguish a series of levels at which the nature of working scientific knowledge can be analysed. The knowledge held at each of these levels is shaped by the cognitive activity of that level. These levels are to be thought of as quasi-autonomous, at least in the short term. They are affected, but not fully determined by, what goes on at other levels.

Discovery of the pulsar

Consider the following story, an idealised version of the discovery of the first pulsar.[1] In July 1967, a newly constructed radio telescope operated by the Cambridge radio astronomy group began systematic recordings of the signals

---

[1] For a fuller account see Woolgar, 1976.

it received from the heavens. It scanned for particular kinds of radio noise in the moving patch of the sky to which it pointed in the Earth's daily rotation.

Perception A research student (Jocelyn Bell) noticed an unusual kind of signal on the recording tape. It occurred for several days at a point in the daily scan corresponding to one place in the sky.

Individual knowledge Perhaps this new perception was illusory, produced by some terrestrial radio noise that came on briefly once a day. However, although it was very regular, the period of repetition matched the sidereal day, which is four minutes longer than the solar day.

Group knowledge Bell reported each new finding to her supervisor, Anthony Hewish. As soon as they could, and after a few tribulations, a recording of the signal was made at a higher rate to capture more detail. It was found that it included a rapid regular pulsation of radio noise that cycled from maximum to minimum several times a second. The new observations were discussed within the Cambridge radio-astronomy group. The first way of annotating the chart recordings was as LGM1, LGM2, etc. – a joking reference to the possibility that these were intelligent signals sent by Little Green Men. Alternative explanations were narrowed down by discussion and by observations with additional equipment, with the aid of other people co-opted into the study. Other similar pulsating radio sources were also found. The group was secretive, in part because the risk of an embarrassing error was initially thought to be great. By the time the first paper was published in *Nature* in February 1968 (Hewish et al., 1968) they had established the region of the sky the pulse came from and the maximum likely size of the source. Because the pulse was so regular, coherent, and powerful, they thought it might be the oscillations of an entire small star, perhaps a neutron star.

Institutional knowledge Once the phenomenon became common property, developments occurred on several fronts, especially after a small star within the Crab Nebula was identified as a source. Visible light from it was eventually found to be pulsating at the same frequency. It was most surprising that a distant bright (and therefore presumably large) object could pulsate so fast. Explanations eventually narrowed down to the view that the source of the pulse lay in some kind of flare on the surface of a rapidly spinning neutron star, so that we were seeing regular flashes as we do from the spinning light of a lighthouse. The star in the Crab Nebula, thought to be the result of a recent supernova, must therefore be such a neutron star. As other pulsars were found and studied, this general picture was confirmed, elaborated, and applied more generally in astronomy.

Common knowledge It was an eminently newsworthy finding and quickly became part of the common knowledge of the scientifically informed after its publication in *Nature*.

Knowledge fit for aliens The plaque attached to *Pioneer 10*, which is leaving the solar system and could therefore conceivably be picked up by extraterrestrials, gives the angular relationship between nearby pulsars in such a way that intelligent beings without knowledge of human language could work out the address of the creators of *Pioneer*.

Hypnotically recovered memories

One area of great controversy in the late twentieth century has been the use of hypnosis to aid in the recovery of memories. Here is an idealised narrative illustrating how the issue involves the interaction of cognitive levels.

The story typically begins with some problem in the personal life of an individual (let us say that she is a young adult woman) which leads her to seek therapy. She meets a therapist who proposes the use of hypnosis to gain access to the suspected cause of the problem.

Group perception Under hypnosis, the subject is seen by those present to remember having been sexually abused by her father when she was a small child.

Group knowledge The subject and her therapist are quickly convinced in the course of their discussion that sexual abuse occurred, especially as it gives a new comprehensibility to the patient's psychological problems.

Individual knowledge As our subject reflects on her life in the context of her newly recovered memories, she is increasingly certain that sexual abuse did occur. She begins to reorganise her life around the blameworthiness of her father. She dissociates herself from her family, and after brooding upon the issue decides to sue her father.

Institutional knowledge In the court case which follows, our subject gives testimony on the remembered sexual abuse, while her father vigorously denies that any sexual abuse occurred and points out that there had been no hint of this until his daughter had gone to the therapist. Each side offers additional evidence to justify its belief. In this case, the jury is persuaded by our subject and her therapist with the aid of their supporting evidence.

Common knowledge In the subsequent public discussion, the case is compared with similar stories. Some interest groups see it as supporting their view that sexual abuse is more common in families than had previously been publicly acknowledged. Other interest groups argue that the use of hypnosis to recover memories is systematically suspect, for hypnotic subjects tend to recover whatever memories are appropriate to the situation in which hypnosis takes place. If the hypnotist had looked for memories of abduction by aliens or memories of past lives, the subject would probably have supplied them. On many occasions, it proves possible to interpret supplementary information as supporting evidence.

More and more issues are drawn in as the discussion continues and no clear consensus appears to be emerging. In this controversial context, the initial conclusions at the lower levels become more ambiguous. What had appeared to be clear knowledge now looks more like opinion. In this case, our subject begins to doubt her own firm memories of sexual abuse, while her therapist remains convinced. The professional organisation to which the therapist belongs declares that, with appropriate precautions, hypnotically recovered memories are reliable. The status of the initial individual and group cognitive judgements is now changed.

In the story of the discovery of the pulsar, the messages at each cognitive level moved into harmony. If harmony is sustained it becomes appropriate to replace the present analysis in terms of knowledge in flux by the older ideal of knowledge as the possession of adequately justified objective truths. When there is no noticed variation in the content of the message with changes in context, attention can entirely be focussed upon its content (in its now universalised context). In contrast, in the case of hypnotically recovered memory, the interaction of the cognitive levels has not produced harmony at the point at which the story is set, so that the traditional analysis of knowledge does not yet apply.

A case can be built for distinctive cognitive processes and indeed for knowledge at each of these levels. In the account offered here, the levels are *quasi-autonomous*. The levels are held separate in the short term, because each must provide a coherent guide to a corresponding form of action, even before information has been fully assimilated from other levels.

There is a pressure for cognitive coherence at each level. In perception, when one pays attention to a scene without being able to work out what is there, there is a feeling of tension, resolved only when a particular thing is seen. The act of seeing provides a sense of order. In an individual's reflective thought, there is an urge to make co-ordinated sense of it all, so that the individual is able to decide quickly what to do. In groups, the need to take co-ordinated action depends on ways of reconciling divergent beliefs. If there has not been time for discussion, individuals tend to follow the suggestion of their leaders. Institutions may sustain the impetus towards co-ordinated answers that can be turned into policy decisions. Above the institutional level, knowledge may fail to cohere. However, in a time of war, national priorities are easier to set than when there is no emergency. A totalitarian society is able to act effectively, while, at the other extreme, a postmodern society has fragmented knowledge with less clear priorities.

The levels are not entirely autonomous because the information processed at each level continually draws upon other levels. In the very long run, all the distinctions tend to blur. Ideally, knowledge at each level would come into correspondence with reality. In practice, time produces its own blurring.

2.2 Sensorimotor cognitive activity

We gain our knowledge of the world we occupy by reflecting upon the combination of our own perceptions and the concepts we acquire from others. Traditionally, the analysis of empirical knowledge begins with perception.

Perception

Modern theories of perception regard it as a complex and sophisticated inferential process. The physical processes by which objects impinge upon our senses are not merely indirect but require complex brain processes. About half the human brain is devoted to visual processing.

One of the issues debated in recent psychology of perception is the relative importance of the approach of J. J. Gibson and that of cognitive psychology. Gibson (1979) emphasised how the perceptual system has evolved into a method of tuning into specific features of the total environment which effectively prepare the individual for action. The perceptual system can be fast-acting and does not always draw upon the slower processes of higher-level cognitive processing (Baron, 1988). Although there is a great deal more to Gibson's view, it is most appropriate to emphasise here that perception can be treated as having a 'separate and significant epistemic function' (Baron, 1988, p. 5). Perception is selectively tuned to external reality in a manner suited to the active functioning of the individual. The perceptual system rubs against reality.

The Gibson approach does not capture all that is going on in perception, at least according to cognitive psychologists. Mental representations of the world are constructed in a cascade of stages of brain-processing, each working primarily upon the output of the lower stages, and also affected by what has previously been learned at higher stages. Only some of the later stages are accessible to introspection. A model of the processes involved can in principle be set up on a computer or any other information-processing system of sufficient capacity. It would be a mistake, however, to follow the information-processing approach too simplistically by abstracting content from context. The higher levels of cognitive activity affect perception in what is commonly referred to as 'top-down processing.' The widest contextual factors affect our thoughts and memories which in their turn affect our perceptions.

Although the nature of our perceptual awareness is partly shaped by our conscious thoughts, the built-in powers of our perceptual system are able to provide us with surprises which give a jolt to our prior concepts, provided they involve the aspects of reality to which our basic perceptual mechanisms are attuned. We do not just see what we expect or want to see; we readily learn to see new things and, in appropriate situations, also see them as significant. Perception is undoubtedly a source of new knowledge.

We do not currently know enough to be able to predict the perception resulting from any particular immediate input of the sensory system of a human being. I claim that at *every* level of cognitive activity the cognitive content of that level goes beyond and is not fully determined by the known inputs to that level. Future research into perceptual processes is likely to provide further understanding of how the initial state of human sensory equipment affects perceptions. However, in the absence of knowledge of these mechanisms, there is an analogy between the way perception works and the actions of a person who tries to produce solutions to problems when not all the ingredients of a complete solution are available. Gregory (1973) suggested that perception is analogous to hypothesis formation in science. The metaphor of an autonomous agent being at work within the perceptual process has some power.[2]

---

[2] Such a suggestion is made in recent philosophical psychology in the view known as 'homuncular functionalism' – our brains work through a set of functioning agents – each rather more stupid than a normal person, the team collectively constituting a person's mind. See for example, Lycan (ed.) (1990); Dennett, (1991).

Single perceptions can easily be illusory. The natural way to eliminate perceptual illusions may be to look again from a different position or in a different way, if possible using additional senses. If we see something unexpected, it is natural for us to keep on looking, or to bring in another sense, such as touch, even before much conscious thought is involved. In more complex cases, especially where there is an incongruity between prior expectation and present perception, conscious thought processes or discussion with other people guide the manner in which we seek new perceptions to act as a check. It is the sustained pattern in perceptions that gains reinforcement at higher cognitive levels.

Empirical science does not build on momentary perceptions alone, but on observations linked to past experience with the aid of rationally constructed systematic thought. When a new observation report is in radical conflict with higher-level expectations, it usually cannot take the strain. (A single reported sighting of an extraterrestrial visitor in a flying saucer is not normally taken as proof of extraterrestrial contact.)

How should we give authority to our perceptions? Should we maximise our receptiveness to novelty in pure perceptions, or should we minimise error? The strategy of welcoming unexpected perceptions is straightforward. We are readily surprised by experiences of kinds to which the perceptual system is attuned, even if we do not have the appropriate concepts. If we are to encourage this aspect of perception, perhaps we should try to experience the world with a heightened sensitivity. Every subtle hint of visual texture, colour, taste, and smell should be noticed because we might learn something from it. In many contexts, this is indeed how we proceed. The strategy is clearest in practical matters. In cookery, for example, we expect to learn at least as much by our senses surprising us as we do from the standard elaborations of how things ought to happen. In science, the same strategy can be used in contexts where a new field of inquiry has just been opened up and needs to be explored. Our first ideas might be inappropriate and perception can provide us with a basis for new ideas.

However, the main trend in science has been in an opposing direction. Rather than maximise the power of perception to inspire our thoughts, we often seek to maximise its reliability. And for that purpose, we need to observe things in trustworthy ways and in trustworthy contexts. We repeat observations and we set up situations in which illusion is less likely to occur. For example, we try to make our observations in situations we understand well, with experience of what can go wrong and of what should be avoided. We also seek to observe in situations for which our concepts provide us with a clearly defined set of perceptual possibilities to choose between. We replace raw observation by measurement, and we replace measurement involving difficult sensory judgement by measurement in which instruments amplify the signal to the point where only crude and undemanding observational distinctions need be made. The case study of N-rays* discussed in chapter 9, tells a story of the failure of an attempt to replace difficult sensory discriminations by reliable instrumental readings. Rather interestingly, it is becoming possible to replace the human perceptual system almost entirely by scientific instrumentation. Observing instruments can now be automated to such an extent that the first direct human involvement in the observational process can be to assimilate instrumental information displayed in an appropriate scientific language. When we minimise perceptual error this way, the information collected can generate knowledge, but cannot so easily give us *perceptual* surprises. The surprises that come are at a higher level, as contradictions between statements in scientific language, or as unexpected results within computer simulations of reality.

It will be argued in chapter 6 that perception does not provide a foundation for our knowledge which firmly attaches it to the world. Rather each perception is like a convenient velcro patch with which we can easily stick the web of thought to immediate reality. As we extend the number of such patches while readjusting our web of thought, we often have to remove and reattach them elsewhere.

Beyond passive perception

We did not evolve our perceptual systems simply as autonomous reality-input devices designed to plug information into our cognitions. We normally experience the world in the context of acting in it and learn from the outcome of our actions. Touching, seeing, hearing, tasting, and smelling are combined with muscle sense to give us feedback on the success or failure of our attempts to turn intentions into action. The process of 'learning through doing' long predated the process of 'learning though matching perceptions to concepts'. The basis of empirical knowledge lies in the way we learn to act in the world so as to generate the expected experience. Science, too, builds on perception linked to action. There is a practical, craft side to science. Often we can make things work without being able to say how we do it. This craft knowledge is a resource for the higher levels of cognitive activity. Even in the most formal accounts of empirical science there is a key place for action. We build and use instruments to enable us to detect some subtle detail. We do experiments in which we try to control events so that they occur as our theory predicts. Properly scientific knowledge builds on actions which push nature to the limits of what we currently think possible, for that way we are able to learn more from the perceptions that result.

2.3 Personal knowledge

The second level of cognitive activity is personal knowledge. The term echoes Polanyi (1958). Polanyi wished to emphasise that not all human knowledge is expressed in publicly shared language; some is contained in the intuitive and procedural skills of the individual. Even in science, Polanyi argued, tacit personal knowledge plays a part. Many of Polanyi's examples would, in the present classification, be called sensorimotor knowledge, the non-verbal knowledge that is tied into personal experience of acting in the world.

Personal knowledge in the present account is held at the level at which perceptions are combined together within the patterns of thinking provided by concepts. Individual experience and social learning are available as resources, giving the individual access to informal and formal reasoning processes.

We have a rich and complex language to refer to such internal states as our conscious thoughts, memories, beliefs, and so on. Wittgenstein (1953) introduced what is now known as the 'Private Language Argument' – the argument that the term 'knowledge' is not appropriately applied to reports of those purely private inner states for which there are no ways of discovering any error and which it is inappropriate to doubt. Knowledge has to be expressed in public language, the conditions for use of which are tied to socially shared forms of life. However, although there may be difficulties about purely private knowledge, there are no such difficulties about the kind of personal knowledge which combines experience and reason in a manner which can, in principle, be checked by others. When an individual uses only his or her personal understanding to act or to make judgements upon publicly accessible matters, the knowledge employed is personal knowledge. I know in this way that I am sitting on a chair and typing at a computer keyboard.

When a single competent individual focusses his or her reflective abilities on a limited problem area, the depth of insight that can be achieved is difficult to match at any of the higher, social levels. An individual can follow through a chain of formal reasoning so carefully and clearly that the same result is reached as would be produced by a machine which turns logical steps into causal processes. An individual can also stand outside the formal chain, and see its potential and limits, which the machine cannot do. An individual can come to an optimal judgement even when there is insufficient information to reach a solution by the formal reasoning available. An individual can consider more than one domain of formal or informal discourse and find ways of making each discourse into a resource for improving the others.

Although the notion of knowledge being held by the individual is a philosophical commonplace, individual consciousness also has its limits. We do not have the insights into ourselves that we think we do. We see in our own minds only what the resources of our culture enables us to see, and, in particular, what the intellectual fads of the age dispose us to see. The notion of subjective certainty is an illusion. Phenomenology of mind is far too fragile an intellectual construction to be a foundation for an acceptable theory of scientific knowledge.

Many of the resources an individual uses are acquired from other people. Although each individual believes that he or she can think freely, it is easiest to think towards socially acceptable conclusions; indeed it may be impossible to move very far away from the horizons of possibility set by institutionalised conventions, just as it is impossible to develop sustained lines of thought far beyond what can be achieved with that most important of social instruments, the current form of language.

There is an argument that because human beings are basically social animals, the intermediate level (between the biological and the social) of the free individual, the possessor of a mind, is a cultural product. According to this view, the concept of mind is a social construction (Coulter, 1979). However, if it is a social construction, it is a fairly stable one, which we have the biological capacities and social interests to sustain. Once society has handed autonomy, freedom, and responsibility over to individual possessors of minds, they are capable of producing knowledge that can be, in the short term, somewhat independent of wider social communities.

When we come across knowledge claims which simply *cannot* be held at the level of the individual mind, we are tempted to question whether the term 'knowledge' should be applied to them at all. A recent example was a computer proof of the famous four-colour problem. This proof, which was produced by an appropriately programmed computer, was so complex that no single individual could work through all of its steps. (It could, however, be checked in many ways – by checking the program, by programming other computers to do part or all of the same task, and by teams of humans splitting up the task of working through the proof.) No one individual could take personal responsibility for the claim that the whole proof is free from errors. If this is so, the argument goes, such a computer proof is not knowledge at all. On the view presented here, all that this argument has shown is that while the computer proof is not *personal knowledge*, it could be knowledge at another level, such as for the group which collectively checks the proof.

The next three levels in the hierarchy of cognitive activity build upon the postulate that knowledge can be social. An incongruous idea in an earlier age of individualism, this notion has been campaigned for in many places and in many ways in the twentieth century.

2.4 Group knowledge

Those social sciences which deal with the social basis of knowledge have built up complex and varied systems. In order to reconstruct some of this complexity in my account, I have chosen to divide interpersonal factors into three levels. The lowest of these, the group, is the one at which it is most plausible to invoke biological mechanisms.

Human beings are naturally social. Indeed, the intellectual life of individuals appears rather like the limiting case of life in a small group. We learn to think by listening to our own and other people's talk. We learn to reason by constructing internal representations of arguments that are successful in persuading others. We construct intellectual edifices in our minds in preparation for impressing others.

A co-operative group shares a coherent body of common knowledge. As conflicts of interest and disagreements of judgement are collectively talked through to mutual understanding, and perhaps to a resolution, group knowledge becomes more highly integrated. Short-lived and unstable groups are likely to have more fragmented group knowledge.

There are quite strong social mechanisms which make individuals behave somewhat differently in a small group from the way they do separately. Social psychology is concerned with the nature of such groups and group processes. There are many mechanisms which facilitate bringing small groups into harmony. In an appropriate context, they will come into cognitive agreement rather readily, though not more reliably to the 'correct' result. Quite often, they agree on an extreme view ('risky shift'). The willingness of groups to share emotional states and to act together in a co-ordinated way could have been important in our evolutionary origins. It is a commonplace that even the most disputatious groups feel the pressure to come together under external threat or challenge. Groups are often more visible and more powerful than assemblages of isolated individuals. In situations in science in which it is more important to show which new ideas work than it is to eliminate every possibility of error, groups can be more effective than individuals.

Although its philosophy emphasised *individual* rational thought, modern science has always involved the interaction of such social groups as university communities and learned societies. In the nineteenth century, science became more conspicuously concerned with exploiting the extra possibilities of cohesive groups in its cognitive processes. There was a very old model of group structure to draw upon – that of the teacher and his students. In the nineteenth century, and especially in Germany, leading university teachers began to teach laboratory and field research techniques systematically. It was found that the collective activity of the professor and his students could be vastly more productive than the professor working alone. Professors could exploit their students in this way – using them as extra pairs of hands with peripheral brains attached. Even in the most exploitative schools, in which all credit for work done went to the leader of the group, the students benefited, for they learned how to do research. The best of them could later become professors themselves and apply the same system elsewhere. The ideal research school was at the time regarded as mutually beneficial rather than mutually exploitative. It was judged that such schools were usually less creative than the best individual geniuses of the time, making up for this by being vastly more productive (Merz, 1965, Part I, chapter 1). The pattern of group activity organised around teaching research was much repeated, especially by those who had been trained in it. It became an institutionalised feature of laboratory-based experimental science, first in disciplines in which it was easiest to apply (chemistry, physiology, physics), and then in many others (like experimental psychology).

A later equally influential model of group activity for science has been the team of researchers who are assigned to deal with a specific problem – the most famous precedent here was such military 'big science' projects as the development of the atom bomb in World War II. This pattern was applied subsequently in peacetime, as in the running of expensive pieces of high-technology equipment such as particle accelerators and radio telescopes. Group structures in science have developed further over the last few decades as part of a wider pattern of development of organisational structures in business.

Groups, then, are important in the cognitive activity of science. Indeed, a number of historians and sociologists of science are now finding special cognitive significance in interactions within small groups of scientists.

Groups like those involved in science may be able to draw upon some of the behavioural traits that small groups are speculated to have had in our evolutionary and social origins. However, unlike the small autonomous communities studied in social anthropology, the members of scientific groups have interests and activities outside the group, and their membership is voluntary.

Small human groups, including those in science, tend to be more cohesive and more powerful than rational calculation of self-interest would require. We gain some of our sense of identity from the groups to which we belong. Although people can learn to be purely selfish, we can also easily learn to be social. (Perhaps our animal origins in small social groups of closely related individuals still has residual effects on the biology of our social nature.[3])

Cognitive activity takes a somewhat different form within groups. The interactions of individuals have an effect on their separate activities including observation and reasoning. It is harder for an individual in a group to move far from the context of shared presuppositions and values. This group context affects the top-down processing in individual perception, and individual cognition in general.

The processes by which knowledge is created and revised are augmented at the group level. Complex tasks can be handled more readily when there is some way of dividing them into complementary parts. Creativity can arise out of discussion and shared activity as well as from individual thought. Scientific discovery moves from being an isolated act of insight which checks out to the negotiation of agreed solutions to agreed problems. Discovery becomes a social process (cf., Brannigan, 1981).

The critical processes that go on in groups are also important. Indeed, it is plausible to suggest that the critical rationality by which an individual constructs justifications of beliefs is the internal representation of the processes of criticism that occur naturally within groups. Isolated individuals often have difficulty appreciating how those who lack their own subjective commitments might see things differently. The task is easier in a group. In a highly cohesive group, an intersubjective viewpoint comes to dominate; in a less cohesive group, some kind of interpersonal market of ideas emerges.

Group knowledge cannot satisfactorily be the highest level of knowledge. The shared values of a group may shape knowledge in a way which makes it acceptable only for that group. A group which mostly looks inward, as extreme religious sects tend to do, risks producing claims to knowledge as disconnected from external social constructions of reality as the most subjective of personal beliefs. Within the tradition of scientific orthodoxy, some

---

[3] This is to take a very small halting step towards the claims of modern human sociobiology, without conceding the rest of the package deal that this controversial discipline offers. It is hard to see any calculation from purely selfish interest which would make having children worth while in present Western society. We have children because we want them, in part as an extension of ourselves, not because as seperate individuals we are better off with them.

hierarchical groups have been totalitarian in their internal social structure. In the case of the school of the late nineteenth-century psychologist Wilhelm Wundt, for example, anecdotes exist that his students were sometimes too frightened to report empirical results to him which were contrary to his current theory; instead they invented alternative, more acceptable results. Internal checks upon the knowledge of a totalitarian regime are notoriously difficult to make.[4]

Insiders and outsiders to groups

One feature of the cognitive activity of the group (and also of larger cohesive social organisations) is the distinction between the capacities and skills of the insider and of the outsider. Is the judgement of experts always privileged, or might such a group merely be defending their collective interest?

A common claim made on behalf of expert groups and disciplines is that only they have the experience and training to make competent judgements about the matters of knowledge over which they claim special territorial rights. This may be an appropriate rebuff to the attack of an *ignorant outsider*. But an outsider may be competent – a student on the threshold of career decisions, arrogant in the awareness of untapped potential, or someone who, like the double Nobel prize-winner Linus Pauling claiming that Vitamin C prevents colds*, aspires to contribute to a new disciplinary area after having reached the top elsewhere. A *competent outsider* is likely to regard specialist knowledge and skill that is not accessible to him or her as defensive ideology serving the self-interest of a closed group. Sulloway (1991) reviewed the arguments for such a view of the psychoanalytic movement, as Freud* created it. In particular, if individuals have to make a major act of commitment to a discipline, before being permitted to study its inner secrets, then defences of the discipline based on the authority of those secrets are suspect.

Freud and his successors offered an additional defence of their own insider knowledge when they claimed a lack of objectivity in untrained outsiders. 'An analyst without proper training will easily fall into the temptation of projecting outwards some of the peculiarities of his own personality.'[5] This defence of psychoanalysis may be plausible in repressed and mentally dishonest societies. Nevertheless, it is distasteful as a general argument to defend

---

[4] The most famous literary formulation of this view is in George Orwell's *Nineteen Eighty-Four* (Orwell, 1949).

[5] S. Freud, 'Recommendations to Physicians Practising Psycho-Analysis' (1912), as quoted in Sulloway, (1991), p. 268.

a system of belief. If the outsider to psychoanalysis can only escape the distortion of sexual repression by committing him or herself to psychoanalysis, and if those who become psychoanalysts and then criticise the movement are excommunicated as heretics, then psychoanalysis has legislated its own immunity to challenge. The outsider rejects this form of defence of insider beliefs.

Should we prefer the cognitive perspective of the outsider then? An argument along these lines can be reconstructed from Marxism*. Marxists have claimed that everyone is affected by the ideology of their class. The outsider will tend to regard the investment of effort of the disciplinary specialist as generating, not merely a commitment to a particular way of seeing things, but also an ideological bias comparable to the ideological bias of the ruling class. The vision of insiders is distorted by false consciousness.

The insider criticism of the uncleared vision of the repressed outsider is comparable to the outsider critique of the false consciousness of the insider. Both have similar weaknesses. Both are circular in that their presuppositions are very close to their conclusions. The resolution of these two views is to admit that either the insider or the outsider view might be correct, but that neither is epistemologically privileged. Science should have an open social system, so the outsider can criticise any discipline that discourages investigation from outside. In science, there is an incentive to make tacit knowledge explicit, rather than keep it as insider secrets.

We should all, insiders and outsiders alike, be suspicious of claims that make false consciousness something from which we cannot escape. Our actions might be driven by sexuality or class interests or insider interests, of which we are not aware, but once this is pointed out to us, we should not be in a worse position than anyone else to see that this is happening. Those who aspire to intellectual honesty should be suspicious of the argument that there are some underlying pressures which make intellectual honesty impossible.

In general, very successful groups tend to grow. Those based on teaching tend to spread to new centres in the second generation, even though the initial teacher grows older and moves to other things or dies. With growth, the cohesiveness of the group declines. Some of the wider social structures of science are the expanded shells of older successful groups.

2.5 Knowledge within a shared institutional framework

Interactions between groups and between individuals who are not part of the same group occur at the next highest level, that of the shared institutional framework. Many social sciences seek to understand, to regulate, and to improve the processes which enable large numbers of people to stay together. For science to be possible, these processes have to allow a high level of stability and continuity. Under these circumstances, complex systems of conventional social practice can build up. Mary Douglas suggests that, minimally, an institution is only a convention (Douglas, 1987, p. 46). She refers to the definition of convention of David Lewis (Lewis, 1968). A convention arises when it is in the interest of all the parties involved to act in a co-ordinated way. No one deviates from the convention, lest the benefit of co-ordination be lost. Science operates within a framework of such conventions already established in wider society and constructs further conventions to co-ordinate its own practices. Suppose that a collection of scientists find that, although they have a lot in common, they risk coming into mutually undesirable conflict. They may well settle upon some rules of conduct. A scientific society, an educational establishment, or a conference might create such rules for itself. And when the results of acting within such rules is seen to be beneficial by other people with a similar problem of unwanted conflict, they too might adopt or adapt the rules.

The institutions of science have traditionally facilitated the search for consensual knowledge. While making it easier for every scientist to broaden the range of information of which he or she is aware, the institutions of scientific communication strive to make it easier to see which information is the most significant. Orthodox viewpoints on new information can build up very easily within scientific institutions.

Science is practised within a complex of such rules, forming an institutional framework. The buildings occupied, the permissible ways of organising the space within them, the organisational structures within which scientists act, the practices approved of as research, the horizons of creative possibility, the symbolic structure of thought, the way conclusions are formulated, the system of communication, the public negotiation of proofs, the conduct of disagreements and disputes, and the way approved work is rewarded are just a few of the many features of the institutional framework of science.

In the early twentieth century, the metaphor of society as behaving rather like an individual was widely used (and criticised). Mary Douglas (1987) suggests that the notion still has some value. The institutions of society, she argues, remember and forget; they do the classifying. They even make life and death decisions, such as who receives the limited amount of food in a time of mass starvation (Douglas, 1987, chapter 9). Her argument is intended to show how highly constrained the actions of individuals are by the institutional context prevailing. These remarks apply to science, just as to any other social institution. And in particular, the role of institutions in the construction of knowledge is a large area of study. Institutions shape the cognitions of all who come within their shadows.

The institutions which are most obviously relevant to the production of knowledge in science are those which provide a framework within which the system of public presentation and public acknowledgement of new knowledge can operate. Its rules have developed over many centuries and have created a level of cognitive interaction more general than those we have already discussed. The rules also provide a barrier partially insulating the internal practice of science from the pressures of the wider society, for the judgements that matter most are those of people whose esteem is most sought by scientists – the other scientists who are competent and well informed on the questions at issue.

In terms of this chapter's argument for a hierarchy of levels of cognitive activity, institutions provide a generalised context in which knowledge can thrive over extended regions and times. Most empirical knowledge is initially constructed by individuals or small groups with some reference to the immediate context. Cognitive conclusions which are well adapted to that local context may not travel well to other contexts. Suppose, for example, that a local laboratory has some special conceptual or experimental skill and it produces knowledge in a way which presupposes that skill, then that work can only be taken up as knowledge by others who accept the expert judgement of the members of the first laboratory, or who find some independent check on it which is under their own control. Unless it can be developed or exploited in some other way, it may not be transferable to new contexts.

However, science in different local centres is often very similar in its institutional form. Most modern science is a professionalised practice, carried out in universities and in industrial and government research and development laboratories. Within such settings, the local organisational forms of science that have recently proved to be most successful at producing the required kind of knowledge tend to get copied. For this reason, if knowledge takes a form which is suited to the institutionalised organisational pattern, it can readily spread from one similar local scene to another. One feature which aids such diffusion is research fruitfulness. In the context of research institutions, knowledge which can stimulate further acceptable research spreads rapidly. Another feature applies to educational contexts. Knowledge which draws upon, or can be applied in, a specific kind of training can get especially favourable attention from all those who provide such a training.

Knowledge which is capable of spreading within a widespread institutional framework does not need the support of idiosyncratic personal or local group characteristics. It may, however, be shaped to suit general institutional constraints such as those mentioned in the last paragraph. Scientific knowledge created within our greatest scientific institutions often illustrates what is conventionally regarded as objective knowledge. Good institutions become invisible from inside the scientific approach. However, institutions always affect the content of the knowledge created within them, and, if we find that we wish to criticise an item of generally accepted knowledge, we are often drawn into commenting on aspects of institutions.

Institutional knowledge is, then, less general than our universal ideal of knowledge. When knowledge is regarded purely as institutional knowledge, it has its own distinctive character, linked to the conventional nature of institutions. It is likely to be loaded with institutional values. For example, consider the notion that only those with a particular kind of training are permitted to make the expert judgements which shape knowledge within an institutional setting. Although their judgements may seem obvious to insiders, to outsiders this may be seen as a preference for matters linked to the training, which looks like an institutional value. This book will consider examples where institutional knowledge diverges from knowledge at other levels. For example, it is sometimes found that institutional cognitive practice in science is resistant to sources of knowledge which are outside collective scientific control. Non-scientific systems sometimes construct knowledge out of private insights or revelation. Such sources are not acceptable in science. Because of the nature of its social system, the knowledge held in scientific institutions must be capable of being checked in a publicly satisfactory manner. An example to which we will return in chapter 10 is whether there could be institutionalised scientific knowledge of psychic phenomena. There have been many occasions on which individuals and research groups have satisfied themselves that a specific psychic phenomenon has occurred, in the absence of procedures to overcome collective organised scepticism.

2.6 Common knowledge

Much of the knowledge contained within our society is linked to practices, expertise and resources in specialised domains. Science is a subset of such domains. Since there are so many forms of specialised activity which lie on the borders of science, the term 'institutionalised knowledge' does not mark the scientific side of any clear boundary with the non-scientific. The most characteristic forms of institutionalised scientific knowledge spread only through limited areas of science, such as particular disciplines or problem areas. They are therefore more limited in range than the whole of science. In addition, much knowledge straddles the science–nonscience boundary. Science-based technologies, for example, may apply long-established scientific results. High technology may employ people with scientific training to do work which is not itself science. Technological laboratories may be encouraged to do some 'blue-skies research', indistinguishable from pioneering science except that it inspires hope of eventual practical application. The aims of seeking knowledge and improving practical technique are not always institutionally separated.

The analysis of science at the level of common knowledge can easily lose the distinctive qualities of the lower cognitive levels. Consider the following idealisation of science made from the point of view of non-scientists.

In an age in which so much of our economic system is driven by the use of technological knowledge, we have constructed a large number of knowledge-using systems. In the desire to realise an aim, a problem arises and we search through available knowledge to find a solution to the problem. For example, the commercial development of new processes and products takes this form. In the large-scale enterprise of finding better solutions to old problems and solutions to new problems, science is a relatively inexpensive overhead on the technological development for which it provides *new* knowledge as and when it is needed.

In this image of science as an augmentation of the way society (or factions within society) use existing knowledge to solve problems, it appears that science is not autonomous. It is produced by the converging interests of the patrons, creators, and users of the new knowledge. In a context in which some sector of society (the underclass, or a minority race, or women, say) is systematically excluded from the commissioning and use of science, then the knowledge created will tend to work against their interests. Since the primary need is not 'truth' but the advancement of a particular aim, what is presented as new knowledge may be ideologically tainted by external values.

On this view, science becomes nothing but another resource for the competitive struggles that go on in our society. It has nothing to offer except aiding the success of factions in such struggles. There can be as many variants of scientific knowledge as there are distinct interest groups who appeal to knowledge. Knowledge becomes just another commodity to be bought and sold in the market-place. Knowledge need only be fitted together into a coherent whole if it is more valuable that way. The processes of accrediting knowledge may add value, but, under some circumstances, contradictory versions of unaccredited knowledge may be worth more. For example, quack medicines can be worth millions of dollars.

The trend in postmodern society is to abandon the search for a single coherent system of truth and to let each fragmentary knowledge claim find its own market value. The Enlightenment ideal image of scientific knowledge is regarded as a hopeless quest, like the holy grail.

However, such a view of science is not universally held. It is often in the interests of scientists themselves to oppose it. Science, the scientists and philosophers tell us, is knowledge objectively established from empirical fact. Only bad science is tainted by the interests of those who commission it. This view of the neutrality and autonomy of scientific knowledge, which has been the orthodoxy since the Enlightenment, has recently come under a new wave of attack. The critics of science note that its very methods are open to contamination by external and other values. Science cannot be value-free; it cannot be objectively neutral. Furthermore, because of the way it is commissioned and used by sectional interests in society, even if it were completely objective, it would still tend to serve these interests, rather than society as a whole. The frontiers of knowledge are not advanced evenly. Feminist writers argue, for example, that an objective science created mainly by men, produces knowledge which is in the interests of men.

The defenders of science argue back. 'What good is science if the answer depends on where you are getting your money?' (Holloway, 1993, p. 11). In terms of the present discussion, one modern reply to critics is that science has long been on such a large scale that its practitioners have their own interests as scientists. They wish to maintain a system which keeps a certain distance between itself and its patrons. Although a specific project might be developed in terms of a convergence of interests between scientists and patrons, that does not mean that the next project must represent the same interests. Science is not, then, completely open to external interests. In general, projects which put science into disrepute are discouraged by scientists. Modern scientists would like to live in a world in which science is given a special place because it is perceived to work so well in its own distinctive way.

This kind of argument will be considered at appropriate points throughout the book. The key point to be made here is that, when we consider science as a cognitive activity from the point of view of wider society, it is difficult to distinguish between the social systems which *use* knowledge and are driven by the values guiding that use and the system of science which *creates* knowledge and endeavours to do so in a way that is minimally contaminated by wider factional values.

Presumably, common knowledge is also limited by being human knowledge. Human beings have all sorts of limits to the possibilities of cognition. An individual has a particular set of senses, can only be in one place at a time, and has limited intellectual capacity. There are clear structural features in human language. Human knowledge caters to such limits. It will always be hard for us to create knowledge beyond these limits.

2.7 Knowledge fit for extraterrestrials

The way we think of knowledge is framed by our human nature. Does it have to be? Can we produce universal knowledge? If we could build into our knowledge claims all the human presuppositions of knowledge, then we could state as universal knowledge the relation between the presuppositions and the conclusions. Some of this has gone on in work on the foundations of mathematics, especially over the last century. The idea of a general a priori framework for knowledge, as envisaged by Immanuel Kant, will be discussed briefly in the next chapter. Presumably one such broader form of knowledge would be knowledge we could share with non-human beings.

There is a relevant problem of interest both to orthodox science and to popular culture. Given the high a priori probability of extraterrestrial intelligence elsewhere in the universe, it would be of enormous cultural significance if we were actually able to make contact. How could we communicate with aliens for the first time? This is the problem of CETI* (Communication with Extraterrestrial Intelligence). A common postulate is that the more universal forms of mathematics will be readily understood by intelligent aliens. The natural number system, it is often suggested, is surely universal, so that a sequence of prime numbers, suitably signalled, should be seen by aliens as a meaningful message, and the task of building up a set of mutually understood signs can begin. In the absence of actual extraterrestrials, we cannot be sure that our speculations are correct. Perhaps a sufficiently different being would have come to the more general forms of mathematics by a quite different historical path than that by which our own species came to mathematical knowledge.

The problem of communicating empirical knowledge to extraterrestrials may be even greater than for mathematics. The problem will be especially great for contact with beings a long way away in an environment unlike our own. We would need to give the aliens the kind of contextual information that human communication normally builds upon. Perhaps they will understand us if they know enough about what we are like and what our situation is. Perhaps we will understand them if we know enough about them.

The exercise in extending the horizons of possible knowledge has the practical effect of showing more of the presuppositions of present knowledge. In the next chapter, I will discuss further how to scrutinise the frameworks of human knowledge.

2.8 Interaction of the levels of cognitive activity

The central claim of this chapter is that an adequate analysis of scientific knowledge must consider the interaction of levels of cognitive activity, and that previous analyses have tended to pay too much attention to an insufficiently broad range of levels. All levels are of theoretical importance and all levels up to common knowledge are of practical importance. Sensorimotor knowledge is at the centre of attention in realist empiricism. Individual cognition dominates discussions of claims for a rational basis for scientific knowledge. Group and institutional processes are of great importance in the vision of the creation of knowledge within the scientific community offered by T. S. Kuhn. Engagement with the wider social world is the key to the claims of social historians of science and macrosociologists of scientific knowledge. In practice, all the levels interact, the lower levels drawing upon the higher ones as frameworks and providing them with more specific content. Only the highest level is different. It is an extrapolation of normal cognitive activity, and (until we meet extraterrestrials) mainly of interest as a check on the most general philosophical claims made about knowledge.

Much can be learned by looking at how information is transformed in the interaction of levels. For example, doubt at one level can be reduced by the pressure for firm judgements at another. Such tensions become acute when some sort of action is needed. A householder cannot afford to dither when his or her senses are unclear as to whether a burglary is in progress. An effective course of action must be followed even if it may turn out to be inappropriate. Nor can a group afford to continue to discuss divergent individual opinions when the first signs appear of an external attack. Each individual appreciates the need to accommodate his or her will to the group's requirement for effective collective action. However, the mechanisms by which doubt is sharpened into conviction can also generate error and illusion. We can become convinced that a threat is present when it is not. Processes like these go on in science as it seeks to generate knowledge that can be a basis for action at the appropriate level.

I do not claim in this theory of cognitive levels that the interaction between the levels must take any very precise and specific form, only that it should be full. When barriers are constructed between the levels, knowledge becomes less trustworthy. (I will discuss this in chapter 9 in terms of 'pathologies of knowledge'.) The relationship between the levels is subject to historical change with developments in science, technology, and society. For example, I think that the relative rate of individual and social processes in cognition is changing. Social communication and social decision-making has speeded up in recent society because the technological means are faster and save money. Modern science, like modern society, hurries individuals along. These changes make social processes more visible and possibly also more important in the construction of knowledge. Traditionally in our culture, individuals have been regarded as having priority over society in conflicts of cognitive value. Perhaps this, too, will change as the relation between the individual and the social levels changes.

2.9 Summary

The decisions of individuals, local groups, or whole nations take place in different ways. The knowledge which guides action at each of these social levels has a distinctive form.

In this chapter, I have shown that the knowledge-construction process does not merely take a different form at particular places and times but also at each of a number of cognitive levels: We can loosely talk of knowledge at each of the following levels of cognitive activity: sensorimotor, individual, group, institutionally shared, and the common or human level. I added a level above human knowledge, one of principle rather than practice (unless we meet intelligent non-human beings), that of all knowing agents.

On a strict traditional standard, we would require the activity of all the levels to be in harmony before describing the outcome as knowledge. In a system such as science, in which knowledge is in perpetual flux, this traditional standard is rarely reached. It is argued here that a useful way to study actual knowledge claims is to analyse the cognitive activity at all the levels and the interaction between them. This form of analysis will be put to work in the chapters which follow.

3

Facts in frameworks

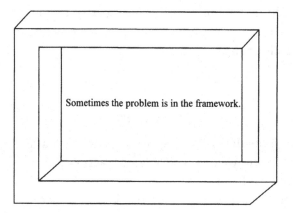

Sometimes the problem is in the framework.

In the introductory chapter, it was suggested that, rather than follow any single disciplinary perspective closely, this book will provide the resources for a kind of cognitive play with the various ideas of science. In this chapter and the next, I will open up the processes of science to scrutiny. I will do it in terms of three aspects of the scientific process:

(1) Frameworks of assumption.
(2) The relevant empirical facts, and how they are constructed and assimilated within frameworks.
(3) The rational procedures employed in adjusting the network of understanding, as in the assimilation of novel facts.

This chapter is mainly concerned with facts and the frameworks within which they are constructed.

At the beginning of this century, positivist philosophy accepted the idea of positive facts that are true because they correspond to the way the world is, and that stand independently of any framework of assumption because the process of their construction is designed to strip away all assumption. Since then, philosophy has identified more and more framework features which are presupposed by any statement of fact. The view that we do not have any brute facts that stand outside all frameworks, which would then have seemed provocative in the extreme, is now a near platitude. This chapter begins by applying the insights of several approaches within science studies to the nature of frameworks and of framework assumptions that affect what we regard as facts.

3.1 Science as the construction and closure of black boxes

A study such as this draws our attention away from the result and towards the process of what we are doing. By looking at the ways in which a scientific procedure can go wrong, we are encouraged to see it as problematic. This is contrary to what is normal in natural science, in which resolutions are sought for problematic issues, so that we may stop thinking about how something was done and move on to use it to produce further achievements. What is being done here leads to questioning what is generally taken for granted.

Latour (1987) used the suggestive metaphor of the 'black box' for the social processes of science. He described science as involving the construction and closure of 'black boxes'. The analogy Latour appealed to was with the sealed automatic workings of complex pieces of modern machinery which ordinary users can put to work without the slightest idea of what goes on inside to produce the observed effects. A laser generator in a compact-disc player is such a black box for most people. As long as it provides its special kind of light as required when switched on, there is no need to understand what goes on inside it. Analogously, behavioural psychologists once referred to the human mind/brain as a black box – again because they did not wish to study its internal workings, only the relationship between the stimuli to which it was exposed and the responses which it produced. In Latour's view, scientists try to convert laboratory findings into such black boxes. The difficult processes of making judgements based on highly fallible laboratory events are eventually hidden behind the façade of the objectively stated 'fact'. Latour also explained that the black boxes produced in science do not always stay closed. Sometimes, as in controversy, our problem solutions fall apart, and we must re-examine the assumptions we made.

The metaphor of the black box is beguiling. It is valuable as a corrective to earlier philosophical images of science, for example, by emphasising that the notion of science being 'founded on a solid bedrock of fact' is illusory. What we regard as facts are constructed, not found ready-made. The appearance that something is a hard fact is merely the result of us looking at the black exterior of the box that hides away its construction processes. A scientific fact is rather more like the 'guilty' verdict of a court. If we accept that there was nothing wrong with the judicial process, we take it as a fact that the person convicted did indeed commit the crime, and act on that basis. Only occasionally, in new circumstances, do we reopen the case.

An alternative metaphor to Latour's was offered by Collins (Collins, 1992). Perceptions are like ships in bottles. Science is in the business of putting ships in bottles, no clue being left behind as to how it was done. Collins presented his own practice as one of exposing the techniques involved.

The black box metaphor for the scientific process has its limits. The box holding a scientific fact is not really black. It has an unrevealing surface texture only with respect to the social process that went into its construction. Every effort is made to make its cognitive significance as transparent as possible. The symbolic meaning of the fact is immediately visible. We can see how it relates to other parts of the cognitive map of science by the connection it reports between its constituent concepts. And we are also given an idea of how it connects to the natural world. It guides us in how nature might constrain further research activity.

In this book we will sometimes open the black boxes which scientists construct and close. This task will be undertaken in a way which meets standards shared by natural science, social science, and philosophy.

When we look at apparent 'hard facts' in science, we immediately see that part of the meaning they carry is dependent on context. Any argument constructed around them will depend on implicit assumptions carried by that context. Sets of contextual assumptions will be discussed in terms of the concept of 'framework'.

3.2 Frameworks

In science, very complex technical issues are considered, often far more complex than the human mind can cope with by explicitly considering all aspects simultaneously. One important strategy for dealing with complexity is to divide an issue under study into those parts that can be taken for granted, and those on which creative work needs to be done to answer questions arising. Those assumptions, both tacit and explicit, which are taken for granted form the *framework*. The framework for scientific reasoning very often treats past conclusions as facts, that is, it encloses them in black boxes. Some framework assumptions are unnoticed and it may take considerable ingenuity to uncover them. We can use historical hindsight, for example, to see how the science of earlier ages was in harmony with cultural assumptions which only became obvious much later. For example, the rationalist presuppositions of the science of the Enlightenment were challenged from within a new framework of assumption in the age of Romanticism. Other presuppositions which may be noticed do not attract critical attention because they are part of widely shared frameworks of assumption, and to challenge them would be to challenge invisible social convention. Religious assumptions about the relationship of the Divine Creator to His Creation helped shape science in seventeenth-century Europe. Other assumptions are explicit ways of making a difficult problem more manageable, as when general theories are applied to complex cases.

The assumptions within the framework range from transient matters of the moment to prerequisites for all knowledge. In order to deal with any problem, we make initial explicit assumptions, which we may later abandon or work into our solution. In addition, an individual's way of working draws upon personal assumptions, values, and commitments. Further features of the framework are distinctive to the local group within which science is being practised. Every institutionalised form of practice has large numbers of implicit and explicit rules which guide, inspire, and limit those who operate within them. Individuals can break away but normally at a cost to themselves. For example, it is harder to gain approval from others for what is done. More general framework features are linked to the wider structure of society and even to the general form of the human species. At the most general level are those assumptions which philosophers identify as prerequisites for all knowledge (in the tradition of Immanuel Kant).

The positivist ideology of the neutral fact which stands independent of all else, long made it easy for people to forget how inescapable framework factors are in science. However, the omnipresence of framework factors is easy to illustrate. Consider, for example, the way science tends to lose sight of the subjective nature of the observer. Every observation requires an observer, whose characteristics are usually buried in the framework rather then being treated as part of the scientific problem. The modern tradition of scientific objectivity obliges the scientist to separate him or herself from the facts of the matter, which should only report relationships among the phenomena that are the same for all observers. When quantum theory found it necessary to put the active interventions of the observer into its equations, modern apologists for natural science were forced to see the observer as an inescapable part of the phenomenon.

New scientific ventures often start with many unnoticed assumptions, which are only exposed in the subsequent development of the activity. Scientists very often presuppose more than the minimum needed to do science, either because they are not aware of a presupposition or because they believe it has heuristic value. The unnecessary assumptions that people make are part of the individual and local setting of science. The historian of science very often finds it helpful to link such features to the creation of scientific novelty – for it may be that without certain personal or local assumptions it would have been difficult or impossible to come to a particular conceptual insight.

Institutional pressures remove superfluous assumptions which inhibit the flow of information. For example, the tendency of scientists to remove the actual observer from accounts used in the negotiation of scientific facts (reporting what happened in the passive mode) is clearly part of the pressure in the institutionalised process to hide personal and local assumptions. It is the appropriate thing to do. To declare one's personal assumptions openly would invite objections from those who did not share them. One person's plausible assumption is another's prejudice.

In the process of making scientific discourse more explicit, objective, and rigorous, we frequently uncover prior beliefs or assumptions. If one such assumption turns out to be avoidable, the objectification process encourages us to eliminate all appeal to it, however important it was in the initial creative process. If, however, the assumption turns out to be essential to the science, it is treated as an explicit assumption, analogous to an axiom in a formalised system of mathematics or logic. If possible, it is derived from earlier objective science, or established by empirical investigation. If that cannot be done, it continues as an explicit assumption, perhaps to be justified by the results of the reasoning based on it. In contrast to positivist dogma, it turns out that attempts at formalisation of scientific discourse typically require many such assumptions.

In the present discussion of science as 'uncertain knowledge', it must be appreciated that the objectification process has not run its course; indeed, it never catches up with the changing frontiers of science. Frameworks on such a vision of science are a mixture of formalised assumption and informal features, only some of which have been identified by the practitioner, more turning up with hindsight.

The task of identifying framework presuppositions is not always easy, even in retrospect. For example, consider the ontological presuppositions of science. Is commitment to realism part of the framework of scientific practice? Immense effort has been devoted by philosophers of science in their efforts to show that realism or such (overlapping) alternatives as idealism, conventionalism, and pragmatism are necessary presuppositions. Practitioners of science are inclined to commit themselves for or against realism. They do not do so in any especially regular way, and may change their minds. My own tentative hypothesis is that scientists, finding that philosophers offer them an open choice, develop their informal discourses in ways which could be formalised in more than one way. When pushed to declare a stance for or against realism, they opt for whatever is most congenial to their current situation. For example, they are likely to be realists when ideas are driving research forward, so that it can be claimed that the helpful ideas correspond to reality. They are likely to be pragmatists rather than realists when several incompatible but equally useful verbal representations are available which cannot all correspond to reality. Of course, what scientists *say* about the ontological commitments of their framework and what they *actually* presuppose in their research may be different. The philosophical questions of the ontology of science remain open.

Implicit personal or local group presuppositions are often made explicit if they are noticed by people in other places who are interested in the knowledge claim. The normal trend is for statements of scientific fact to lose locally held framework features. Scientific facts are reconstructed to build only on widespread framework assumptions, such as those diffused through institutional traditions or more generally through whole cultures. We can see that such facts still depend on framework assumptions by looking at those extreme individuals and groups who try to set up alternative frameworks which reject the orthodoxy. Such extreme outsiders are labelled as revolutionaries (if successful) or cranks (if not). (See, for example, the story of Velikovsky* in 3.3, below. Velikovsky rejected the general assumption of uniformity in astronomy and geology over the timescale of recorded human history.)

In order to develop the discussion of frameworks in science we should divide the object of study into manageable constituents. A valuable way to do this is in terms of the now familiar hierarchy of levels of cognitive activity. Let us consider the framework concept as it applies at each level.

The framework of sensorimotor cognition

Perception is the least explicit and most context-dependent of all the levels of knowledge. Our sense-organs and the associated parts of the brain evolved to meet a specific repertoire of challenges. Whatever we learn from experience becomes part of the context of new perceptions. The structures of thought we develop to anticipate new situations affect how we perceive them. In general, the context of perception is shaped by innate factors, by immediate context, by past experience, by social constraints, by our general culture, and by our theories. The top-down processing theory of perception shows how such factors can make a difference.

The attempt to eliminate such framework factors from perception encouraged philosophers to talk of raw sensation, of sense-data, or of immediate appearances. However, the certainty of a sense-datum, such as that I see a white patch, is in its closeness to the verbal statement it legitimates rather than in its attachment to the outcome of the perceptual process. Furthermore, context-free perception statements are no basis for science. In general, a perception, like seeing an instrumental reading of '167.39', gains its significance in the context of understanding the instrument used in context.

The framework of individual thought

When an individual reflects upon an issue, the mechanisms of perception are of no concern, unless tension between the content of the perception and the context leads to doubt about how it was generated. Normally, by then, the sensory processes have become enclosed in the 'black box' of perceptual facts.

In science, as in other human activities, when we look at people some distance from ourselves, we notice how external factors are working on them. When we look at a situation from the inside, the framework factors virtually disappear, and we focus upon the explicit problem and its desired solution.

When we think, we can only manage to hold a certain amount in our immediate consciousness at any moment. There are ways around this – by using headings which refer to remembered details, and often also external cues that can prompt memory when it is appropriate, such as writing on paper. The details we momentarily put aside become part of the framework of thought at that moment. They may be picked up and reflected upon later. If our thoughts are to be reliable, it is often helpful to work through tasks with some kind of system or structure. We rely on memory of what worked before, on sensory cues immediately accessible to us, and on idealised conceptualisations of the problem. The assumptions all of these carry are part of the framework of individual thought.

An especially important way for an individual to simplify a task by hiding some of the complexity is to reduce a problem to a clear and coherent conceptual representation, and then think in terms of that. Although the effectiveness of such a strategy is obvious (indeed science and many other intellectual activities crucially depend on it), it has limitations. To think in terms of a particular conceptualisation of the problem actually makes it harder to look at the problem in other ways. Perhaps there were other conceptualisations we could have chosen; as we work at developing the preferred one, the others become more difficult to gain access to. Perhaps we preferred one framework because of its links to our wider external interests. Thoughts about framework choice are not readily conducted within a framework once we have begun to structure our thought in its terms. Perhaps the conceptual orderliness of our preferred way of thinking was initially difficult to impose upon the problem domain. As we succeed in coping with this problem, we become locked into a way of thinking imposed by our concepts. Our pattern of thought imprisons us in its own framework.

There is always an escape from imprisonment in the present coherent framework of our concepts, though it may only lead to some other prison. The easiest way to discover the limits to our own thought is to interact with others. Science also gives special importance to the method of seeking perceptual surprises. Within our own thought, we may also move to some other conceptual framework. Sometimes, by jumping back and forth between several alternative viewpoints, we can see things that cannot be seen in any one of them. If it seems worth it, we can then construct new or extended frameworks to express the new insights. Another technique to liberate our thought is to construct conceptual frameworks with wide horizons of possibility going far beyond what we hold to be true. We are then better able to envisage new possibilities of truth, although we must then maintain a clear distinction between possibility and truth lest we introduce new errors about what the truth is. Another technique to enlarge the prison of the frame of personal thought is to develop not one but several ways of thinking about our system of thought. There is as much tyranny in having only one form of meta-discourse as there is in having only one frame of discourse.

A conceptual prison that is well suited to our purposes only looks like a prison from the outside. The framework of an individual's thought is not so much a rigid prison as an inertial system, holding him or her back when (s)he tries to escape too quickly from the limits of past thoughts.

In science, as in the rest of life, no man is an island. A vast amount of effort in the social sciences is devoted to displaying the ways in which our every action is affected by our social situation. Even though social factors do not completely determine our actions, they are undoubtedly there, a part of the framework in which we think and act. Even when we choose to rebel against orthodoxy, we cannot help taking social guidance, though we choose different authorities. The ways adolescents find to shock their parents are boringly familiar to their grandparents. With greater insight into an individual's situation more possibilities for autonomous individual action become visible. These possibilities are themselves constructed out of the resource kit society provides.

The group framework

When a group of scientists engage in cognitive activity, attention is concentrated upon language-based discussion, social interaction and collective action. Many details of perception and individual reflective judgement disappear from view, whilst they are not thought to be problematic. Many features of the wider context are also taken for granted by the group.

The cohesive social groups of science are likely to construct frameworks which are fuller than those of an isolated individual, and so harder to escape from. Frameworks which may be implausible to outsiders become self-evident to group members. Cognitive conflicts can arise between rival groups because the commitment each gives to its own conceptual framework does not allow for alternatives. In science, such conflicts, whether between individuals or groups, are conducted at the higher level of disciplines.

The levels of the group and the institution were blurred together in T. S. Kuhn's idea of a paradigm-sharing scientific community. Kuhn (1970a; 1970b; 1977) argued for the distinctive nature of a community of researchers who communicate primarily with one another, sharing training, research skills, procedural assumptions, and standards. Kuhn elaborated some of the features of community frameworks in terms of his concept of 'paradigm'. Kuhn identified a very large number of presuppositions shared by members of paradigm-sharing communities which may not be held by other people. His account emphasised the self-confident dogmatism of such a community, which no longer devotes critical attention to its procedural assumptions, getting on instead with the task of putting them to work. In Kuhn's account, the willingness of insiders to stop questioning features of the disciplinary framework and get on with agreed puzzle-solving activities became an identifying feature of normal science. The framework features that Kuhn identified can be divided between those which involve direct personal interaction, as in such cohesive groups as academic schools and interdisciplinary teams, and the more impersonal features which are shared by dispersed communities, which have a high level of internal communication and share institutional values and practices.

Institutional frameworks

Disciplinary frameworks include procedures for handling cognitive conflicts. They standardise frameworks to facilitate collective discourse between individuals and/or groups. As a result, institutions are often not ready to consider viewpoints formulated in terms of other frames of discourse – those of other disciplines or of outsiders.

It is useful to explore the tension between two metaphors for the institutional framework of science. One is the comparison with a literary genre. In any form of literature, a kind of bargain emerges between authors, publishers, and readers, of what can be taken for granted in a literary production. The cover, the title, and the first line of a story are usually enough to mark out the genre to which it belongs, and the reader can settle down in the expectation that the culprit will be found in a detective story, or that love will conquer all in a true romance story. Occasionally that expectation will be toyed with, as when writers try to stretch the creative limits of the genre. How that may be done is itself a genre feature. The institutionalised constraints on individual sciences are like genre rules. They too are dynamic conventions which can be modified with good reason, but which persist as long as they are unchanged by effective challenge.

The second metaphor is that science is like an organised game. Every such game has rules. It may even have an authority structure to reconsider the rules. The players operate more or less within the rules. In science, the winning career strategy is repeatedly to present new findings as knowledge and to have one's claims accepted and rewarded. As in all such games of repeated plays, an individual's reputation of being successful makes winning easier, and a little harder for opponents. While science grows in society, the game of science is not a zero-sum game, for the rewards for those who win by creating new knowledge are not entirely at the expense of losers.

In the metaphor of science as a game, it becomes plausible to develop an analogy in which the lower levels of cognitive activity are playing the game with the institutional rules as framework, and the higher levels involve setting the institutional rules under which the game should be played. For example, the individual graduate student might see him or her self as playing a game in which gaining a Ph.D. is the prize. The graduate studies committee of his or her university might see itself as setting the rules for such a game. Decision-making bodies for science, such as those which control research grants, sometimes present themselves in terms of such a rhetoric. They have their own forms of rationality, their own forms of evidence, and their own framework assumptions. The power to change events that is carried in such internal managerial politics may sometimes be considerable and at other times a mere illusion. It is a noticeable part of the total activity of science.

These two metaphors suggest that the institutional framework for scientific activity is at once a set of implicit genre rules and the rules of the game of science. The two metaphors are each helpful, and diverge most clearly when we think about the significance of breaking or changing the rules of science. In literature, there are standards of behaviour in breaking the genre rules. If a book seems the better for having broken a rule, a change in the rule can result. In a game, changing rules is an activity with its own rules, which are kept clearly separate from actually playing the game.

The framework of human knowledge

The positivists' image of objective scientific knowledge was constructed around the idea that no arbitrary assumptions or values are present in positive science. In recent decades, historians and philosophers of science have shown that we cannot make science independent of non-science. Even the kind of metaphysics that positivists liked least can turn up in framework presuppositions (cf., Burtt, 1932; Bartley, 1968). In this book it is suggested that presuppositions, once identified, are regarded with suspicion in science. Sometimes it can be demonstrated that the alternatives to the presupposition are not viable. If, however, a recently identified presupposition cannot be given adequate support, it generally comes to be seen as non-scientific and the related science is reconstructed so as not to presuppose it. So science tends towards being based upon factors we have yet to articulate, or to which we can imagine no alternatives. This is not a stable equilibrium state, as creative processes may open up new horizons of possibility. A real or imagined extraterrestrial visitor, for example, may show us new forms of arbitrariness in some of our cognitive assumptions.

Presuppositions of all knowledge

A traditional concern of philosophy, especially in the Kantian tradition, has been to identify necessary presuppositions for all knowledge, and for all empirical knowledge. Kant denied the possibility of knowledge of *noumena*, things in themselves, and constructed an elaborate scheme of the categories of knowledge of *phenomena*, things as they appear. I do not think that the Kantian enterprise adapts well to subsequent developments in science. Kant had, for example, thought of space as necessarily being three-dimensional and having properties which mark it as Euclidean, and had a fairly restricted conception of causal relationships. Non-Euclidean geometries were soon discovered. General Relativity Theory attributes a non-Euclidean form to physical space. In these changes, concepts of space have increasingly been informed by developments in mathematics. Some modern theories seek to unify physics within ten or eleven dimensional space.

Just as we have moved away from the Kantian conception of space, we have also revised our notions of causality, especially in the wake of quantum theory. We now find possibilities conceivable which Kant explicitly denied. Furthermore, there is no clear indication that the issue will ever settle down. Are current speculations in physics about the possibilities of certain kinds of time travel incoherent nonsense, or a sign that the boundaries of conceptual possibility are forever being extended along with the boundaries of empirical possibility?

Just as there are no absolute facts which stand outside all frameworks, so there is no absolute framework, within which all facts can be fitted and in which they take their most general form. In particular, I offer the challenge that for any feature of a postulated absolute framework for facts, it will be possible, with some ingenuity, to imagine non-human beings operating in a contrary framework in which the postulated absolute feature does not hold.

3.3 Interlude: Velikovsky – science outside the normal framework?

It is part of the general strategy of attempting to cope with excessive complexity by regarding some issues as closed black boxes that should not be reopened. There comes a time in any group of scientists, or of true believers, when it seems more productive to stop arguing with non-believers about fundamental questions, to treat the orthodox answers dogmatically as part of the framework, and to get on with developing the consequences of the specific beliefs held. The more inward-looking the group, the more dogmatic they can be about framework assumptions, and the more outward looking they are, the less dogmatic they can afford to be.

Some individuals and groups aspire for their work to be given the status of scientific knowledge, but reject shared features of relevant disciplinary frameworks. They are likely to find that the scientific establishment responds negatively, as it regards its own interests as under threat. Work which conflicts with dogmatically held framework assumptions can only be given rational consideration by disciplinary insiders at the cost of admitting that there is something arbitrary in the disciplinary assumptions under challenge.

A famous case which illustrates this point was the Velikovsky affair.* Velikovsky had claimed in a series of books, beginning in 1950 (Velikovsky, 1978, originally 1950), that catastrophic interactions between the Earth and other planets could explain some of the more amazing stories of ancient history and legend. A number of social scientists complained a few years later (De Grazia, 1978, originally 1966) that Velikovsky's revolutionary theory was not being considered openly and fairly, as might be expected according to the ideals of science, but had been rejected in an intolerant way. For example, when Velikovsky's first book was about to be published, a boycott was organised against the initial publisher, Macmillan. Attacks on the theory by scientists also appeared in the popular press. (For a history, see De Grazia, 1978.) The reasoning in these attacks appeared to be that Velikovsky (whose formal training and career were in medicine and psychoanalysis, not in the physical sciences) was making claims about recent astronomical and terrestrial catastrophes which were contradicted by scientific knowledge of the stability of the solar system and the uniformity of geological processes over many millions of years. However, this reasoning appealed to the very dogmas that Velikovsky was explicitly denying. One way to explain why the scientists appeared to be so unfair in their peremptory dismissal of Velikovsky was that the uniformitarian dogmas challenged by his theory depended on the successful closure of sustained debates up to the mid nineteenth century. To take Velikovsky seriously would have required the enormous effort of unwinding relevant parts of science by more than a century. Such an action did not seem justified. M. Polanyi (1969), for example, discussed Velikovsky in terms of the tacit knowledge on which the judgements of the scientific community was based. He defended the functional nature of the judgements of scientists on such matters, even when the judgements were not fully backed up by clear reasoning. Those tacit judgements, were, in the present vocabulary, judgements based on implicit framework assumptions.

The Velikovsky affair illustrates how, when individuals challenge disciplinary framework assumptions from outside the discipline, without demonstrating a preferable disciplinary perspective, it is quite easy to reach a point where rationality fails and the individual is abruptly dismissed from serious consideration.

It should be added that, although Velikovsky's theory was constructed outside the framework of orthodox science, it was constructed within a well-established non-scientific framework. Velikovsky's theory gave substance and reality to dramatic details in the myths and legends of many cultures in which the heavenly bodies, with which gods were associated, did some of the things the gods were remembered for doing. Most persuasively of all, Velikovsky provided a way of taking literally some miraculous events in the Old Testament. The Red Sea really did part for Moses and the departing Israelites in the Exodus, and manna really did fall from heaven in the desert. These were effects of the near collision between the Earth and the comet which eventually became Venus. Velikovsky turned out to be building on framework assumptions taken from religion (Dolby 1975).

The religious framework to which Velikovsky's non-scientific readers related his work provided an alternative to the orthodox scientific framework. Similarly, Creation science* builds on the presuppositions Fundamentalist Christians bring to their reading of Genesis in the Old Testament. In other cases, frameworks of unorthodox science derive from special views in politics (e.g. anti-fluoridation*), in the search for health (e.g. Mesmerism*), and in the search for self-understanding (e.g. transcendental meditation*). These views ignore disciplinary frameworks of current scientific orthodoxy. Instead, they are framed by pre-scientific traditions, or by discredited science, or by the pioneering intellectual constructions of individuals who operate outside science. In general, we will see that thinking about framework assumptions is important in relating unorthodox popular science to the mature institutionalised scientific orthodoxy, with its positivistic cult of the fact and its confidence in the paradigms of its specialist communities.

3.4 Facts

A fact is a specific assertion that some statement is true, some state of affairs is the case, or some action has been done. Science as a systematic form of knowledge, is built out of facts, rationally connected.

Empirical science makes use of two kinds of facts, old facts, produced by others and accepted because they are adjudged true (on the authority of the individual or tradition from which they came) and new facts, normally established with instrumentally aided observation, often in the context of an experiment.

Old facts

Pure science has some concern with the replication and application of old facts. Textbook traditions, for example, recycle old material through many intellectual generations. Disciplines outside science are often more concerned with old facts than new. Applied science, for example, *uses* knowledge rather than generating it. In traditions which apply knowledge it is not to be expected that the skill and effort required to check for oneself the observational basis of old facts will be sustained through long chains of transmission. Nevertheless, all scholarly disciplines impose standards in which one is responsible for the integrity of one's sources. The requirement to cite one's source of all old facts also makes it possible for others to make further checks by moving back up the chain. An important aspect of the higher levels of cognitive activity is the establishment of procedures for giving authority to old facts.

The importance of these procedures can be dramatised by seeing how they can go wrong. For example, popular and unorthodox traditions such as astrology* often sustain traditional facts by handing them down through long chains, in which the same thing is repeated without any attempt being made to check it against some source earlier in or outside the chain. Orthodox knowledge is not immune from the tendency for corruptions to cumulate in such chains of repetition. For example, Einbinder (1964) analysed how encyclopedia entries could become corrupted and out of date in successive American editions of the *Encyclopedia Britannica*. The problem of corruption of traditional knowledge is often serious in popular traditions. L. Kusche (1975) in his critical examination of the literature on the Bermuda Triangle* illustrated how apparently significant phenomena were generated out of the normal uncertainties of original news reports in a tradition of popular journalism which imaginatively embroidered its 'factual' material without adequate checking procedures.

New facts

In pure science, the emphasis is on generating new facts. Traditional philosophical conceptions of science assumed that scientific facts report observations which correctly represent local reality. However, reasons have cumulated for doubting the closeness of the connection of factual claims to the world they represent. I will summarise that view in terms of the argument of the first part of this chapter that facts are relative to frameworks.

It has become a commonplace of modern philosophy of science that the kinds of observational facts which science employs are theory-laden. The process of observation has become less direct in modern science. The world that scientists experience directly is the artificial world of written messages and of manufactured instrumentation, hidden within and behind which are small carefully chosen and controlled chunks of the natural processes under study. Less reliance is placed on direct observation of the object of study and more on observing the (what they hope is illusion-free) read-out of instruments. Observation becomes inferential, presupposing, among other things, the theory of the instruments used. Direct observation still plays a role in some sciences, but, in effect, human observers are reduced to an especially flexible and also highly fallible kind of data recorder, mainly useful in preliminary stages of inquiry.

The arguments of N. R. Hanson (1958) and others sought to persuade philosophers of science of the English-speaking world that science is not founded upon raw sensations (pure sense data), but upon holistic construals of situations. What we *see things as* in science is dependent on innate factors, on context, on culture, and on the background theories. As the psychologist R. L. Gregory (1973) argued, perception is analogous to constructing a scientific hypothesis. To a considerable extent it is through our discussions with others that we build the expectations which structure our observations.

In the tradition of scientific empiricism, it appears trivially obvious that scientific observation is a uniquely individual act – each of us is responsible for his or her own sensory judgements. In principle, if a new empirical fact builds on other empirical facts which are not yet universally accepted, the presenter should also take responsibility for the facts which are presupposed. It may even be appropriate personally to make confirmatory observations (or to set a student to doing them). So all the facts used in a work of natural science should have an observational basis for which the author can take responsibility. However, in practice, our sensory knowledge is constructed by individuals drawing upon a complex network of interdependent social interactions. The theory of a hierarchy of cognitive levels suggests that it is not always sensible to make individual proposers completely responsible for new facts. In a chemical laboratory, a researcher may take reagents from labelled bottles and combine them in the way he was taught, as confirmed in a textbook. But the reagents might be mislabelled, or he might have been taught wrongly.

Very often the notion that a single person can be responsible for a fact breaks down. The complex findings of team research may involve collective responsibility. When one fact is closely dependent on others, there are often good reasons for sharing responsibility for the collective outcome. If the research involved in generating a key fact is expensive or difficult, then those who have been given funds and have the skill substitute for those who have not.

Modern sociology of knowledge has developed accounts of the generation and nature of new scientific facts. In such microsociological studies as Latour and Woolgar (1986) and Knorr-Cetina (1981), accounts are given of how scientific facts emerge from laboratory practice. These emphasise that science does not begin with reliable observations, but with tentative and provisional laboratory results which come to be regarded as reliable only as the outcome of an ever-widening chain of negotiations between those whose expert opinions count. Direct replications of results are not often published, if only because results become more readily publishable when they introduce original variations on existing work (Collins, 1992). The negotiation of a new fact is concerned with winning over other individuals and with finding a generally acceptable way to represent the result in the relevant network of understanding.

Negotiations are much easier to bring to a satisfactory outcome if all stages of the production of a fact are demonstrably under professional control. The scientist making the claim is responsible for all that it involves and professional colleagues have the right to challenge any point. Although they may choose to ignore a claimed fact, they take some of the responsibility when they give credit for the work done, repeat its claims in publications, or use it in further research. The negotiation process requires the claimant to establish that (s)he has set up the observational situation appropriately, avoiding known 'pitfalls' (Ravetz, 1971, chapter 3) – a demanding task in experimental science. People must be persuaded that the new observations are not subject to unreported error or illusions – this normally requires the production of instrumentally recorded data which others can check or replicate if they choose. The scientist must argue for the meaning attributed to the data that has been recorded. This requires the display of relevant competence in the processing of concepts in the network of understanding. (S)he must be able to argue persuasively that (s)he is aware of, and has eliminated, rival interpretations, especially those which would enable critics to dismiss the work as fatally flawed.

Microsociologists of scientific knowledge have found that proposers of new scientific facts do not negotiate their claims by addressing the whole of the scientific community simultaneously. Typically, the first stages of negotiation are with immediate colleagues in the same laboratory, and then larger, less friendly, and more competitive groups of fellow researchers based in other similar laboratories are addressed. Finally, more intellectually distant groups are brought in. In these later stages, the people involved have less intimate knowledge of the particular fact-producing process involved; they have to take more on trust, tending to accept the authority of those who have already been won over. They can still be critical, especially if the new result is difficult to harmonise with their own work. In the cases studied, informal communication tends to dominate earlier stages; formal publication is slow, its impact being mainly on the later stages. To a rough approximation, the process of negotiation can be schematised as first convincing oneself, then one's immediate colleagues, then other members of the same specialism, then those scientists of other specialist communities who care, and finally those in the non-scientific world to whom the matter is relevant. If the process is successful, the new scientific fact becomes like a closed black box, which no one bothers to look inside when it is used. The institutional context of normal science reduces the negotiation of facts to a highly structured ritual process.

New scientific facts are, then, the outcome of complex processes in which socially embedded forms of rationality are employed to negotiate new claims of knowledge. Such a view has come to be linked with the conception of knowledge as relative to its social setting. This book will stop short of complete relativism. It will be argued that something can, after all, be made of the claim that scientific facts correspond to reality. The task of evading relativism is difficult, especially since we change our world with the aid of our knowledge. The conclusions reached in chapters 4 and 5 in discussion of the problems of realism versus relativism will undermine the idealised simplicity of the present chapter.

3.5 Summary

Frameworks All cognitive activity is conducted within frameworks of presupposition. It is usually easy enough to consider a framework assumption explicitly, but we cannot pay attention to all of them at once and some assumptions may never attract enough attention to be exposed.

When we engage in cognitive activity at any of the cognitive levels, the other levels are typically part of the framework, implicitly and indirectly affecting our judgements.

Facts Facts are constructed within frameworks. There is no simple way of accessing neutral facts which stand outside all frameworks and correspond directly to the way things are. Statements of fact are, therefore, complex cognitive constructions.

Within its frameworks of assumption, science involves the construction of new facts, using old facts as its key resource. However, unless old facts are used critically, errors cumulate in fact-building traditions. The process of constructing new scientific facts has been studied by microsociologists of knowledge. They argue that it is only after a succession of negotiations that initially tentative and provisional laboratory results can come to be regarded as clear and uncontroversial facts.

4

Rationality, irrationality, and relativism

4.1 Rationality

One of the intellectual skills we have inherited from the ancient Greeks is a way of formalising arguments by laying down explicit rules which ensure that a conclusion reached from true premises is also true. These rules of valid reasoning were elaborated by the ancient Greeks into deductive systems such as Aristotelian syllogistic logic and Euclidean geometry.

The success of this tradition made explicit formalised deductive reasoning into a model for all reasoning. It made argument the primary form of rationality and explicit argument according to universally agreed rules became a skill inculcated by education and applied widely in systematised learning. Our ideals of knowledge came to be shaped around this model of rationality. We should strive to formulate our ideas in statements, beginning with assertions about which there is minimal room for doubt, building on this firm foundation with valid reasoning that preserves truth at every step. Because such reasoning is explicit, it can be publicly scrutinised. Because it is valid, it does not depend on who produced it or where it is produced, but holds for anyone who is prepared to accept the premises.

Aristotelian syllogistic logic (which renders reasoning as triads of statements, in which the conclusion follows from two premises, according to a limited set of rules, e.g. 'All men are mortal' / 'Socrates is a man' / Therefore 'Socrates is mortal') was the main model of deductive reasoning for more than two millennia. However, it was increasingly clear that it was too limited for the grand aims described.

Mathematics was the other ancient model for this dream of fully rational knowledge. The greatest successes of natural science have been in finding ways to describe the world so that mathematical reasoning can be applied and new descriptions reached deductively. There is a rich philosophical tradition, going back to Plato, which has sought to base *all* knowledge in self-evident or a priori truth, of which mathematics was the primary exemplar. One of the opposing views originated by the ancient Greeks was empiricism, according to which experience rather than pure reason is the source of knowledge. In the seventeenth century, René Descartes was inspired by mathematics in his project to construct all knowledge from pure reason within his own consciousness. He hoped to derive all of science in his a priori deduction of knowledge. (His own contributions to natural science used less rigorous methods.) However, the empiricist philosophers of natural science of his time were not persuaded. They insisted that there is no innate knowledge adequate to serve as axioms for knowledge derived from experience. We must build scientific knowledge out of the facts we seek and discover in our experience. Deductive reasoning has a major place in a mathematically formulated empirical science, but empirical science cannot be purely deductive.

Modern logic underwent a radical enrichment with the development of mathematical logic around the beginning of the twentieth century. Logical positivists were among those who seized upon the new developments and dreamed of reconstructing all normally accepted reasoning processes within the powerful resources of mathematical logic, so that their precise logical status would always be clear.

Within a broadly empiricist framework, deductive reasoning alone is insufficient to produce science. The twentieth-century logical positivists hoped to build a fully rational reconstruction of science on statements of immediate experience called 'protocol statements' (either in the mind of a single individual in the phenomenalist versions or in the public domain, available for anyone to check, in the physicalist versions). However, for reasons developed throughout this book, their programme failed.

The logical positivist dream was most dramatically frustrated by the contemporary publication in 1931 of Gödel's famous Incompleteness Theorem, which showed of any axiomatic system adequate for the construction of arithmetic, that if it was consistent it must be incomplete. In any axiomatised formal logic, it is possible to establish the truth of a statement which cannot be derived formally deduced from the axioms. Axiomatised formal logics cannot represent all rationally reached conclusions.

The old arguments between rationalist idealism and empiricism still reverberate around philosophy. The modern variants of idealism do not normally base knowledge on a transcendent world of ideas from which our souls come, as Plato argued. More often, they try to base knowledge on our direct awareness of what is immediately experienced in our own consciousness. However, phenomenology and related movements do not at present provide an adequate foundation for science, even in their own terms.

Although a satisfactory rational account of science cannot be developed on a purely deductive basis, science is full of fragments of explicit formalised deductive or quasi-deductive reasoning. Scientific arguments very often hinge upon persuading others to accept conclusions on the basis of such fragments. Efforts are continually made to systematise the fragments into more coherent wholes. Systematised bodies of formal fragments are well suited to being taught, and so the didactic image of science emphasises explicit semi-formal accounts of large bodies of scientific knowledge. Our image of scientific rationality is centred upon such extensions of formal deductive reasoning. Some of the most admired scientific achievements are cast in the form of systematic bodies of deductive reasoning, in the manner of Isaac Newton's 'Mathematical Principles of Natural Philosophy' (the *Principia*: Newton, 1968, originally 1687).

An important part of the extension of deductive reasoning in empirical science has been the construction of quasi-deductive ways of dealing with incomplete knowledge, for which unaided deduction would reach no conclusion. Philosophers of science introduced the concept of induction, which seeks to give a rational basis to the process of extrapolating from a run of facts to a new case or to a general law. If all the swans ever seen by Europeans before the eighteenth century had been white, surely it was reasonable for them to expect the next swan they saw also to be white. If all the ravens we have ever studied have been black, perhaps all ravens are black. Such inferences are not valid deductions, but might be made deductive by adding further premises, such as that what we have previously experienced is representative of what goes on more generally. Such an added premise might turn out to be wrong, (as turned out to be the case with swans when black swans were discovered), but, while it is assumed valid, deductions can be made. Similarly, scientists and applied mathematicians have developed statistics and probability theory to deal with uncertainty. Again, the valid deductions are based on probabilistic assumptions and the conclusions reached may turn out to be false if those assumptions are themselves wrong or incorrectly applied.

Explicit rationality and informal rationality

The discussion so far in this chapter has been constructed to ease the introduction of a fundamental distinction in the notion of rationality. I wish to contrast *explicit rationality* and *informal rationality*. By explicit rationality I mean reasoning based on the ideal models of valid logical inference and mathematical proof. Even when such reasoning is not fully deductive, it aspires to be. Every effort is made to render each step explicit and open to rational scrutiny. Such reasoning has proved highly effective in science. It has worked best within a rigorous analytical approach that anchors its chains of deductive reasoning in mathematically idealised accounts of the entities and processes that cause appearances together with precise measurements of the succession of observed phenomena. As we try to make such reasoning more rigorous, we construct coherent formal systems, in which all the starting-points for reasoning and the rules for reasoning are contained.

In the ideal limit, we can write text which contains such formal argument, in the sense that we can reconstruct the argument from the sequence and arrangement of symbols together with the rules for generating and interpreting concatenations of symbols. We can also program a computer to model such a system. Then the succession of the computer's electronic states can be made to implement the logical rules of the system so that its causal processes allow it only to arrive at states logically derivable in the system. In the ideal state of the formal text or the computer program, the explicit rationality of science has completely escaped the messy human context of its construction and become an objective entity in its own right in the material world. For many people, scientific rationality is precisely this ideal type of rationality. However, this kind of rationality can never completely represent human reasoning about the world. In particular, it requires a context if it is to be given significance, and in science, the processes by which an explicit system of rationality is attached to the world (or even to another explicit system) are not at present capable of being completely represented by explicit rationality.

By *informal rationality*, I mean rationality in a broader sense, the power of adapting thought or action appropriately to the achievement of some end. When we make a judgement, solve a problem, or perform an action, we may sometimes be said to have done so rationally, even if we have not supplied any explicit valid reasoning in justification. In sequences such as that in which experience leads to thought which leads to action, formal reasoning can only be involved in part of the process, for not all the stages involve explicit statements.

The extended notion of informal rationality is not just about the relationship of statements, but also about how such statements are used in human life. Informal rationality is very often holistic. That is, it takes account of the total situation, not just what has been represented in the current explicit system.

Informal rationality in the broad sense applies to individual thought and action slightly differently to social action and interaction. I will introduce these two applications in terms of individual and social rationality. And I will describe how explicit rationality and informal rationality are tied together in scientific rationality. The general argument offered in the account of individual rationality which follows is that it is more wide-ranging than formal rationality. We can only rely on explicit routines of formal reasoning to a limited extent.

Individual rationality

Individual rationality appears to be a slightly narrower term than thinking or acting reasonably, which in turn appears to be a slightly narrower term than adaptive behaviour. The broadest concept, adaptive behaviour, applies to animals as well as humans. The domestic animal that runs back and forth when blocked from its goal by a wire fence is more likely to find a way round the obstacle than the animal that stays at the closest point to the goal. So we say that its behaviour is adaptive. Its behaviour may only be adaptive because of the peculiar nature of such a fence. Perhaps the animal that stays still is accustomed to barging through obstacles it can see through and will go on to try to tear through the fence or to dig under or jump over it. So we may be mistaken in what behaviour we say is adaptive, given the mechanisms of adaptation available to the animal. (Different animals may have different adaptive behaviour.)

In humans, we are most likely to describe adaptive behaviour as reasonable if we can find supporting reasoning for it, particularly, if we can show that that reasoning was employed by the individual involved. In the limit, we might just say, 'well, it worked, didn't it?' pointing out that the behaviour was in fact adaptive. To describe what is going on as rational (rather than merely reasonable), we appeal to the thinking of the agent involved. The thinking should follow an appropriate systematic pattern, for which a case can be made that it is adaptive. It may well be that Western rationality is not as adaptive as the thinking of people in another culture for the particular circumstances of the other culture. Perhaps Western rationality is not even as adaptive in Western culture as we are taught to believe. Western societies try to generate forms of life which make appropriate the repertoires of explicit reasoning in which people are trained. But the matching of life style and methods of coping is never perfect or complete.

Consider the child's puzzle: 'Which is the heaviest, a pound of bricks or a pound of feathers?' When asked in a context in which a rapid answer is required, the naive respondent will very often say, 'A pound of bricks.' And that is not, I suggest, an unreasonable answer. Of course, it is the wrong answer, as you can see if you process the explicit question fully and correctly. It is only wrong because the respondent is being tricked. The context appears to require a choice in a hurry. The fast comparative method of the naive respondent is based on our normal experience of bricks being heavy and feathers being light. So to answer 'bricks' is the result of well-adapted behaviour that only fails because the respondent has not realised that an attempt is being made to trick him or her. Similarly, it is well established that even sophisticated statisticians do not always spontaneously apply the complexities of statistical inference in their ordinary thinking and can be caught out along with the rest of us when ordinary thinking would lead to the wrong result.

The contrast between rationally defensible adaptive behaviour and explicit reasoning can be highlighted by the functional advantage of a number of psychological states that are common symptoms of psychiatric disorders when they become dysfunctional. Examples are high anxiety levels, depression (and its opposite extreme, euphoria), paranoia, and phobia. In the discussion of these terms which follows, I will attempt to supplement their customary psychiatric meanings by introducing an evolutionary way of thinking about them.

In a desperate situation requiring a response within seconds, it is not rational to sit down and think out all the alternatives in a formal risk analysis. We should take the first answer that occurs to us that survives preliminary scrutiny and set about making it work. In contrast, if we have all the time in the world, it is not rational to jump panic-stricken at the first possibility that offers itself. Natural selection has prepared us to think differently when we are pushed into one of these two states. There is an appropriate formal training for each state, which only works when it is appropriate to the specific new situation. The fighter pilot responds to a known danger with lightning reflexes, but faced with a different kind of danger, his trained reflexes can be non-adaptive. The academic responds to a new problem by drawing richly upon available knowledge. In an urgent crisis, that response only overloads the complexity of the situation. In general, we can make use of our training; we should also appreciate its limits and when we must go outside it.

Suppose that a person is depressed for some reason. The task of seeking a solution to a problem now involves suspicion of easy answers. The subject no longer has the self-confidence to trust an initial judgement. He or she will be reluctant to start and to sustain the task. If, in contrast, someone approaches a task in an elated mood, then, with high self-confidence, (s)he will seize upon every impulse and attempt to work it into the solution. Ideas will be given a try which would never even occur to the less exuberant.

It can be adaptive to respond to challenges in a depressed way. If one is physically ill, it is unwise to take on new challenges. It is in one's interest to have a high threshold to taking any action. Similarly, if the people around are in a subdued state, it may be unwise to go against the trend. (My dog similarly matches my mood when I am ill with a fever.) There may even be an advantage in Seasonal Affective Disorder (SAD) in northern regions in discouraging imprudent behaviour during the long winter. If one is trying to survive on hoarded food until spring, it is not wise to take unneeded risks. When spring comes, and the sun shines, the time for such extreme caution is over.

For most of the time, and for most people, it is reasonable to allow emotional and mood states to affect our judgement. In a world full of problems, the clinically depressed may be more realistic than normal people. In a world full of opportunities, those whose enthusiasm borders on mania may take advantage of chances the rest of us waste. We each move towards these extremes in terms of our assessment of our situation.

Paranoia has its origins, I suggest, in extreme forms of what can also be highly adaptive behaviour. An animal, and especially a vulnerable young animal is easily pushed into behaviour appropriate to the slogan, 'Watch out, there's a predator about!' This is different from the fearful behaviour when the predator has been spotted and there is immediate danger. In humans, the corresponding adaptive slogan may well be, 'Watch out, there's an enemy about!' for the main threat so often comes from our own species. We feel the need to identify and locate the enemy and gain a clear idea of what he or she is doing. This natural and adaptive reaction can become highly developed in certain cultural elaborations, especially in conflict situations, of which spying on the enemy is perhaps the most obvious. It can be appropriate action, and therefore rational, to look for and find enemies who are acting against one's interest. The paranoid individual is being unreasonable if, in ordinary life, other people have no interest in conspiracy. The paranoid delusions are linked to an inappropriate sense of self-importance. A much disliked political leader or a rich and elderly eccentric who has deliberately alienated his or her family would, in contrast, be unreasonable *not* to expect to be conspired against.

There is a problem with the interplay between our defensive mode of nervous circuitry and the positive feedback we give ourselves through the use of restricted modes of language. Emotions typically build up with the repetition of appropriate internal and external cues. A specific kind of language-based thought may sustain a particular mood even after it has stopped being adaptive. If we give ourselves (individually or collectively) paranoid fears they can be very hard to shake off. You can not be sure that other people, no matter how open and friendly they appear, are not just putting on a show of friendliness and actually harbouring ill intentions. Behind every corner, there may be an aggressor hiding, ready to spring out. If a quick look shows no one to be there, that merely shows how well hidden the enemy is. This kind of reasoning is not, however, a universal form of irrationality; it is pathological behaviour only in contexts where other factors give reason to believe it is ill-judged.

When paranoia is judged to be a symptom of a psychiatric condition, there are likely to be additional factors at work. For example, a significant minority of people hear external 'voices' which are not heard by others. If such a person's voices are saying nasty things, or if mention of them makes other people suspicious and potentially hostile, that individual may, appropriately, react in a paranoid manner. While the hearing of voices in itself may sometimes be harmless, the addition of paranoid behaviour makes a diagnosis of schizophrenia even more likely.

As the hearing of voices illustrates, an individual may show heightened suspicion of others which is appropriate to his or her subjective condition but which other people do not easily appreciate. The paranoid subject may fail to take what observers judge to be a balanced account of all relevant considerations, either because there does not seem to be time for slow and careful consideration or because obsession with some problem results in every relevant thought delivering the same paranoid answer. In contrast, the observers may judge that the subject's thought and behaviour is irrational, being poorly integrated and inappropriate for the world as they understand it. The discrepancy between internal and external judgements of the irrationality of the paranoia is, then, as much a function of the thinking of the observers as of those to whom the term is applied.

Similar arguments can be mounted about the adaptive origins of phobias. There appear to be some kinds of things we can rather easily learn to be fearful about, while it is harder to become fearful of others. Although such a phobic response may have had survival value in our animal origins, it is more often maladaptive than adaptive in modern society, where the objects we should be most fearful of take different forms from snakes, spiders, and other easily aroused fears.

Similarly, agoraphobia can be seen as a maladaptive form of development. The young of many species are bred to be fearful about moving away from the conditions in which they are initially raised. They keep close to their mother, or stay in the nest or lair. Developmental and environmental factors eventually result in the loss of this infantile behaviour. In humans, the developmental process takes place in less stereotyped circumstances. If residues of infantile fearfulness remain (for reasons which psychotherapists seek to understand), it is quite possible for an individual to revert to anxious fear of less familiar surroundings.

In the remainder of my discussion of individual rationality, I will introduce an introspective account of my own informal reasoning, in terms of the idea that it seeks an optimised solution, and can function at a range of levels of generality.

Individual rationality cannot be as closely focussed on explicit reasoning procedures as formal rationality. It is more holistic. I find that in much of my own thinking there is something like an optimiser at work. In each situation, the optimiser acts by taking account of all the factors it can cope with, both explicit and implicit, and producing the best answer available. There are formal rational systems which can function as optimisers. My personal optimiser differs from these in that not every factor has to be rendered explicit before account is taken of it.

Suppose I have to solve a practical problem. Sometimes such a task can be solved (optimised) by direct scrutiny. It may, however, require more extended thought. Then my optimiser may draw upon the heuristic problem-solving strategies of my culture. I might first judge that an optimal strategy of solution is to follow through a particular series of steps. I may then separately consider the first step, and optimise the solution to the task it involves. That output is made one of the inputs into the second step, and so on. If at some stage, the optimal way of dealing with the step is to question the output from an earlier step, I go back and reconsider the earlier step in the new context. (Since I can do this, I can also begin by working back from the solution.) If enough steps have to be retraced, then I may have to go back to the problem-solving strategy and reoptimise that in the new context. By such iterative methods, I can often come to a solution to the original problem. In the context of resources provided by explicit formal reasoning, the optimiser's task is simplified, for some intermediate steps may be purely deductive, but its nature is not changed.

The process is well illustrated by a task that has never been performed before, so that there are not enough familiar routines to slot into standard places in a quick and easy manner. Student learning often has to face such challenges. Then the initial conceptualisation of the task may not suffice, and an optimisation process may have to be gone through repeatedly until a satisfactory result is reached. A good student may then go on to try to find the very best solution.

My own optimiser does not reason in explicit sentences. Its processes draw upon symbolic as well as sensory resources. Language is only constructed by a subsidiary optimiser that constructs speech or writing as best it can to suit the output of the main optimiser. I do not know what sentence I am going to utter until I have said it. I can not remember the words I used a minute or so after uttering them, though I can generally produce the same optimised thought in different words. If I take notes from a lecture, I stop using the lecturer's own phrases once I slip more than a few seconds behind what (s)he is saying. My own internal rationality only crudely approximates explicit rationality.

I remember that my personal optimiser sometimes tended to cheat when I was a mathematics student. I had noticed that solutions to certain complex types of mathematical question tended to be very simple. That insight into the pedagogic strategy of the teacher simplified some problems in ways he had not taken into account. I could often guess what the simple answer might be, confirm it, and then work backwards to construct a more acceptable problem solution. I was told off for doing this – I was not learning the intended problem-solving methods. In general, as this anecdote shows, my optimiser functions holistically, being quite good at finding widely disparate material which can aid in the solution. For example, it is very good at picking up material in one mode of discourse and using it metaphorically in another. It can draw upon preverbal perceptual experience and upon wider social issues. I find it hard to make a purely aesthetic judgement about a creative work without being affected by how other people regard it, especially at the early stages of examining it, when I have little to go on.

My optimiser does not have to function holistically. I can focus my attention on a narrowly defined problem and try to deal with it entirely in the terms provided. There is a 'zoom function' on the optimisation process. I zoom in on a topic of thought while studying it intensely, and then find that in a subsequent unhurried moment of relaxation that more broadly defined synthetic insights come to me which had eluded the narrowly focussed concentration.

The zoom function in the optimisation of thought is a feature of others as well as myself. The psychiatric condition of schizophrenia can leave the zoom function drawn right back from the target, so that the problem gets lost in the buzzing confusion of the rest of the individual's private thoughts and sensory experience. Obsessional attention can zoom the optimiser very close into the target so that awareness of the wider context is lost. Our society encourages obsessional behaviour, especially when attention is devoted to socially valued goals. In this respect, our whole culture distorts informal rationality.

I suggest that the rationality of other individuals is at least sometimes like my optimiser. It is capable of holism, it can zoom into narrow issues, it can settle upon trained methods of procedure, or it can see everything in the world as reflected in the topic of study.

One other example of where individual rationality goes far beyond our present models of explicit rationality is worthy of attention at this point. Individual thought can, with practice, do a trick which we cannot yet make computers do, and that is to process more than one logical system or framework of semiformal discourse at once. Those who learn a single framework for processing a particular kind of problem soon find that their thought is locked into that framework. They cannot question its rules, for that would be to question the very structure of their thought. In contrast, in the kind of higher education in which I am employed, great attention is paid to teaching people critical thinking and critical comparison. Students are taught to confront one set of intellectual resources with another that takes a different approach. In social life and social science, we greatly develop our capacity to understand other people's ways of thinking without confusing them with our own. The capacity to do this may have some genetic basis, for it seems to be lacking from birth in the children we call autistic. In fiction, we learn to deal with imaginary events according to narrative rules without confusing them with reality. Our creative thinking makes use of analogy and metaphor, in which we bring an idea from one context to make it work in another. According to Koestler (1964), creativity often comes from using a second, apparently less appropriate, framework to overcome difficulties that are not being solved in an initial framework.

We have, then, very powerful reasoning systems unavailable to present computer systems. One way to look at this skill is to imagine that your thought, like mine, works through an optimiser function. If the optimiser is set to operate within one coherent system of discourse, it will behave like the autistic thinker, locked in a single self-oriented perspective. If the optimiser is allowed to operate while zoomed out further from its target, it can just as easily optimise over two systems of discourse, following the rules of each when that helps, and jumping from system to system when that is more helpful. The optimiser may run into capacity problems in such tasks, and therefore such thinking can be difficult. But it is often worth the effort. For example, the reflexive process of thinking about what one is doing rather than doing it unselfconsciously requires operating simultaneously a base-level discourse and a higher-level discourse designed for talking about the base level. The benefits of reflexive thinking come from optimising such levels.

Even though explicit formal reasoning is widely appreciated to be important in science, the broader processes of individual rationality are no less important. The concept of the optimiser process serves to illustrate some of these wider rational processes.

Such wider rational processes are essential for science. How do scientists decide which is the best available theory when there is more than one available? T. S. Kuhn (1977, chapter 13) discussed this issue in the context of developing his concept of paradigm, as the broad and coherent frame of discourse of a scientific community. Kuhn listed the rational considerations like accuracy, consistency, scope, simplicity, and fruitfulness which might enter our judgement in theory choice. These turn out to vary with the paradigm which frames the analysis. If there are two theories embedded in two paradigms, then there is no single point of reference from which measured judgements can be made. And yet scientists make such choices. In my terms, they are making decisions which require them to optimise for several competing considerations, while simultaneously dealing with two reference frameworks.

The question of what is optimised in such scientific choices is probably more multi-factored than Kuhn allowed for. The scientist has to carry internal representations of the various levels of cognitive activity (as explained in chapter 2). The levels internal to science may diverge from those involving wider society. He or she may have to optimise with respect to the *rapidity* with which decisions must be reached (for complex matters can take a longer time than is available). Simultaneously, consideration has to be given to the *rigour* of the science and the *reliability* of its application. Some attention must be paid to the *richness* in range and detail of the results promised. And it is also important to take into account the *relevance* of the work to making further research fruitful and further significant applications easy to make. When such optimising judgements have to be made in choosing between whole paradigms or rival theoretical frames of discourse, there is no single formulation of explicit rationality that can manage the task, and yet the optimisation processes of individual scientists cope with it.

Finally, in this discussion of individual rationality, I will end with some comments on the relationship of individual rationality to explicit rationality. There is psychological evidence that we think with greater facility and make less mistakes in reasoning about concrete cases than about abstractly formulated problems to which logic can be applied more directly (Atkinson et al. 1990, p. 329.) The relationship of abstract systems of rationality and correct thinking is not simple. We naturally learn to use limited pragmatic rules rather than the most general rules possible. Furthermore, some of the more contrived systems of abstract reasoning are highly artificial, in the sense that though they can be made impeccable in their internal logical workings, they make strained and unnatural assumptions about their subject-matter. Ordinary thinking does not always bother with artificial restrictions. In some learned fields, the subject-matter may be constrained (as in experiments) to fit the artificial limits of the system of formal reasoning. Under these circumstances it can be hard to relate the reasoning employed in that field with what goes on elsewhere.[1]

Social rationality

In introducing informal rationality, I distinguished individual rationality and social rationality. The two blur into one another when individual thought approximates to internalised speech. That is, one trick of thought is to listen and react to one's own statements. This trick may be fundamental to the process of learning language. It also has the feature that in the early stages, before a child has learned to think in a fully concealed manner, others can monitor the process and encourage the child to think in socially acceptable ways. A properly socialised individual can eventually develop ideas conversationally and present arguments to support particular conclusions, whether or not other people are present.

It is this kind of discourse, whether public or internal, which is at the heart of social life and the life of the socialised individual.

Social rationality is, among other things, about how people come to co-ordinate their actions within groups, as by discussion and argument, and how groups come to act in co-ordinated ways, as in the establishment of agreement. In these processes, a key role is played by verbal exchanges. Other factors are also important, for example the social relationship of the language users. The reprimand of a child by an adult would often be a rude insult if said by the child to the adult.

Any form of argument might be said to be reasonable if it accomplishes the required purpose. If an individual has systematic means of winning arguments which rely on social inequalities between speaker and audience, then he might be said to be using a form of social rationality. It would not, however, be the explicit rationality of the Western tradition, for that relies

[1] These remarks apply equally well to examples as diverse as mathematical physics, economics, or the theory of risk assessment. They are espceially well illustrated by paradoxes of probabilistic thinking.

on reasoning which does not depend on implicit features of the specific situation, only on matters which hold universally, together with factors covered explicitly by the initial assumptions of the reasoning.

Group consensus might be obtained in a manner that approximates to explicit rationality by arguing objectively to a conclusion that all must accept, by the very rules of argument. Consensus could also be obtained by some other socially effective means which is adaptive for the group. For example, many groups require everyone to accept the judgement of the leader. That can be quite appropriate, as when there is no time to argue before collective action is required.

We see, then, that social rationality is open to more influences than explicit rationality. It is more holistic, just as individual rationality is. Because of this, social scientists in general and sociologists of knowledge in particular, have devoted a great deal of effort to finding systematic social factors which might shape social rationality. Suppose, for example, that men employ different patterns of thought from women simply because they always occupy different social positions in present patriarchal society. Perhaps the social rationality of women leads them to different conclusions, and groups of women to construct different viewpoints. Perhaps the very idea of objective depersonalised abstract thought is a male construct, abhorrent to women. This kind of argument is offered by writers on feminist epistemology. The implications for feminist science are examined further in chapter 10.

Social rationality will take somewhat different forms, depending on the level of social generality one addresses. Within one's own group, interactions are with other individuals one already knows and partly understands. At higher social levels, an individual is likely to show solidarity with his or her own group and to interact with others in terms of their relationship to groups not one's own.

The concept of the optimiser process can be used to advantage in discussing social rationality as well as individual rationality. A well socialised member of a community will tend to optimise judgements and solutions to particular problems, in terms of the collective interest of the group rather than purely selfishly. The zoom function on such an individual's optimiser is set at the level of the whole community. Many such individuals can reinforce one another in paying more attention to group considerations than to issues which involve single individuals or larger social domains. Collective discussions then optimise output to the requirements of the communal level. In science, there is evidence that social rationality takes such a form at the level of the specialist community. To the extent that individuals optimise their discourse in this way, they tend to suppress idiosyncratic factors if they are not optimal for the community as a whole. In a learned community, for example, creative efforts tend to be focussed upon shared problems, building upon shared resources and using shared methods. Poorly socialised individuals may, however, be more capable of producing intellectually radical work than the well socialised.

The zoom function in the optimisation of individual social action can be set at more than one social level. We can see ourselves as members of a small group, of a nation, of the human species, or even as representatives of life on Earth. At each level a different set of values apply. Similarly in science, all the levels of cognitive activity discussed in chapter 2 play some part in the social rationality of the knowledge system. At the level of groups, the informal social processes just described co-ordinate social rationality. At the level of scientific institutions, formal rules are established to sustain approved forms of social rationality. At such levels, abstract systems of formal reasoning serve as competence indicators within disciplinary practice at least as much as aspirations towards the ideal rigour of explicit rationality.

One form of social rationality at the level of institutions is the procedure of arguing an issue out. While an isolated individual can only optimise a problem solution in terms of factors impinging upon him or her personally, a community can set up rules which allow a greater diversity of factors to emerge. Each participant can offer arguments based on a different perspective. In this public context, individuals can produce their own optimisation of the discussion. In democratic contexts, the further optimisation process can be employed of putting the matter to a vote. In science, cognitive issues are not decided democratically. Those who take part in discussion are nevertheless affected by it in their subsequent actions.

An important ideal of social rationality in science is to make scientific knowledge as rigorous as possible. Modern science seeks the minimisation of error and maximisation of trustworthiness. However, it is interesting to note that science does not typically use the *most* rigorous forms of reasoning possible. (It would seem that mathematical reasoning in physical science is less rigorous than in pure mathematics, reasoning in pure mathematics is less rigorous than logical treatments of the foundations of mathematics, and logical treatments of the foundations of mathematics are always turning out to be less rigorous than the pioneers had thought.) Stopping short of perfect rigour is very sensible. The rigour of the formal system used covers only *part* of the scientific process, other parts remain informal and intuitive.

Rigorous systems of formal reasoning play a functional role in the social rationality of science. Competent use of technical forms of reasoning is a valuable indicator in a disciplinary field of whose work is most worth taking seriously. Just as with failure to gain the appropriate academic qualifications, or to have the right institutional affiliations, or to make copious and appropriate use of technical jargon, it aids in the elimination of work which does not deserve serious attention. This kind of selection at the entry points of the communication system is known in sociology of science as *gatekeeping*.[2] If a contribution is offered to a scientific discipline which fails to use technical reasoning correctly, then it is easy to judge that its authors do not have the wit or the education to do better, so they are judged incompetent. Gatekeeping is especially important in any field in which ideas are in plentiful supply, but *good* ideas are rare, and it is difficult to work out which ideas are good. Under these circumstances, the limited resources are best applied to the most promising candidates. It is a key practical problem to identify the most promising candidates at as early a stage as possible. At the stage of deciding which research to fund, it is a problem of science policy. At the stage of evaluating work that has been done, the main path to serious attention is to get the work presented at important meetings and published in elite journals. At the various steps in this process, the correct use of technical rationality is a valuable competence indicator.

In general, social rationality includes, but goes beyond, the rationality of formal systems. It is context-dependent and is sensitive to social relationships.

Explicit rationality is an idealised intellectual construction from the social rationality of the tradition of Western philosophy. Scientific rationality, which is sometimes seen in this image, is also the rationality of a real social system made up of real individuals. And the success of science has as much to do with the individual and social rationality of scientists as it has to do with rigorous following of the principles of explicit rationality.

Social rationality should be a tool for thought rather than a prison for thought

Finally in this section, I will try to show how we can avoid being ensnared by the coercive appearance of explicit rationality. We can make it a useful but optional tool for our thought rather than a prison for thought.

Explicit rationality is not merely an idealised form of social rationality; it is also possible for an individual to internalise and apply it to his or her own concerns. What gives it its special powers is still subject to philosophical investigation. This power comes in part from the conventional nature of

[2] Merton (1973), p. 521, explains the term, attributing it to K. Lewin.

the principles of reasoning involved. The coercive power may be more than mere convention; it is at least this. In an explicitly formulated argument, we follow through a chain of steps which we have found do not give people opportunity to argue back. The starting-points of the chain are included as premises.

What is the nature of the agreement about the logically valid steps? Rational assent has a voluntary rather than a coercive nature. Suppose you (perversely) wish to defend the view that $2 + 2 = 5$. If, in discussion, it turns out that you are using the symbols in a standard way, then your failure to follow the rules can be represented as the holding of an inconsistent position. You can be said to have broken the rules of deductive reasoning. Not wishing to be inconsistent, you may, rationally, abandon your claim. If, instead, you declare that you are not bothered about being inconsistent, that you do not want to play the rationality game, people are likely to stop reasoning with you. Rationality will have failed, but you will not have been forced to change your mind.

For permanently settled knowledge, ideals of rational rigour have considerable attraction. In this book's conception of science as knowledge in flux, adapting in the longer term to a changing environment, the crystalline perfection of extended structures of formalised rationality is too brittle. Science, like species, must be able to meet changing demands, including the challenge of the totally unexpected. A little vagueness, and the coexistence of alternative approaches can help in an evolutionary epistemology. As an example, consider how one should face the demonstration of inconsistency in a system of science. Clearly, inconsistency is a major rational failing, if only because normal forms of logical inference allow one to deduce any statement at all from premises that include a statement, A, and a statement, B, from which the contradiction of A can be derived. Nevertheless, something very close to inconsistency can be built into a scientific system merely by claiming that the science has not reached the final truth in a particular context, and that A and B are provisional approximations to the truth, each useful in its own as yet imperfectly defined context. Such a move might be useful for the 'instant rationality' of a system which optimises present knowledge. Clearly it has a built in flexibility to cope with unexpected possibilities if A alone or B alone turn out to be well suited to the needs of a new context. We may find it convenient to use several rational discourses as resources and optimise their applications on each occasion.

It is for reasons such as these that explicit rationality is a tool for scientific rationality rather than its definitive form. Science strives for explicit rationality for such social reasons as the maintenance of its authority at the same time as it tries to make itself into the ideal type of explicit rationality. This aim is self-limiting. If scientific investigation is repeatedly done according to a particular ideal, a residue of problems which are not tractable to this approach will cumulate. A balance has to be struck between maximising the explicit rationality of a momentary state of science and giving science the capacity to survive and adapt to the unexpected challenges of the future.

In the rest of this book I will employ the 'composite materials' view of scientific rationality. The brittle fibres of explicit formal reasoning are embedded in an amorphous plastic matrix compounded from individual and social rationality. The composite is stronger and tougher than the brittle fibres or the soft matrix would be separately.

4.2 Interlude: is the search for perpetual motion irrational?

In modern scientific orthodoxy one idea that is regarded as especially foolish is to search for machines which will run forever without an external source of power, as in a perpetual clock, or, better still, provide an inexhaustible source of energy. Nevertheless, the quest for perpetual motion has had continuing fascination for inventors since the Middle Ages.[3] Perhaps, after the development of machines producing usable power from the movement of water or wind, it seemed reasonable to look for a way in which a rotating wheel could power itself. Some of the earliest schemes used a complex wheel, the parts of which were forced to move so that it was always out of balance. As it rotated in order to restore equilibrium, parts would move so that the new position was still out of balance and it would rotate for ever. Almost as old as the overbalancing wheel is the water-wheel which returns some of the water to feed its own mill stream, and the rarer self-blowing windmills. Some important intellectuals of the seventeenth and eighteenth centuries dallied with the idea of perpetual motion (Dircks, 1861, p. xv); by then the idea mainly attracted less literate practical inventors. The rate of patent applications for such devices rose rapidly after the industrial revolution, when the need for new energy sources was widely appreciated. There was a further rise in such patent applications in the second half of the nineteenth century.[4] One might wonder how there could be so many patent applications for devices that we believe could not possibly have worked. Typically, the inventor had a

---

[3] The richest source of historical material on perpetual motion is Dircks (1861, 1870). For an extended recent account, see Ord-Hume (1977).
[4] Dircks (1861) reprints virtually all British patent specifications for perpetual motion up to 1860. He lists 3 patents before the late eighteenth century, then 4 per decade until the 1850s, for which 52 patents are reported. Hering (1924), p. 76, reports a different count by F. Charlesworth of British patents up to 1903. Of over 600 patients, all but 25 were after 1854.

plan, and needed funding in order to build a successful working model. There were, however, stories of successful machines, the secret of which had died with the inventor, which may have motivated others to produce their own devices (Dircks, 1861, introductory essay). However, stories of past successes were commonly rejected by establishment figures as apocryphal and current working models were generally rejected as frauds.

In the absence of unambiguously successful examples of perpetual motion, the scientific orthodoxy increasingly accepted that perpetual motion is impossible. P. de La Hire in 1678 produced an argument for the impossibility of the commonest type of perpetual motion machine, the overbalancing wheel (reprinted in Dircks, 1861, 514). When, in 1775, the Paris Academy of Sciences decided not to consider any proposals for the construction of perpetual motion machines, the only scientific reason given was based on the (now unacceptable) idea that *vis viva*, the Leibnizian force of motion, cannot be created, but only conserved, or destroyed by friction or by being put to use. Their main justification for the ban was that many mechanics were ruining their families by undertaking costly but illusory projects when their talents could have been put to better use (translated in Elkana, 1974, p. 30).

Although there were no general *proofs* of the impossibility of perpetual motion, the orthodox view is made clear by the increasing use of proofs of other matters in science which *presupposed* the impossibility of perpetual motion. The first famous argument of this kind was Simon Stevin's use in 1586 of the idea of a 'wreath of spheres' looped around two inclined planes. His analysis of the forces acting on the spheres assumed that they must be such that the wreath does not tend to revolve perpetually (Dijksterhuis, 1961, Part IV, sections 60-70). I will shortly give an account of how famous arguments in early presentations of each of the first two laws of thermodynamics employed the assumption that perpetual motion of particular kinds was impossible (cf., Zemansky, 1964, 147; Angrist, 1968).

Is the search for perpetual motion irrational then? That judgement has certainly been made. For example, Hering (1924) regarded it as such. Hering quoted a nineteenth-century editor of *Mechanics Magazine* on the idea that the British government might have offered a large reward for the successful invention of such a machine.

No reward has been offered by government; it has done many foolish things, but none so foolish as this. Before our correspondent wastes any more time on his schemes, let him first seat himself on a three legged stool, and try to lift himself by the legs of his stool. If he succeeds in that he may go on – the want of government reward notwithstanding.[5]

[5] *Mechanics Magazine* for January 29, 1848, as quoted by Hering (1924). pp. 73–4

In Hering's view, this cranky view was so popular because the economic benefits of an unlimited source of free power became an idea which infected ardent enthusiasts. As they made the idea central to their very being, they dared not risk shattered illusions, and criticism and ridicule seemed to increase their ardour rather than quench it. Even when it was too expensive to make working models, the dream could feed on plans and half-completed practical projects the failure of which was easily explained away. Hering also pointed to the irrationality of those who put money into such schemes, including fraudulent working models. One working model, for example, was detected as fraudulent by the irregularity of the sound it made, indicating to a trained ear that it was powered by a hidden crank. This turned out to be operated by an ancient wretch imprisoned in a nearby loft (Hering, 1924, pp. 87-8).

A less extreme dismissal of the rationality of the quest for perpetual motion is to see it as having changed its nature in the course of history. Angrist (1968) suggests that there is nothing inherently irrational about perpetual motion, but that science, and, in particular, thermodynamics, has shown its impossibility. Therefore, the early searchers after perpetual motion were not irrational, while later seekers are either irrational or ill informed.

Angrist's view, though widely held by scientists, is not entirely plausible. Early formulations of the ideas underlying the first two laws of thermodynamics *presupposed* perpetual motion to be impossible, rather than proving it. Helmholtz used the non-existence of working perpetual motion machines to support the idea of energy conservation in 1847 (Helmholtz, 1853), but did not consider the reverse possibility: machines which conserved energy and yet ran perpetually. Ideas were soon being devised for perpetual motion machines which did not *create* energy, but perpetually *recycled* it by creating temperature differences which could then be used to power a heat engine.[6] The impossibility of there being machines working in closed systems which reuse the same energy repeatedly was made the basis of arguments for the second law of thermodynamics. Because the second law of thermodynamics came to be statistical in form, physicists in the later nineteenth century entertained themselves with the idea of a minuscule demon, Maxwell's demon, which could open and close a trapdoor between two chambers to let gas molecules through selectively and so build up a heat (or a pressure) difference in the gas in the chambers. This could power a heat engine indefinitely. Arguments over whether or not the demon's actions could be per-

---

[6] For example, a machine was proposed which would use liquid air to liquefy larger quantities of air, by perpetually reusing the heat in atmospheric air (Hering, 1924, pp. 89-90).

formed, allowing perpetual motion, have gone on well into the twentieth century. They relate to the quantum interference and the energy costs in finding the information the demon requires.[7]

If one accepts the two laws of thermodynamics as being without exception, one cannot show that all perpetual motion is impossible, only that motion which violates these two restrictions. It is possible to imagine perpetual motion machines which work on some other principle. It is conceivable, for example, that the universe is infinite and contains infinite stores of potential energy which could be tapped by a self-contained machine interacting with the energy fields within itself. Other scientific possibilities put such limits upon the applicability of the laws of thermodynamics. For example, in the days before the Big Bang theory took firm hold in cosmology, Fred Hoyle and others constructed a rival steady state theory in which matter-energy is continuously being created in an expanding universe. In a sufficiently large closed system, according to this theory, the total matter-energy grows, and this process might be harnessed by new kinds of perpetual motion machine.

What survives from this discussion of Angrist's view of the increasing irrationality of the search for perpetual motion is that although perpetual motion has not been shown to be forbidden by science, it has been shown to be unproductive to search for it in well-explored scientific pathways. Dircks (1861) and Ord-Hume (1977) consider that modern seekers tend to be irrational in their repetition of earlier confusions and errors when they should be learning from them.

A third approach to the irrationality of the search for perpetual motion is to defend the rationality of the pursuit and to agree that many in the past tradition deserve heavy criticism. Smedile (1962) suggested that it is a pity that competent scientists and engineers are not attracted into a quest that would be so rewarding if successful, while the fools who are not so readily discouraged are most unlikely to succeed.

All these approaches have tended to judge harshly those who have sought perpetual motion in a way which fails to appreciate the pitfalls of past failures. However, if our task is to understand the motivation and thinking of an individual in the manner normally employed by an intellectual historian, we may find it more appropriate to reconstruct the reasoning which guided the action of the individual we study. There have, for example, been many occasions through the centuries in which men of science developed ideas which seemed to presuppose the possibility of perpetual motion. For example, Charles Lyell's uniformitarian system of geology in the early nineteenth

---

[7] For historical accounts see Klein (1970); Brillouin (1962); Feyerabend (1966).

century postulated an indefinitely long geological history of the Earth. Charles Darwin's theory of the evolution of species by natural selection built on this idea that the Earth has been inhabited by our animal precursors for hundreds or thousands of millions of years. A famous attack on this view, led by Lord Kelvin, argued that the energy sources available to provide subterranean heat and to power the sun had a limited capacity and were running down. The period of the inhabitability of the Earth could have been a few tens of millions of years at most. Lyell rejected this conclusion, even though he appeared to be invoking some unknown perpetual source of energy. His reasoning, that the problem comes from the limits of our present knowledge, may have seemed implausible to his critics. With hindsight, we do not wish to say his defence was irrational, if only because Lyell turned out to be correct, as was generally admitted with the discovery of the energy of radioactivity at the end of the nineteenth century, and the subsequent appreciation that radioactive decay of uranium and thorium can keep the Earth's interior hot and that nuclear reactions can power the stars for billions of years.

The conclusion I draw in this case is that the search for perpetual motion cannot be demonstrated to be objectively irrational. We can present the actions of a particular individual as rational or irrational depending on the requirements of our evaluations.

4.3 Irrationality

Some of the more romantic and mystical variants of modern culture proclaim their irrationality. I take these to be forms of social rationality which wish to distance themselves from the ideal of explicit rationality. Those who ask you to trust your own emotions, or to seek to enrich experience even if it is dangerous are asking you to conform to a carefully constructed social standard. Informal reason can be asked to serve different values.

In the discussion of this section I will regard deviation from explicit rationality as remaining informally rational when some kind of considered defence can be provided. On this view, irrationality is the abandonment of both explicit and informal rationality.

In terms of the discussion of 4.1, it may not always be reasonable to expect individuals or groups to follow a trained pattern of explicit rationality. For there may always be a defence that the judgements made are being optimised in a way which does not require the high level of objectivity and transferability that explicit rationality achieves so well.

Individuals can, of course, fail to optimise their decisions even from their own point of view of that moment. If an individual later regards a decision as a mistake, then that is what it is, and probably not an example of irrationality. Mistakes are especially likely to occur when an individual optimises for speed of response, as when anxious or in a panic. An individual may optimise decisions in a considered and sustained way which others regard as wrong but the individual does not. This may be a discrepancy between the individual's rationality and social rationality. The individual rationality involved may be quite defensible when it is better understood.

A common move in the discussion of irrationality is to regard it as the failure to make full use of the resources of explicit rationality which are available. I argue in this section that such a view is mistaken. Explicit rationality is a limited tool; in practice the rational forms available are inevitably incomplete. They are as much an indicator of intellectual competence as a way of establishing that conclusions are beyond doubt. Explicit rationality is normally supplemented with particular forms of social rationality. These vary with setting and with time, and it may be quite reasonable, though unsociable, for an individual to refuse to conform to them. In particular, the creative process very often involves going beyond the prevailing social conventions for rationality in order to construct new more optimal ways of collective reasoning.

In general, the attribution of irrationality has a lot to do with one's wish to distance oneself from the person or reasoning process involved. To say of other people that they are acting irrationally is to declare the inappropriateness of their actions to the circumstances and to what they might be expected to know of them. To say of one's own past actions that they were irrational is to indicate that something prevented the full application of sound reasoning.

To say that an agent is being irrational may be part of an attempt to persuade one's listeners that such actions do not optimise the agent's interests as well as the use of trained reasoning procedures. This kind of persuasion is not a conclusive argument, for there is always the counter argument that the social rationality being appealed to is itself imperfect rationality in the context.

I argue that the kind of sympathetic investigation of cognitive activity appropriate to the case studies in this book makes the attribution of irrationality extremely difficult. We are not dealing with people's hasty mistakes but their considered actions, often backed up with supporting argument. Where a pioneer failed to produce adequate supporting reasoning, we will often find a champion prepared to provide it. This reasoning may not gain our approval, and we may judge it to be in error. It will rarely be the case that we are forced into the judgement that it is irrational. All considered attempts to produce knowledge claims are, prima facie, rational. If we also think that not all knowledge claims can be accepted, this will lead us towards the issue of relativism, which will be explored more fully in section 4.5.

The argument that the fair-minded investigator can always find a rational basis for what are sometimes presented as irrational ideas and actions will now be developed through a series of examples which move us away from intellectual orthodoxy. The movement is from modern science to past science, to magical explanation in primitive societies, to magical explanation in Western scientific society, to phenomenological theories of madness.

In the ideal of explicit rationality, logical arguments show reasoning at its best. To the extent that logical principles can be applied to science, this is our ideal standard of rationality. However, since there are no rational methods which produce completely rigorous empirical scientific knowledge, we accept that the best combination of formal and informal reasoning of the great scientists of our day is also to be understood as rational. Our approval may be conditional upon lasting results being obtained, but it is usually unstinting.

The errors of 'great' scientists of the past

What, then, are we to say about the rationality of the 'greatest' scientists of the past? In their more carefully considered ideas and actions, they are likely to have been superior in creativity and persuasive power to many of their modern commentators. Where we still accept their conclusions, they appear as exemplars of scientific rationality. What about where they are now thought to be wrong? Consider, for example, that leading figure of philosophy, mathematics, physics, and physiology, René Descartes. In the early seventeenth century, Descartes formulated seven laws of (elastic) impact, relating the initial and final velocities of two bodies in collision. All but one of his laws are wrong, and the errors seem glaringly obvious to a modern eye. For example, he imagined that a small moving body when in collision with a larger body at rest would not make the larger body move, but would bounce off it. This result was to hold even if the small body hit at high speed, contrary to our childhood experience of marbles and our understanding of the effect of high velocity bullets on their living targets. Although Descartes was soon criticised and his laws improved upon, the counter-intuitive nature of his results was far less obvious in his time. Historians of science have sought to reconstruct Descartes' reasoning so that it can be seen why he was so confident of his results (e.g. Dijksterhuis, 1961, Part IV, sections 194-220, especially 205-8).

Although modern historians do not all agree, the general view is that it is appropriate to see Descartes' treatment of impact as quite reasonable in the context of its time and not as irrational. It required new insight (based, for example, on the idea of the relativity of motion) for his immediate successors to represent his results as being in error.

Azande belief in witchcraft and oracles

Let us next consider the rationality of the witchcraft and magic of the Azande as described by the anthropologist, E. E. Evans-Pritchard (1937) on the basis of his studies in the 1920s. This tribe believed that when misfortune occurs to someone it is because another person who possesses a witchcraft substance bears him or her ill will. Suppose the supports of a granary are eaten away by termites and it falls on someone sitting underneath. The Azande understand that termites have caused the fall of the granary, but invoke witchcraft to explain why it should have fallen on that individual. A Zande cannot identify a witch who is the cause of misfortune simply by looking at him or her. The presence of the witch substance in a person can only be proved directly by a rather messy autopsy after the witch's death. In practice, oracles are used to establish what is going on. The Azande use many oracles in day-to-day decision-making. The most powerful is the 'poison oracle'. Having worked out that someone, Ndoruma, say, might be the source of the witchcraft, a small dose of a carefully prepared substance, *benge*, is administered to a fowl while a question is asked that permits the answer 'yes' or 'no'. 'Was Ndoruma responsible for the granary falling on me?' If the chicken dies, the answer is yes. A check is then made by giving the *benge* to another fowl and asking the question the other way round, so that confirmation requires that this fowl survives. It is possible for this check to fail – if both fowls die, or both survive. This is construed as showing that the *benge* was improperly prepared rather than that the oracle method of decision should be rejected (Winch, 1964).

The entire pattern of magic and witchcraft of the Azande involves many more features than this. Witch doctors and sorcery are also important aspects of the culture, for example. The claim that this elaborate pattern can function as an effective explanatory system was made by Evans-Pritchard in terms of the adequacy of the beliefs in each context to which they are applied. Evans-Pritchard told us that he found it quite straightforward to run his household, making all difficult decisions by consulting an oracle. We should not think, then, that our science provides a successful way of managing our affairs in the world, while Zande magic does not. In both cases, the beliefs are adequate for the survival of the culture in its own natural setting. We can formulate more detailed standards to show up the contrast between Zande beliefs and our own, but they inevitably presuppose features of our own culture. In contrast, the Azande themselves were not in a position to question their practices. Winch (1964) quoted the following remarks by Evans-Pritchard:

Let the reader consider any argument that would utterly demolish all Zande claims for the power of the oracle. If it were translated into Zande modes of thought it would serve to support their entire structure of belief. For their mystical notions are eminently coherent, being interrelated by a network of logical ties, and are so ordered that they never too crudely contradict sensory experience but, instead, experience seems to justify them. (Evans-Pritchard, 1937, p. 319)

In this web of belief every strand depends on every other strand, and a Zande cannot get outside its meshes because this is the only world he knows. The web is not an external structure in which he is enclosed. It is the texture of his thought and he cannot think that his thought is wrong. (Evans-Pritchard, 1937, p. 194)

Although Zande beliefs hold together, it is in a far looser way than in European systems of abstract thought. Apparent contradictions are not noticed because the beliefs function in different situations and are therefore not brought into opposition (p. 202). The rationality of the Azande is not the same as scientific rationality (Bloor, 1991, chapter 7). It is quite easy to criticise Zande beliefs by the standards of modern explicit rationality. Evans-Pritchard (1976, pp. 201-4), for example, notes twenty-two points on why the Azande do not perceive the futility of their views on witchcraft. Although we can criticise their beliefs, we can see the point of saying that in their culture, with their way of life, there is no better way to proceed. Their thinking is not our thinking; it is not appropriate to call it irrational as long as it is adequate for their purposes.

It might seem that even if it is acceptable to understand and appreciate commitment to magic within a closed traditional African society, in our culture there are superior frameworks of belief available, and only the peculiar or the perverse among us would be attracted towards magic. However, it is not clear that we should automatically judge the Western pursuit of magic as irrational. As J. G. Frazer showed (1911-15), magical ways of thinking have been widespread in European culture; there is still a natural plausibility in the assumption of sympathetic magic that it is an effective procedure to work upon something which is similar to or has been in contact with the thing we desire to influence. This kind of thinking has never died out in the backwaters of our society, and thrives in the cultic fringe of Western society. Modern Western groups who practice magic find no difficulty in attracting and keeping recruits, especially from among those who judge that the social orthodoxy has in some way failed them. In their particular situation, magical ways of thinking are not merely plausible, but a powerful ingredient of the fabric of social life (see, for example, Roszak, 1975; Tiryakian, 1972-3; Truzzi, 1972).

R. D. Laing's view of schizophrenia

Perhaps the furthest limits to which models of rationality have been applied is in certain theories of schizophrenia. R. D. Laing in the 1960s and 1970s argued that 'the experience and behaviour that gets labelled schizophrenic is a special strategy that a person invents in order to live in an unliveable situation' (Laing, 1967, p. 95). On Laing's view, schizophrenic behaviour may not be rational, but it is intelligible, an understandable effect of the family situation, especially if the family also tells the child that he is mad. An important feature of Laing's approach was to try to show that the bizarre language of schizophrenics could reflect a different structuring of experience, often expressed in a highly cryptic form. Laing took the side of the patient, claiming that patients may even have insights missed by the sane (e.g. Collier, 1977, chapter 5; Boyers & Orrill, 1972). His work was drawn into a movement critical of psychiatry.*

Even though Laing's championing of the intelligibility of the schizophrenic is becoming a shipwreck of psychiatric history, stranded on the reef of added knowledge of abnormal brain-functioning, the notion survives that rationality can be found wherever you look for it.

Models of irrationality

Just as the apparently irrational can be studied as a rational process, so other people apply models of irrationality to apparently rational processes. There are many accounts of the limits to, and illusions in, our sense of self-control. The proponents of operant conditioning say that we tend superstitiously to repeat what happened by chance to work in the past, so that we build up complex rituals of preparation for the performance of important actions such as competing in sport. Our mental outlook can be abruptly changed by people who deliberately exploit the techniques of brainwashing, as in recruitment into religious sects. We can acquire self-legitimating views such as the paranoid fear of persecution, in which even the lack of evidence of conspiracy can be taken as evidence of how thorough the conspiracy is. We can acquire a subversive set of values, such as those held by a charlatan, who spins a self-serving web of deceit until the risk of being caught in his lies makes it advisable to move on.

All these models have a certain plausibility and some supporting evidence. None seems entirely persuasive to those who try to use them critically in detailed cases, such as in science in which carefully considered judgements are being proclaimed. The irrational models seem to contain mere grains of truth. They work best for the unthinking errors of people who do not have the wit or wisdom to see things more clearly. They are mostly applied to individuals failing to perform optimally in contexts where there is little time for deliberation. If someone is very competent in what they do, it is much harder to make the attribution of irrationality stick. It is much easier to display the irrationality of the casual remarks of an illiterate buffoon than of some clear-headed intellectual who has thought long and hard about an apparently foolish idea. In his defence of the medically unorthodox idea that Vitamin C prevents colds*, the *competent* medical outsider, Linus Pauling (1976), was able to pull to bits the claims of his opponents that early experimental studies had already shown that his claims had no statistical significance. He could show how the conventional use of statistical argument failed to *prove* that he was wrong, and that the balance of the evidence was actually in his favour. It is hard to make the charge of irrationality stick to Pauling's clear and carefully reasoned case for a non-standard viewpoint.

One form of abnormality which is a feature of apparent failures of scientific rationality is that they are very often produced by people with an obsessional concern with some object or issue, perhaps pursuing it at some cost to their social image. However, there is no simple connection between being obsessional and being irrational, as the very best science is also most often done by individuals and organisations which concentrate obsessionally on some specific target of study.

There is a tension between the attribution of a model of irrationality and the opposing temptation to look for rationality as long as there is a chance that it is there. The use of either option suggests that some prior judgement, favourable or unfavourable, has already been made. Rationality and irrationality, it would appear, are rival evaluative labels the use of which is dependent on the cultural context. This leads us into a discussion of relativity.

4.4 Interlude: women's brains

Consider this story: physical anthropologists of the mid-nineteenth century made detailed measurements of the size of human brains. After some argument, it became a commonplace to say that in general there is a clear relationship between intelligence and the volume of the brain. Mammal species, like other animal species, generally show a clear relationship between brain size and body size. The most striking exception is the human species, for which the ratio of brain size to body weight is much larger than the mammalian norm. Human beings are also exceptionally intelligent among mammals. This provides a prima-facie connection between intelligence and brain volume. The anthropologists went further. The superior races of man, they decided, had larger brains than inferior races. A mature adult has a larger brain than a small child or the atrophied one of an elderly person. And a man generally has a larger brain than a woman. The evidence seemed clear, and the conclusion obvious. Let us consider one aspect of this conclusion – that men are more intelligent than women because they have larger brains. How objective was this conclusion? One of the more important investigators was Paul Broca, the leading figure in French physical anthropology. He had personally measured some hundreds of human brains at autopsies in Paris hospitals, and found that the male brains averaged 1325 g, and the female brains 1144 g, the difference being 14 per cent of the male weight. This difference exceeds what would be expected from the greater average physical size of males.

We might ask if the small size of the female brain depends exclusively on the small size of her body, Tiedemann has proposed this explanation. But we must not forget that women are, on average, a little less intelligent than men, a difference which we should not exaggerate but which is, nonetheless, real. We are therefore permitted to suppose that the relatively small size of the female brain depends in part upon her physical inferiority and in part upon her intellectual inferiority.[8]

Modern feminists regard findings like those of Broca as prejudiced pseudoscience. I think that many people would agree that, in mid-nineteenth-century European culture, it was rather easy for the conclusion to be drawn that women were intellectually inferior. The culture was male-dominated, with very few women making significant intellectual achievements. The men who discussed such matters were not aware of the cultural barriers in the way of the more able women who wished to contribute to intellectual issues. Today, we would say that the conclusion that Broca and others reached was not justified by the evidence. Intelligence is not simply related to brain size.

---

[8] P. Broca, as quoted by S. J. Gould (1978, pp. 364–5.).

For example, although brain size is statistically correlated with physical size, there is not much connection between physical size and intelligence among comparable human beings who have had adequate nutrition. Even allowing for the fact that women are physically smaller than men, there is a residual difference in Broca's data. Gould suggests that Broca was wrong in his interpretation because he had omitted to note that the women were, on average, older at death. Possibly more of the women had suffered the degenerative diseases of extreme old age in which the brain tends to shrivel (Gould, 1978, p. 365).

By the twentieth century, Broca's conclusions had come to be rejected by the intellectual orthodoxy, and attention was paid to another hypothesis, that males are biologically more variable than females, intellectually as well as physically, so that we may expect on biological grounds that the most intellectually outstanding individuals are mostly men. This hypothesis, too, had more to do with what men found plausible in their culture, rather than with what followed inexorably from biological fact (Shields, 1982).

The story of women's brains has been interpreted as illustrating how science is inescapably affected by the social context in which it is produced (Shields, 1978). The idea that science is autonomous, protected by its institutional setting to follow its own rational principles is, on this view, mistaken. Every society shapes its knowledge in a manner affected by the specific social circumstances. The line between rationality and irrationality is relative to the culture.

My own view is that the men did not have to be prejudiced to reach their conclusion. It was correct. In terms of the concepts then employed, women *were* intellectually inferior. It was clear for everyone to see. The women of that society were inferior in intellectual accomplishment and this was the primary indicator of intelligence at the time. However, it was not biology that made it so, but the actions of the male-dominated society. It was not a general fact that women were intellectually inferior, but an historical fact about Victorian society. An alternative judgement was possible that Victorian women were not permitted to achieve their potential. In any section of the culture where such a view prevailed, a different judgement could have been made about gender differences in intellect.

4.5 Relativism

The case of women's brains introduces the idea that social factors affect the way attributions of rationality and irrationality are made and also the content of what is asserted as knowledge. It seems to me that the case for the social determination of knowledge is most persuasive for whole societies in which individuals only have access to a limited range of viewpoints. And the case is most interestingly provocative when it is extended to the idea that alternative claims to knowledge match divisions in society. In this section, the general issue will be applied to science in its most plausible area – societies which produce science but put explicit social controls on what is accepted as science. Special attention will be given to totalitarian societies. After a brief discussion of science in the equally closed world of the religious sect, the discussion will turn to claims for relativism in more open societies.

In section 4.3, some attention was given to the world view of the Azande. Evans-Pritchard was quoted as saying that the Azande are imprisoned in the web of their own thought. In this context, it is worth noting the discussion by Horton (1967), which looks at a number of points of similarity and of difference between African traditional thought and Western science. In particular, Horton develops Popper's contrast (in Popper, 1966, originally 1945) between the open predicament and the closed predicament. Horton emphasises that traditional thought is typically closed – unable to consider alternative viewpoints – while Western thought in the tradition of ancient Greece is open – in culturally heterogeneous communities, it is appropriate to recognise that the locally prevailing way of thinking may be wrong and to appreciate the existence of alternatives. On this view, a key difference between traditional African thought and ours is that we have as a cultural resource more powerful ways of representing sympathetically the views of people whose form of life differs from our own.

Perhaps the relativity of knowledge is mostly a feature of the closed predicament. In contrast, science seems to be a product of the open predicament. Although there is considerable appeal in the view that every culture generates something equivalent to science that works satisfactorily in the context of its form of life, science emerged in a distinctive form in Europe. It did not emerge in a form capable of explosive self-sustained growth in ancient China, even though the cumulated learning and the technological level of ancient China was higher than that of medieval Europe. Nor were there close analogues to science in ancient India, for all that sub-continent's cultural accomplishments. Science is best regarded, not as the explanatory understanding of any society, but as a much more specific kind of understanding. This book will investigate in Part II the postulate that there is something distinctive about Western science.

Science and pseudo-science in totalitarian systems

Within European and Europeanised society, there have been a number of occasions in the twentieth century on which the national orthodoxy in a field of science was labelled as pseudo-science by the international main stream of science. Two examples I have especially in mind occurred within totalitarian regimes.[9] They are firstly, Nazi science, especially its racist theory and its rejection of some disciplinary areas as defective because of Jewish involvement; and secondly, science in Stalinist Russia, including psychology and social sciences, and Lysenko's agronomy.* Both these societies feared contamination by foreign political systems, and developed forms of science that were regarded as being more in harmony with local political and practical needs than was the international scientific orthodoxy. Their social optimisations of rationality were within their own culture rather than across all European cultures.

In Nazi Germany, the abstract and theoretical parts of physics, such as Einstein's Theory of General Relativity, were criticised (most notably by the physicists P. Lenard and J. Stark). Nazi science was to keep close to the concrete and practical. Jews and those who supported them and their achievements were obliged to leave the country or to abandon academic life as soon as the Nazis came to power. Psychoanalysis was among the doctrines that suffered. Nazi racism* was the most extreme deviation of Nazi science from the scientific orthodoxy elsewhere.

The Soviet Union and the case of Lysenko At the beginning of the twentieth century, Russian science was modest in scale in comparison with that of nations like Germany, France, or Great Britain. Science was regarded as a liberating factor within social circles of the Russian intelligentsia and was often linked to ideas of social and political reform. There was some state repression of particular ideas, such as those materialistic and anti-religious views which seemed to be most subversive of the existing social order. Then came the revolution and associated disruptions. The new regime was materialistic in its philosophy, so it tended to encourage science, especially those parts of science which had already been invoked as supportive of materialistic movements.

In the decades after the Bolshevik Revolution, the country was striving to remain true to its (changing) interpretation of Marxist principles, to protect itself from the corrupting effect of foreign influences and to transform its agricultural and industrial base. All this was to be done within a system of

---

[9] A third, which I shall not discuss here, is Maoist science in the 1970s

central planning. Scientists tried to anticipate the needs of the state and of its ideology.

Soviet science was able to grow under a number of constraints. The political leaders of the society were intolerant of anything explicitly opposing their ideological stance. Although most natural science was not directly affected by this, the rhetoric of the state ideology was a resource that could be called upon by anyone within the normal cut and thrust of scientific debate. It was always possible for an opponent to argue that some branch of inquiry or of applied science implicitly presupposed ideas which conflicted with the orthodox political line. If the political leaders could be aligned with this view, then the offending branch of science could be suppressed. Most often, the connection of ideology and science was tenuous and the political leaders did not act directly. There was room for dispute. It was up to the scientists themselves to work out what conformed to the official view. In Stalin's time, this was an especially serious matter, as the purges he instigated eliminated people who were considered a potential threat to the regime – for intellectuals and scientists, this happened especially in the late 1930s. Fear made mere accusation a powerful weapon.

In general, the influence of political ideology on science was as much positive as negative – it encouraged scientists to try to develop science supportive of the approved ideological stance or at least to dress their ideas up in suitable form. For example, if one could show that an idea was closely connected with, and supported by, something that had been said by one of the great Marxist leaders – especially Marx, Engels, or Lenin – then that gave it greater prima-facie plausibility. The idea still had to be defended against other forms of criticism.

A second general constraint on the Soviet science of the time was the need for the nation rapidly to implement an immense programme of change in social organization, in the pattern of agriculture, and industrialization. These were urgent practical problems. The state scorned interesting lines of inquiry that might produce a return in twenty or thirty years, preferring to support projects that could give results, if not straight away, then within two or three years.

A third general factor was the xenophobic atmosphere – a distrust of the ideas of foreigners. The science of the West, for example, was often claimed to be contaminated by Western ideology (for example, behaviourism in psychology). Western science needed to be sifted and transformed if it was to be used in the Soviet Union. Soviet arguments were developed, especially in the late 1940s, that science is best understood as a product of the social context in which it emerges. Much of the science of Western Europe and America is

merely capitalist bourgeois science that props up the social order of decadent
Western societies. It should be rejected in the true socialist state of Russia.

These three factors interacted. An approach which claimed to apply to
approved Marxist theory within science, which promised practical results,
and which gave a basis for criticising Western ideas and preferring Russian
ones, was obviously well suited to the context. And that is what Lysenko's
agriculturally oriented biology was. In such a context, a science could grow
rapidly on the basis of its promises as long as it could distract official attention from its lack of positive achievements.

T. D. Lysenko (1898-1976) was a poorly educated agricultural worker who
was able to exploit the desperate need of the country to produce increasing
quantities of grain with a lower input of labour to feed the new industrial
workers taken from the land. Traditional peasant practices were unreliable
and inefficient. However, the remedies promised by the few scientific experts
were likely to take many years and to be expensive to develop. For example,
plant breeders offered a programme over decades of collecting breeding
populations of many potentially useful pure strains of grains, breeding
more productive strains, building up seed stocks of the best, and encouraging
peasants to change to them. The process would be slow, and would be
delayed further by the fact that improved varieties would require new knowledge and techniques in their cultivation, and the peasants did not readily
change their farming practices. Lysenko was one of those who offered alternative methods nearer to the traditional practices of the peasants which
promised immediate practical results. In the context, scientific experts who
rejected wild and improperly tested schemes, without themselves offering an
alternative with quick practical results, were open to the charge of
unconstructive criticism, or 'wrecking'.

Lysenko's own practical projects changed over the years, as earlier proposals did not turn out as he had promised. His schemes began in 1927 with the
'vernalisation' of winter wheat to enable it to be planted in the spring in areas
where 'winter kill' led to loss of winter wheat. His pilot study had shown a
higher productivity from vernalised winter wheat than from spring wheat.
Lysenko went on to propose the 'vernalisation' of spring wheat, which meant
little more than soaking the seeds before planting. He also proposed
'vernalising' potato tubers, in effect a particular way of allowing them to
start sprouting before planting.

In the mid-1930s, Lysenkoist attacks on orthodox genetics built up.
Lysenko did not favour the orthodox practice of inbreeding strains until
they bred true and then looking for the most productive hybrids. He argued
that environmental pressures can change the genetic form of species. He

promised to develop new more productive types of crop seeds in this way from the impure strains bred by peasants. Lysenko was never very clear about his theory of plant genetics, but soon supporters were elaborating a version of Lamarckian theory for him. At this time, Stalin again emphasised the urgency of effective procedures. He offered the slogan, 'practical success in agriculture is the ultimate criterion of truth in biological science' (Joravsky, 1970, p. 95). The timescale on which practical success was to be judged was so short that all that could be compared was the bold promises of opportunists like Lysenko and the cautious promises of orthodox geneticists. In the late 1930s, when the purges ravaged the intelligentsia, Lysenko and his supporters, who tended to be young, poorly educated, anti-intellectual, and in subordinate positions (Joravsky, 1970, p. 126) more often survived the purges than those who remained neutral or opposed Lysenko. The most eminent geneticist of the time, N. I. Vavilov, who was vaguely supportive of Lysenko, was accused of also supporting too many 'wreckers', was arrested in 1940, and died in prison.

During World War II, Lysenko's new practical schemes were listened to, and increasingly put into effect. Opposition from orthodox geneticists was very occasionally heard until 1948, when their views were officially discouraged. Lysenko and his followers were then supreme in the Soviet biological sciences. In the years which followed, Lysenko stayed on top by directing public attention to new schemes and discouraging negative comments about old abandoned ones. In 1950, Stalin began to withdraw from his attempt to direct the course of science from the top. He signalled the withdrawal by denouncing tyranny in science. Science was to be allowed self-government, subordinates were to be allowed the freedom to be creative (Joravsky, 1970, p. 155). By this time, Lysenko and his supporters were the old guard. Criticisms again began to be heard, as for example of the immense wastefulness of Lysenko's scheme to change the climate of large areas of the Soviet Union by planting many trees. The trees were to form shelter belts, and were to be planted in dense clusters on the principle that competitive struggle does not occur within species. The closely spaced trees would not compete with one other, but the weaker plants would sacrifice themselves for the stronger. However, most of the trees that were planted this way soon died.

After Stalin's death in 1953, attacks on Lysenkoism grew. Kruschev, more sympathetic to those with a peasant background than to trained scientists, was initially reluctant to force change. An undercurrent of criticism grew until the official change away from Lysenkoism in 1964.

Let us try now to assess in general terms the extent to which the Russian experience showed that science is or is not culture-dependent. On the one hand we have Soviet arguments that Western science is infested with decadent capitalist and bourgeois ideology. On the other hand we have the knowledge that Soviet attempts to develop a specifically socialist science within a totalitarian system produced some notable failures.

In terms of my discussion in the first part of this chapter, we may see the Soviets as constructing a social rationality which sought to integrate scientific thinking as fully as possible into the thinking of the state as a whole, including its ideology. They were encouraged in this by their conception of science as primarily a resource for solving immediate practical problems. However, the state ideology was not as effective as the main stream of Western science at providing a framework for generating sustainable answers to scientific questions, in particular because Stalinism encouraged judgements in terms of who made the boldest promises.

In spite of the Soviet rhetoric about Western science being merely bourgeois capitalist science, an important feature of Western science was that its rationality was optimised by a new kind of interest group, the scientific intellectuals, who were concerned with the integrity of science, and not (directly) with its ideological value to the capitalist system as a whole.

The lesson that was drawn by apologists for Western science was that good science needs freedom from external interference to sustain its qualities of neutrality and objectivity.

The argument from totalitarianism for the need for science to be autonomous In the 1940s, in reaction to their perception of Nazi and Soviet science, a number of Western commentators developed arguments that science could only flourish properly when allowed to develop autonomously without external interference. When totalitarian societies sought to make science take a desired form, they either inhibited its development, or produced such pseudo-sciences as Nazi racism* and Lysenkoism.*[10] The discussion by R. K. Merton led to his account of science as being based on institutionalised norms, that scientific truth should be universal (universalism), that the findings of scientists should be shared (communism or communalism), that scientists should behave in a disinterested rather than a partisan way (disinterestedness), and that judgement should be based on empirical and logical criteria (organised scepticism). Merton was concerned that Nazi society did not recognise the same science for Aryans and Jews (which is contrary to universalism), did not wish to share knowledge with non-Aryans

---

[10] For example, R. K. Merton, 'Science and the Social Order' (1938) and 'Science and Technology in a Democratic Order' (1942), reprinted in Merton (1973) and elsewhere; M. Polanyi (1951).

(contrary to communism), and wanted scientists to be guided by the state's political dogma on racial matters (contrary to the values of disinterestedness and organised scepticism). In terms of the Mertonian ethos, the social stability of science can be ensured only if adequate defences are set up against changes imposed from outside the scientific fraternity itself (Merton, 1973, p. 259).

Science in non-totalitarian society

The case of women's brains discussed in 4.4, was introduced along with the suggestion that every society shapes scientific knowledge in its own image. On such a view, our own society is just as likely as a totalitarian society to affect what is accepted as scientific knowledge.

The sociology of knowledge takes the view that it is not just closed societies in which knowledge is shaped by social factors. Classical sociology of knowledge argued that most beliefs of a society are causally dependent on the particular social group or class out of which they emerge.[11] Thus, the political ideas of the ruling class are to be understood as part of the ideology which that class uses to justify its dominant place in society, and by which it maintains the compliance of the lower classes. Left wing critics of the establishment may identify such ideological effects when they occur in science.(eg Rose & Rose, 1976) The easiest examples come from the social sciences, for they can most easily be related to sectional interests in wider society.

Intelligence testing and race Consider the controversial example which has long muddied the intellectual waters of our own society: the question of differences in intelligence between races and other major social groups. This discussion has very often been conducted in terms of IQ scores resulting from standardised intelligence tests.* One can pose questions about IQ which seem suitable for empirical scientific investigation. It is easy enough to measure the IQ of two socio-cultural groups and to draw statistical conclusions about the difference. Of course, there will be problems about what the IQ score signifies, and whether it is measuring something which is dependent on the culture of those tested, or on the culture of the testers, especially if they are linked to the eugenics* movement.

We quickly come to realise that any conclusion drawn is loaded with political and moral implications. These are so sensitive that to draw a scien-

[11] The pioneers of sociology of knowledge included Max Scheler, Karl Mannheim, and Werner Stark. E.G. Mannheim (1936). For a modern presentation see Berger and Luckmann (1971). On science and the sociology of knowledge, see Mulkay (1979).

tific inference is also a moral and political act. For example, trying to use the IQ results to argue some thesis about the heritability of IQ scores involves using dubious approximations as assumptions. They are not, however, obviously more tenuous than those which are made in many other areas of social science. In this case, if you say that Americans of African origin do worse on such tests than Americans of European origin, the political implications of inferring that this is due to the different genetic make-up of the Blacks are culturally provocative.

Although hereditarian views of intelligence seemed plausible in the more racist culture of the early twentieth century, since the mid-century, merely to make the inference is to risk being labelled as right-wing and racist. The problem in this case seems to be that drawing such a conclusion is *not* the result of rigorous science, and yet many people do draw it, and are ready to apply it politically. A typical gap in the reasoning is illustrated by the striking way that East European (Jewish) immigrants were getting very low IQ scores in the early twentieth century, but the generation of their children obtained higher average scores than the American norm. (The third generation moved nearer the average scores.) In that case, at least, environmental factors related to immigration and cultural assimilation were at work. The left-wing critic naturally infers that the thinking about the hereditary base of Afro-American IQs was not so much an imaginative leap to open up new questions for research as an attempt to legitimate political prejudice by calling it science. Any further research on this topic is, according to the critic, likely merely to be the development of the same racist ideology.

All this is well known.[12] This case shows that, because of its controversial nature, it would take an incredibly rigorous combination of argument and evidence before the bulk of intellectuals in our society would allow themselves to be convinced of any conclusion on the hereditary basis of racial differences in IQ. In other words, this is a bit of potential knowledge which is not now readily acquired by our culture. It does seem that a culture can more readily accept as knowledge those conclusions which are compatible with its general outlook.

Another example of the way social changes have led to changes in scientific plausibility comes from the secularisation of society and of science. The scientific revolution occurred in the context of deeply and unreservedly Christian thought. Seventeenth-century science was saturated with religious attitudes and assumptions which are no longer universally shared. An important item on the scientific agenda of the time was to reconcile God's Word

---

[12] See, for example, Kamin (1974); Block and Dworkin (1977); Blum (1978); Montagu (1975).

(the Bible) with understanding of God's Work (nature). Modern Creation science* foregrounds that problem. Because it wishes to adjust science in order to sustain a particular way of reading the Bible, it appears as an anachronistic pseudo-science to the evolutionist scientific orthodoxy of the present day.

Science and the sect The discussion so far has looked at societies as a whole, examining how beliefs constructed as science are sustained in the common context. With changes in the make-up of society, knowledge itself will change. The same argument may be applied within our own society to such sub-communities as religious sects, which erect barriers of protective social practices around their own belief system, so that outsiders can not easily give informed critiques and insiders are taught how to protect their beliefs from self-doubt and from external rational challenge. Some religious sects retreat into closed communities; others, however, recruit their members from city dwellers who are exposed to a wide range of alternative beliefs and sustain their sect beliefs by compartmentalising their lives and avoiding situations which would put them under challenge. Occasionally such sects proclaim some of their views as science. Scientology*, for example, even incorporates science into its name. It is not difficult to see how non-standard beliefs can be regarded as scientific within a setting as defensive as that within a sect.[13] For example, many such sects show symptoms of paranoia, in the sense that they represent the social orthodoxy as tending to conspire against them. Very often, the antagonism displayed towards such groups provides clear supporting evidence for an establishment conspiracy. Religious sects are often described as ensnaring vulnerable people as recruits, exposing them to intensive conversion programmes that look like brainwashing, and often moving their new recruits out of the reach of any relatives who object. Those who criticise the sects are often said to be subject to abuse or worse by sect members. Such a hostile view of a sect can appear to its members as persecution. It is rather easy for the sect to respond suspiciously, and for its external relations to be antagonistic and conspiratorial. Paranoia may well be a rational response to the enemies of the sect, rather than a sign of irrationality. Such a response to external criticism makes it easy for members of the sect to discount and deflect conflicting viewpoints.

---

[13] See, for example, Wallis (1976), chapter 8, which applied to Scientology the question of reality maintenance in a deviant belief system.

The relativity of knowledge within the open society We see, then, that different conceptions of reality can be sustained within different sectors of our society, at least for those groups which defend their views in the manner of religious sects. Many people would think that such sects are unrepresentative of modern society. A common view of our society is that it is more 'open' than traditional societies or totalitarian systems or closed sects. For example, we allow our rival forms of belief to confront one another in open competition. This is said to be true of science in particular. However, in the discussion of Velikovsky,* Ralph Juergens (in De Grazia, 1978) gave an account that showed that the scientific orthodoxy had not fairly and dispassionately considered this revolutionary alternative. It would seem that the scientific orthodoxy operating within conventional scientific institutions is open to scientific alternatives only on its own highly restrictive terms. More radical variant views get little respect or consideration.

Within the sociology of scientific knowledge, the view has emerged that, even within modern science, local groups following distinctive forms of local practice will construct their own variants of knowledge out of claims that are successfully negotiated in the local context. On such a relativistic view (e.g. Collins & Cox, 1976), even the observational and experimental 'facts' around which the web of scientific understanding is woven are nothing but social constructions. We are all trapped in the web of our own thought as surely as the Azande. There is no way of demonstrating unambiguously any particular effect of the external natural world, and therefore claims for realism in science are meaningless.

These issues have stimulated much discussion and much research in recent sociology of science. Evidence has been gathered from comparison of science in different cultures, from the effects on science of cultural change over time, from the impact on scientific controversies of conflicts and tensions within the culture, and from the way the practical activity of scientists in the laboratory goes through stages of discussion and negotiation from uncertain experimental readings into firmly accepted scientific facts.

If the reality we construct and sustain is entirely relative to the creative limits of the social system in which it is constructed, then it would be wrong to say that good science corresponds in some approximate way to reality, while bad science has been distorted by external factors. There is no longer any obvious difference between good and bad science. We should try to understand the social construction of belief symmetrically, noting only that some beliefs held to be scientific knowledge are socially orthodox, while others are marginalised as deviant.

The main categories employed by the sociology of knowledge are social. The shared system of understanding that a society identifies as science is to be understood in terms of how individuals interact in that social system. When competing systems of understanding emerge, perhaps to confront one another in controversy, we are to understand their cognitive contrast in terms of differences in their social setting. Case studies of scientific controversies show that social interests often impinge upon the rival cognitive judgements being made (Collins (ed.), 1981; Engelhardt & Caplan (eds.), 1987). It is not rare for the rival parties in a heated moment in a controversy to accuse one another of incompetence or irrationality, and to go on to label opposing doctrines as pseudo-science. It is also not rare in such controversies for participants to identify the social processes that appear to be the cause of the error in their opponents' actions and assertions (e.g. Dolby, 1976). Links between cognitive claims and perceptions of the social mechanisms at work can thus be based on the perceptions of the protagonists. If one side triumphs in such a controversy, its labelling of the other as pseudo-scientific may stick. Those who have been marginalised and discredited either give up the fight, or continue as a form of unorthodoxy. There is something in the idea that a scientific orthodoxy is defined by those who have won all relevant scientific controversies. However, more often, natural scientists try to avoid controversy. They are likely to think that if an issue cannot readily be settled by appeal to mutually agreed facts, then it is not a properly scientific issue at all. So controversial science may not be fully representative of normal scientific knowledge or of the social processes of its production.

The cases discussed in this section have been ones which have attracted the label, 'pseudo-science' from their critics. We should note that, within the sociology of knowledge approach, 'pseudo-science' is a social label used for those who have been excluded by a boundary-construction procedure. The judgement that the beliefs are false or unreasonably arrived at is a social construction (without a privileged status) as much as the beliefs themselves. On the symmetry thesis of the Strong Programme in the Sociology of Knowledge (Bloor, 1991), there should be no fundamental difference between our explanations of how such cases arise and of how the science that is generally accepted is produced.

Let us consider the plausibility of such a claim of symmetry in the difference between orthodox Western genetics of the 1930s and 1940s and the alternative offered by Lysenko. If we say that each society produced the product appropriate to its practical and political needs, it would seem that the term 'pseudo-science' should only be used to describe what was locally rejected. On this view Lysenkoism was pseudo-science in the West and genetics was pseudo-science in the USSR. But the situation was not quite so symmetrical. Among the factors shaping Soviet agricultural science were external political considerations, while the factors shaping Western (agricultural) genetics were more often internal to the institutionalised scientific practice. The nature of the demands in the Soviet system meant that only modest technical training and technological competence were required for those who used political routes to scientific influence. Lysenko could flourish there even though it is unlikely he would have done enough to have been heard of in the Western system. This difference of technical ability might itself be treated symmetrically – as the special skill required to negotiate one's claims in a particular social system. Lysenko may not have understood orthodox genetics or statistical arguments very well, but he was most astute at operating within the Soviet political system. Both Lysenkoist and orthodox genetics were, of course, affected by the success of the practical agriculture which applied their principles. The practical constraint operated on a different timescale in East and West. Stalin insisted on practical results within a year or so and created a political system which did not allow criticism of the past actions of those in power. On a longer timescale of decades, however, Soviet agriculture was perceptibly falling back in comparison with Western agriculture, making increasingly incongruous the claim that the 'science' on which it was based was superior in practical terms to Western science. In addition, Lysenkoism's reliance on external political factors in the deflection of criticism weakened it when the Soviet system changed, especially after the death of Stalin. We may, therefore, conclude that the symmetry of short-term comparisons between Eastern and Western agricultural science should be contrasted with the greater difficulty Lysenkoism had in the long term in sustaining local judgements of its plausibility.

Conclusion

In describing the society in which science is embedded as open, I suggested that the various levels of individual and social activity nested within it are not completely directed from the top. The interests and practices of the constituent parts and levels of such an open society are able to develop away from the dominant interests, and occasionally to pull the equilibrium of society after them. Change can be driven from below as well as from above. In the case of science as a knowledge-producing system, although it is inevitably affected by the society in which it is practised and by the interests of the dominant class within that society, it is also affected in other ways. In the current form of capitalist society, that involves producing economically valued solutions to the problems of a market economy. The practitioners of science also have an interest in making a living from the close study of physical reality, producing knowledge as a cultural adornment for its creators. The rhetoric of objectivity, reliability, and progressiveness are an important part of the scientific interest. Within science, individuals and local groups have further interests in improving their own intellectual standing in the competitive struggle for scientific honour. Scientists may also have purely personal motives for what they do. The entire nested system sustains a diversity of motives and interests, in competition at a number of levels.

The very heterogeneity of science in such an open system makes it more difficult for it to resonate with the rhythm of any particular social pressure. Of course, a local manifestation of science can resonate to the social vibrations of local circumstances, and critics can point this out. However, if the science of other places and other cognitive levels is being shaken differently, the local variant produced there will be the more difficult to assimilate more widely. To become recognised science, a particular accomplishment must be capable of being appreciated by local members of a scientific community spread across a range of (European or Europeanised) societies, some of which may be unlike that in which it emerged.

An approach like Lysenko's, which depended on such local conditions as a particular ideology and a feeling of national chauvinism, could not thrive in other countries. For example, Lysenko's ideas never got established in Eastern European countries other than the USSR after World War II – countries which were under Soviet dominance, but had different practical problems and their own national pride. Their criticisms did not exactly help Lysenko in the USSR. In many cases, local factors are important in the origin of an idea in science. What gets internationally transmitted is not that idea, fully informed by the local issues which inspired it, but a version of the idea which has been reworked elsewhere to suit the needs of other contexts. The transmissible idea is that which survives on its own merits because of its appeal at the levels at which it is transmitted.

In my view, then, what flourishes in the geographically dispersed multi-levelled cognitive system of science is what can survive in many contexts. A society which isolates itself from the international community, as a totalitarian society does, immediately eliminates a source of many good ideas. And then it is only those few ideas which are especially likely to emerge within the constraints of that society which will grow. Thus a society which isolates itself tends to impoverish its science. Similarly, the society which has the greatest diversity of alternative contexts in which a scientific idea can emerge is the one which is the least narrow in the production of science. This is an open society.

This view, however, does not give a complete answer to relativism. After all, the international culture within which local variants of science interact is itself a social context. The knowledge resulting from negotiations in this wider social setting can still be regarded as a social construction. It would appear that there is no escape from social relativism except more social relativism. The way out must surely be that when the diverse social domains of science all confront the same external reality then the science which flourishes is that which is most in tune with it.

In the next chapter I will offer a route from relativism back to a form of realism which allows external reality to play some role in the determination of our ideas, in such a way that the key concessions made to the importance of the social setting may be maintained.

4.6 Summary

Adaptive behaviour is that which is conducive to survival. If performed by a human and given an acceptable retrospective justification, it is *reasonable action*. Any preparatory reasoning for an individual's considered reasonable action is *individual rationality*. Optimising judgements over choices taken in complex situations are good examples of individual rationality.

The reasoning shared by groups of individuals is *social rationality*. It is normally expressed within a common language. Socialised individuals internalise social rationality and use it in the absence of others. *Explicit rationality* is a particular cultural form of social rationality as idealised within the Western philosophical tradition. *Formal deductive reasoning*, as practised in mathematics and logic is a particular ideal type of explicit reasoning.

Although traditional accounts of science emphasise the importance of explicit rationality – usually in the form of deductive reasoning perhaps enhanced by inductive logic or probability theory – it is argued here that science *also* employs less explicit forms of social and individual rationality.

The view taken here makes the concept of *irrationality* difficult to apply to considered actions, especially those which, if subsequently challenged, continue to be defended by the actor in a manner acceptable in the immediate cultural context. As long as a form of social rationality can be found which shows the reasoning to result in adaptive behaviour, it would appear to be rational, in its own context at least. Irrationality becomes an evaluative label, telling one as much about the labeller as the labelled.

A closed group may sustain its own form of rationality and even its own substitute for science by locking out other cultural viewpoints. If we accept that any such group is, in its own terms, rational, we are led to an especially clear form of *relativism*. Extreme relativism is rejected in the chapters which follow.

5

Knowledge and reality

5.1 From relativism through scepticism to a modest realism

At the end of the last chapter, some of the ideas that pull us towards relativism were discussed. Perhaps we are locked in the world of ideas, from
which we can never escape, because whatever we claim wider reality to be,
we can never get beyond the web of symbolic representations to that which
our language seeks to represent. Relativism is a kind of scepticism about
what we can know of objective reality. In its extreme form, ontological
relativism, the claim is that reality is nothing but our shared social construction. In this chapter, a correction to this view is proposed. It is argued that
there is a route from relativism to realism. By concentrating on active rather
than passive experience of the world, the reader will be led to a minimal
realism that is very hard even for sceptics and cynics to dismiss persuasively.
On this basis, bolder realist positions may be constructed, even while conceding the argument that what is conventionally called knowledge is the
outcome of construction processes that are social as well as individual.

The core of the argument of this chapter is that the inclusion of a minimal
conception of reality (linked to experience of our own actions and interactions) makes a great deal of difference to the normal kinds of analysis in
science studies. Most such studies, and especially those after Kuhn (1970a,
originally 1962) clearly separate reality as constructed by the scientific practitioners under study from reality as constructed by the commentator. The
history of science shows the indirectness of the connection between sciences
and their ultimate subject-matter, for every age construes scientific reality
rather differently. It is assumed by modern commentators that to project
the commentator's beliefs about reality on to the scientific actor can be
extremely misleading, even if the commentator is reflecting on his own
actions at an earlier stage in his life. In the history of science, for example,

past scientists cannot be expected to act in terms of what is thought to be the case in the late twentieth century, any more than we can be expected to conduct our research into (say) genetic engineering in terms of the insights into reality of scientists of the twenty-second century. But if we leave reality out of our science studies, and merely comment on what the scientists under study collectively hold to be the case, our discussions tend towards relativism. I hope to show how some of the significance of reality can be brought back into our understanding of the scientific process, while still admitting that what we say about reality is constructed through processes that are subject to change as society changes.

How can we successfully make our concepts correspond to the world? It appears inappropriate in the philosophy of science of the modern age to accept the old magical, religious and mystical answers, which would attribute a symbolic character to reality itself. Modern orthodoxy is that concepts are human constructions and humans have an insignificant place in the material universe. In a secular age, we can no longer presuppose a correspondence between God's Work (Creation) and the symbolic forms in which God's Word is expressed (whether the Bible or science). Nor can we be confident about modern echoes of the Pythagorean idea that the world is constructed out of numbers, or the Platonic idea that the world has a mathematical form. In the present age, we are persuaded that we should not confuse causal relationships of things with symbolic relationships of ideas. Symbols only stand for things. With such a sharp separation of concepts and things and so much room for doubt about the relationship between them, it would seem that we are indeed trapped in our own concepts. Our only hope would be to extend the web of words by matching concepts to intuitions based on pre-linguistic forms of experience. Science is, it seems, vulnerable to scepticism.

One application of the argument developed in this chapter is that this distinction between concepts and things is rather doubtful in everyday life. Everyday knowledge is of a social reality produced by individual action and social interaction, and made out of thoughts turned into deeds.

Questions about our knowledge of reality trouble those with a philosophical inclination. A commonly stated view, which philosophers raise in order to reject, is that the world is as it appears (naive realism). Generations of philosophical argument show that this view is untenable, for the world is not immediately present to the senses, but is inferred indirectly from sensory information in a manner subject to errors of illusion and judgement. An alternative common view in the present age is scientific realism – it is our scientific knowledge which gives us the most reliable account of what the world is really like. Although this view is intuitively obvious to some people, it is implausible to others. Many in our culture claim that science is very far from capturing the way things really are. The scientific world picture is, they say, blind to what it cannot deal with, and many aspects of life are not represented adequately by science. We would do better to seek reality through religion, or some other cultural resource. Another alternative to naive realism is that we can never be sure that we have caught any of reality in our network of understanding. Perhaps reality is something quite different from what is proposed in any theory and the only remaining constraint on our knowledge is our personal efforts to make our thoughts cohere with one another.

A modern way to set up the sceptical problem of the last view is to imagine a living brain severed from its body by a mad scientist. It is stored in a vat of nutrient fluid. The nerves from the brain are not connected to sensory organs but instead to a computer in such a way that everything that the brain experiences is an illusion constructed for that purpose by the computer. How do we know, the sceptic asks, that *we* are not such brains in vats? (e.g. Dennett, 1991).

Much discussion has arisen out of variants of this modern version of the argument by Descartes that we should be prepared to doubt the evidence of our senses, because a demon might be deceiving us. I will not go deeply into such issues. Perhaps, when confronted with such scepticism, we should concede that all we know is the pattern in our perceptions rather than their causes. The pattern of successful perception creatively integrates our immediate experience with what we expect and what we already believe. If we are brains in vats, the mad scientist's computer must sustain the pattern which our past experience leads us to expect, plausibly augmenting it with perceptual surprises.[1]

The computer's task will be enormously more difficult if we start performing (what we think are) actions. Whatever action we choose to perform, the computer must provide the appropriate feedback to our senses to make us think that the world is changed by our actions as it appears to be. Keeping the illusion of reality going for an observer who is active rather than passive poses severe computational challenges to the computer (Dennett, 1991). If the computer anticipates possible actions, it must evaluate an explosion of contingencies. If it responds in 'real time' it is possible to force it to do forbiddingly complex calculations very fast. (The constructors of virtual-rea-

---

[1] Of course, the mad scientist's task will be much easier, if some of the mental checks we can make on the consistency of our esperience are themselves under her control. If there is only part of a brain in the vat, and memories too are actually supplied by the computer, the illusion of consistency will be more readily be maintained.

lity systems have taken on these challenges, and have so far succeeded only in producing grossly oversimplified reality representations in response to simple movements of head and glove.) If, further, we do things such as subtle social interactions, which require very precise and complex connections to be made through the reality which connects the action and consequent perception, the computational task set the computer will escalate still further. One might say that it will need a whole world in which to simulate the reality we believe we are reconstructing in our understanding.

5.2 Immediate reality

The detailed argument is more complex, but let us take up the hint from this discussion that each of us actually does know more through experience than awareness of the immediate contents of our own mind. We know that our minds are locked together with whatever it is outside the mind (and outside the web of our concepts) that links action and perception. I wish to introduce a new technical term here, *immediate reality*, by which I mean the immediate local linkage between our actions and our perceptions of the effects of our actions. Immediate reality is only directly known by its function. The conceptual representation we construct of the causal nexus in which we are embedded *can* leave room for philosophical doubt. We have less room for doubt about what immediate reality *does* to us, for that is directly experienced. Immediate reality can be defined in terms of a single individual, or in terms of the collective actions and interactions of many individuals. It can extend as far across the universe as we can perceive our actions having an effect, but is very far from capturing everything within that range. Immediate reality is restricted to the short timescales on which we act and perceive. On a realist view, to perceive the real world is for it causally to affect our senses, and to act in the real world is causally to affect it by our actions. The notion of immediate reality ties these two concepts together to define that part of external reality in which we have the greatest confidence. Through the perception of the consequences of our actions we rub our ideas against reality. The intimacy of the contact is the greater because when immediate reality is rubbed against our ideas, it too is changed, very often into a more readily conceptualisable form. The term 'immediate reality' applies readily in everyday life to episodes like the treatment of an illness, and in science to the construction and use of physical models, the performance of scientific experiments and the controlled operation of scientific instruments.

The concept of immediate reality is superfluous in a philosophy (or, indeed, in everyday life) that presupposes that our concepts already adequately capture the nature of reality. For then, we can simply *describe* reality. I will loosely refer to conceptual representations of the causal processes involved in our actions and their perceptual consequences as *local reality*.[2] It will sometimes be helpful in the discussion which follows to refer to the scientific descriptions of the underlying processes as local reality, in order to show what the ontological discussion is about. Because local reality is constructed out of concepts, it varies with those concepts from region to region and over time. The historian's local reality will often be different from that of the historical actors (s)he studies.

In a philosophy which is troubled by the kind of scepticism that easily leads to relativism, that is, by the ontological question of what there really is, it is helpful to introduce the new concept of immediate reality as revealing what is most immediate and basic in our knowledge of the external world. It is limited to the functional connection of action and perception, which even the brain in the vat can know.

If we admit immediate reality into the realm of what we can know, a first hesitant step towards realism has been taken. If we are to make immediate reality the basis of scientific realism, a key task of science is to use immediate reality as a window on the world. We should try to situate ourselves and to act in such a way that immediate reality can reveal wider patterns in reality.

For example, when an experimental scientist conducts an experiment, he or she shapes a bit of immediate reality so that it can model wider reality while it simultaneously feeds back the perceptions that theory predicts.

The use of models in science is intimately connected with this idea that in immediate reality we can put our ideas in intimate relationship with the world. Models blur the line between thought and thing. For there are conceptual models of physical processes and physical models of reasoning processes. Since the beginnings of arithmetic, we have constructed physical models which behave in accordance with the conceptual process we are struggling to operate. It was difficult for the ancients to do arithmetic on large numbers with the notations they used, so devices like counting boards and abacuses were favoured. The placing of pebbles on the counting board could be used to check the arithmetical thought just as the thought could check the use of the counting board. The same processes now underlie our use of calculators and computers. Physical systems, electronic systems, and human thought can be made to follow the same pattern of change so that

---

[2] This usage approximately follows Jardine (1991), who uses it to refer to what is real in a community of practitioners of a given discipline at a given time.

we can use any one to model another. We model our thoughts in the world and we model the world in our thoughts.

The difference between this story and an older empiricist view, that we learn of the world by finding regularities in our perceptions, is that the immediate reality we assume to be representative of the wider world is being changed by our actions. This new foundation for empirical knowledge has the advantage that by acting upon our more complex thoughts in immediate reality, we experience them more fully and persuasively. The feedback loop of listening to our own thoughts blurs into the feedback loop of experiencing our own actions conducted in terms of our theories. Along with these advantages there is the disadvantage that the immediate reality might be more responsive to our actions than we realise. It might merely be reflecting our own expectations back at us in a way that does not correspond to wider reality as we think. There are, then, added possibilities of error. Not only may our senses or reasoning be in error, but immediate reality may have adjusted to our actions in ways not taken into account by our concepts. If our error is not purely subjective but the result of unintentionally changing immediate reality, scrupulous checking of perceptions and reasoning may leave us all the more convinced that our misleading impressions are correct. I will apply the concept of immediate reality to practical situations in which our confident attribution of our concepts to the world seems to be leading us into difficulties.

There are some kinds of immediate reality which are especially amenable to reflecting back our own expectations. If it is not appreciated that this is happening, there can be problems for any attempt at producing science. Such an immediate reality changes to fit the local context of culture and expectation, rather than to model the wider material world. Our ideas appear to work, but for the wrong reasons. Examples of such *excessively congenial immediate reality* are easy to find. The human mind examined introspectively, many areas of human interaction, and human society in general, frequently display this kind of fitting in with our expectations. In all of them it is rather too easy to discover empirical support for one's initial ideas. Prescientific culture finds such excessively congenial immediate realities very attractive, and world views are constructed which do not get into difficulties as long as they are only put to challenge within the congenial immediate reality preferred. As a result, highly diverse views of wider reality thrive across human cultures. For example, because our brains are predisposed to find patterns in whatever we study, when this predisposition is applied to itself, in producing theories of the mind, the brain does its best to generate an appropriate pattern. Looking for support for a pet theory of the mind by the introspective study of one's own mind succeeds rather too easily. Recognition of this as a problem was one factor in the behavioural psychologists' rejection of introspective methods (e.g. Watson, 1913). Freud's critics, for example, were inclined to think that if Freud believed that we are driven by sex, that is what he would tend to find in the minds of those whose thinking he influenced. Similarly, if a cognitive scientist thinks our minds process information like computers, that is how he finds himself thinking, as his brain does its best to oblige him.

Especially *un*congenial areas of immediate reality, which shape themselves under our actions so as to defeat our ideas, can also give problems to realist science. This can happen when the immediate reality in which we act is part of some competitive process in a wider ecosystem or in society. Then other entities in the system tend to react to our actions so as to recover any lost advantage. For example, if we develop an insecticide to destroy an undesired type of insect, the insect may well develop immunity. Similarly, if we seek to learn more about a human enemy and to exploit the advantage gained, the enemy may act so as to defeat our efforts. Knowledge is not impossible under such circumstances, but it is certainly more difficult to produce and quickly loses its relevance as reality changes with the effects of the competitive struggle.

Natural science thrives best in areas in which our actions allow or facilitate correspondences between immediate reality and the wider reality we wish to learn about. The idea that we can know the function of immediate reality in linking our actions and perceptions and can use this knowledge as a window on wider reality will prove to be of great value in the rest of this book. However, once we take the idea seriously, new issues arise. These issues, I argue, are of wide significance, and their investigation is of value in advancing our understanding of knowledge more generally. Three classes of problems arise immediately. These are reliability problems, reflexivity problems, and the special complexity of everyday reality.

5.3 Problems of knowledge based upon immediate reality

Reliability problems

The introduction of immediate reality makes it easier to identify new kinds of cognitive error. In addition to perceptual illusions and faulty reasoning, we should now appreciate that immediate reality may fail to correspond to wider reality as we think, if it is especially congenial or uncongenial. Consider, for example, the way the placebo effect shaped medical ideas throughout the history of medicine, especially before its existence was fully appreciated. We are now able to give more sophisticated critical accounts of what was going on in the immediate reality of specific episodes of medical treatment. A doctor who treated a patient according to a theory of the nature of the illness was likely to find his treatment successful. In some cases this might have been for well-understood reasons. The patient may have recovered naturally, or deceit may have been involved, as when the patient was fooled into thinking that the illness was originally present or that recovery had occurred when it had not. But treatments could appear to work, even when such factors were recognised and taken into account. The mere suggestion that a cure will be effective can *make* it effective. This is now known as the placebo effect. Another treatment, based on a contrary theory, would have worked just as well if it also invoked the placebo effect. The patient's immediate reality tended to match itself to the doctor's suggestion. In the history of medicine, medical theories have tended to work well in the immediate reality in which they are created and firmly believed in, and have not worked well enough beyond that immediate reality, in circumstances in which the placebo effect did not operate. Eventually, less direct techniques allowed progress in medicine – the scientific study of anatomy and physiology, and the use of animal models to investigate human illnesses. These indirect methods helped make clear the existence of the placebo effect. Investigative techniques were modified so as to minimise the problem, such as statistical techniques based on blind and double-blind trials.

The errors that immediate reality makes possible can occur anywhere in the hierarchy of levels of cognitive activity. I will discuss each level in turn, with an illustrative account of a theory of dreams at the level of the individual.

At the level of sensorimotor cognition, we are familiar with the kinds of error in which our sensory processing is misled and we experience illusion. We should now add the perceptual error in which immediate reality misleads the senses because it tends to adapt itself to what is expected. Some of the best examples are from the study of ourselves. Consider, for example, individual memory under hypnosis. It is a well-known problem that when hypnosis is used to try to elicit hidden details of an elusive memory (perhaps from a witness to a street crime), the slightest hint by the hypnotist of what is needed tends to be incorporated into the memory. Even the suggestion that there is more to be remembered is obligingly satisfied by the addition of extra detail. Faint residues of the past blur into wishes and hints extracted from the present to produce augmented memories which cannot be distinguished from genuine ones in later recollections. In this case the problem is that the immediate reality of the memory under hypnosis adapts itself to what is expected of it.

When undemanding observation and loose forms of reasoning are combined with a congenial immediate reality, the people involved can persuade themselves of the truth of almost any belief about which a superficially plausible story can be told. Many of the beliefs that flourish in the cultic fringes of society are sustained by an enthusiasm that overcomes any transient doubt about the reliability of its evidential base. My present point is that such extreme ideas as the memory of past lives or of having been kidnapped by extraterrestrials are not always simply a lie or faulty observation and reasoning. They are very often aided by the realisation that the feedback from perception is just as the stories says they should really be. In the cultic fringe of modern society, for example, hypnotism is a powerful tool in the social construction of all sorts of extreme claims, because hypnotic phenomena in general and the hypnotic reworking of memory in particular tend to turn out the way one expects, whatever one expects. If we are to identify and eliminate such problems we must deal with the pitfalls of excessively congenial reality, just as we must take account of perceptual illusions and errors of reasoning.

Personal knowledge which is based on reflection upon repeated experience should be able to correct instantaneous illusions, but may still suffer from systematic mistakes. Among the systematic illusions are those produced by focussing attention upon an immediate reality which increasingly shapes itself to match expectation, as a result of one's actions.

Dreams as an excessively congenial immediate reality I have a speculative (partial) theory of dreams which holds that they constitute a misleading immediate reality of this kind. I presume that dreams have some as yet unclear physiological function in the mammalian nervous system. I presume, too, that their content is due to the amplification of whatever is passing along nervous pathways of the brain which link sensory inputs to the processes of conscious attention. These pathways are not tuned to sensory reality during sleep, but continue to provide contextual significance for whatever stray signals are passing through them, even those which are generated within other parts of the brain.

As they occur to the dreamer, dreams seem to have a life of their own. Those who take dreaming seriously regard them not merely as real in themselves, but as a window on a wider reality. Some see them as containing symbols of repressed desires, others as giving hints of the future, others as allowing a form of action in which the dreamer can affect other people in their sleep. My theory is that because dreams are not completely independent of our waking thoughts, people come to dream in ways which fit their expectations for dreams. At the least, waking memories of dreams selectively highlight the aspects of dreams already regarded as most important. Perhaps also, people are more likely to wake up immediately after a dream that they have reason to find significant so that it is such dreams they most often remember. They may actually come to have more dreams linked to questions which concern them (including worries about the nature of dreams). This could happen if, for example, a preoccupation with a particular conception of dreams was worked into the context the brain provides for the internally generated pseudo-sensory messages of the sleeper. On such a view, dreams form an immediate reality which shapes itself to ideas to which the individual in question has become sensitised. Because this immediate reality was partly produced by such ideas, it tells one much less about wider reality than the dreamer suspects. At a higher level of cognitive activity, if a culture or a subculture shares a particular conception of dreams, collective experience of the dream process will tend to conform to that conception.

Although the collective activity of the group is able to expose and eliminate the idiosyncratic errors of perception and perceptual inference of an individual, the group, too, may misjudge how representative the (interpersonal) immediate reality they study is of wider reality. Groups can, for example, build up a body of confidently held collective insight based on shared experience of congenial immediate reality, which they subsequently find is regarded as illusory by outsiders. Many such examples have occurred in the history of psychical research*. Not all are based on pure perceptual illusion or mistaken reasoning. Sometimes, it is simply because the immediate reality they study is plastic enough to be reshaped by their actions into a form congenial to their wishes. A trivial example would be a group who investigate a fraudulent medium. What the group wants, the medium learns to provide, within his or her capacity for persuasive trickery.

Our institutions seek to make a world in which our ideas work. Sometimes the settled patterns we construct for ourselves can be misleading. For example, given the problems previously noted of the placebo effect in medicine, the institutionalisation of methods of arguing for the efficacy of treatments based on case studies perpetuated errors that were only reduced with the modern statistically based methodologies of blind and double-blind trials. When the use of animal models to test the scientific basis of modern medical treatments was institutionalised (in part for ethical reasons), it was soon realised that the new immediate reality involved was less misleading than direct studies of human patients. However, the immediate reality of medical research on animal models has its own pitfalls. Successful animal tests do not always transfer well to the human case and failed animal tests might not always have failed in the human case.

At the level of common knowledge, society seeks and partially succeeds in making its own immediate reality according to its current needs. A cohesive society can collectively produce a immediate reality that gives an external validity to the ideas prevailing. The harmony between the prevailing ideas and immediate reality can be disrupted if either is forced to change, leaving the impression that the old ideas were illusory. In a less cohesive, more combative society, the struggle to create a immediate reality in harmony with prevailing ideas is inevitably less successful. In such an uncongenial immediate reality, ideas which would have worked well elsewhere are found to fail. I ask myself, for example: can any particular economic idea ever work in a non-totalitarian society as long as there are groups around who are disadvantaged by it so that they have an interest in making it fail?

We see, then, that the concept of immediate reality offers a new kind of certainty, the functional connection between action and perception, but also introduces new kinds of uncertainty in that the connection we produce might have come about for other reasons than those we are conscious of. It is only by systematically looking for failures in our understanding of how our actions lead to the resulting perceptions that we can reduce this source of added unreliability.

On the view offered here, there is no specific form of experience adequate as an autonomous foundation for reliable knowledge.[3] Knowledge requires the integration of all sources, reliable and unreliable, at all cognitive levels. Very often we generalise from knowledge claims which work locally. The main check on such knowledge claims is that they may be sustained satisfactorily in other local contexts. Reliability comes from sensitivity to the possibility of error and immediate response to errors as they are found. Although it would seem that reliability problems are reduced in the long term when short-term local failures have been exposed, we will see that another class of problems, reflexivity problems, are actually increased in the long term.

Reflexivity problems

The human ability to know the world has long included the capacity to represent knowledge of individuals like ourselves. It is a small step within such a cognitive system for it to move from representing the knowledge of

---

[3] In contrast to the view implied by the title of Ziman (1978).

others to representing itself to itself. Reflexivity is a feature of conceptual systems which must include some account of themselves if they are to deal with their subject-matter fully. (For example, the sociology of knowledge is normally thought of as reflexive, as the sociological account of knowledge it provides must apply to its own claims.) Problems can arise from self-reference in logical systems if the knowledge asserted is self-negating (as with 'This statement is false'). Problems can also arise if present cognitive claims can causally affect the future about which the claims are being made. There can be a causal analogue of self-negation, as when social predictions of the 'greenhouse effect' in global heating are believed, leading to large-scale technical measures which prevent the greenhouse effect from occurring. The problem can also arise in interacting systems, the state of each system being logically or causally affected by the representational output of the other. It would be a silly game in which one player guessed how many fingers were kept straight on the other player's concealed hands, if the second player could take into account the first player's guess before allowing his hands to be seen. The problem is not inherently about free will, as the issue can arise in a fully deterministic causal system, as long as it has a representational function which is required to represent its own future state.

Reflexivity often helps us. If we are self-conscious about our knowledge, we can use the resources of our thought to improve our thought. In particular, our knowledge might not be completely trustworthy when we apply it in actions which will change the situation we claim to know. In softer sciences, such as social science, a certain amount of reflexivity is always well advised, because such factors apply. In natural science, it may be advisable to pay attention to our own knowledge when it is new and uncertain, or incomplete, or applied in a new way. Reflexivity is less of an issue in well-established natural science, in which the required precautions and pitfalls are well known and sign-posted. In well-established research patterns of pure science, similarly, scientists tend to be fairly unselfconscious about their knowledge system, at least until their research gets into difficulty. The present book, by calling science 'uncertain knowledge' is emphasising the appropriateness of reflexivity.

I apply the term 'reflexivity' to immediate reality because it tends to be affected by our current cognitive representation of (local) reality. Therefore, a complete specification of the knowledge involved would have to include an account of the effects of itself. Problems arise in any case in which the rest of the system is changed by the process of making or communicating a representation of reality so that the system no longer matches the representation. These problems are least in the very short term, when there has not been time for our new knowledge to make a difference to what we do, and greatest in the long term as we use our present understanding to try to remake the accessible parts of the world.

Immediate reality produces reflexivity problems for knowledge because it facilitates the construction of loops in which the reality about which knowledge is claimed is changed by the knowledge claimed. Traditional conceptions of empirical knowledge concentrated on how we reflect rationally upon sensory information. The notion of objectivity presupposes an ideal state in which the relevant attributes of the object of study are unaffected by the process of studying them. However, my argument is that our knowledge of reality is most informative if we *act* on reality. By our actions, we tend to change the aspect of reality we are trying to understand. The presumption of objective natural science, that our actions should make no difference, can be inappropriate, except in rare favourable circumstances, such as the observation of distant stars. Ideally, any form of cognitive inquiry which changes its immediate object of study, should take into account the effect of its own intervention.

The point is widely appreciated in the social sciences that if a social theory is sufficiently successful to change people's beliefs and actions, then it should include an account of the changes it produces. The longer the timescale on which the social system is considered, the more important this kind of self-modifying feedback loop becomes. It is very often acute in social science, for the timescale on which issues are studied can affect the relevant social action. Political opinion polls (especially of those who vote strategically on the basis of information about how other people will vote) only evade the problem by distinguishing between opinion at the time of the poll and subsequent voting behaviour, and limiting the knowledge claimed to the former.

A weak form of the reflexivity problem is the norm in social science. The knowledge we seek can, in the short term, be kept separate from the human activity under study. But it is always the case that the background understanding, in terms of which the new knowledge is sought, helps shape the subject-matter. In the longer term, research findings will affect this background knowledge.

Reflexivity problems are less acute in natural science (outside quantum theory), but they can occur there too, for example, in contexts in which reality is changed in the long term by human technological action. Long-term predictions (for example, about the ozone layer) have to be worded conditionally in a way which takes into account any effect of our action as a result of the prediction.

Problems of reflexive knowledge are often soluble. If we think about such issues quantitatively, we can see that there are two extremes in feedback loops that change the object according to beliefs about it. It is more serious when there is a major change as a result of the feedback and less serious when the feedback makes only a marginal difference. Knowledge contained within systems that are close to some critical threshold can show a substantial effect from such feedback. A small change in public judgements of the likelihood of panic withdrawals from a bank in the 1930s, could, for example, make a great difference in whether a run on the bank occurs. In contrast, subjectively distorted estimates of error in part of a scientific experiment may not matter at all if they are outweighed by objectively specifiable errors in the rest of the experiment. The presence of a 'correction' factor that favours the experimenters' pet theory only begins to matter as other sources of error are successfully reduced.

Scientific knowledge can reduce reflexivity problems by studying only domains beyond the influence of our actions. For then, only the last stages of the causal processes by which the external world affects our scientific perceptions are open to the kind of interference that is shaped by existing knowledge, as in the design, construction and use of instruments. The behaviour of distant stars is presumed to be oblivious to human concerns. However, immediate reality still *mediates* between ourselves and the wider domain of reality being investigated. In studying a star with a telescope, we go to great trouble to make a locally produced image correspond in some clear way to the star. We have to know how our adjustments of lenses and mirrors are affected by our belief-guided actions in order to know when the image we construct tells us most about the star. The local reality which we *conceptualise* as made up of light and lenses, is the one we *experience* in terms of how our actions change our perceptions. In order to know objectively about stars, we have to know reflexively about how to produce good images in telescopes.

This kind of reflexivity is not normally a problem in natural science; on the contrary, it is an important component of the process of learning about the world.

In science, as in the rest of life, we learn by building upon our successes and avoiding our mistakes. We can learn more effectively if we have a clear idea of what produced the successes and where the errors lie. This requires a degree of reflexive awareness of the learning processes. There should be some kind of internal representation of perception, of individual reflection, of group discussion and of institutional rules that can guide repetition of success and modification of error.

The special complexity of everyday reality

Everyday reality is a special case in which interpersonal immediate reality is continually being made and remade by our actions and interactions. Its complexity has no obvious limit, even in quite simple forms of social life, for there is no limit to the extent to which we can try to make life more difficult for one another, perhaps in order to gain a competitive edge. The problems of reflexivity pervade knowledge of everyday reality. Consider, for example, a game, a war, or another competitive struggle in which two players or teams are opposed and each can benefit from knowing something of the plan of play of the other. Every effort is normally made to ensure the inadequacy of knowledge of one's own intentions by the other side. The very idea of such a competitive struggle presupposes imperfect knowledge. Power and epistemology are connected questions.

Everyday social life is co-operative as well as competitive. We hope to surround ourselves with people we can trust. We often co-operate with some people in order to compete with others. In co-operative groups our ideas succeed more often than they would otherwise. In consequence, the immediate realities we collectively operate are at once both congenial and uncongenial.

Reflexivity is a feature of everyday life. Even when individuals do not appreciate the reflexive features of their situation, the effects of beliefs on actions is, on a longer timescale, changing the situation. Knowledge is one of the constituents of immediate social reality.

Because the complexity of everyday reality comes in part from deliberate human actions, the most obvious tools to cope with it are the resources of human symbolic systems of representation. In everyday reality, we are often not much bothered about what lies outside our network of concepts, for we are operating in a world made by humans. Our problems are in finding out the secrets we do not know and in combining them with what we do know. When we act in terms of partial knowledge, the consequences of our actions are not always as we intend, and further thought and action may be required. The computationally extreme demands of local social reality may well be one of the factors that the human brain evolved its exceptional size to cope with.

In everyday life, most of the situations we find ourselves in and most of the challenges we face are indirectly of our collective making. Many of the limits to our powers arise because what we want is at the expense of others who also try to make immediate reality accord with their wishes. The greatest challenges we face in everyday life are competitive struggles for success and for power. Everyday reality is less congenial here, because one person (or group) gains only at the expense of another. It can be more rewarding, at least for the losers in power struggles, to make life easier by seeking out more congenial forms of immediate reality. Educating one's children is easier than seeking personal fame. Even though our educational theories are very often seriously inadequate, the educational process still works, and we find it rewarding to invest effort in it.

Everyday social reality in comfortable western societies operates largely insulated from the rest of reality. The artificiality of our lives, in which our concerns are mostly directed towards invented goals can continue indefinitely, as long as its material base can be sustained. For example, rather than being what we eat, it matters little what we eat, as long as our food meets minimum nutritional standards. Our worries about food are, within this minimal constraint, largely of our own making.

Knowledge in such a context has to meet rather different challenges from those we set for natural science. Knowledge of matters outside the domain we construct for ourselves can be approximate, or even wrong, as long as the local social consequences of the errors are not serious. Knowledge of matters within the domain of everyday life can also be minimal for those who do not set themselves demanding challenges. Unambitious people can cope well enough with unselfconscious everyday knowledge. But in more extreme circumstances, and for the more ambitious, the requirements of survival can be very demanding indeed.

For these reasons, everyday knowledge of everyday life is not like science, but is a mixture of the simplistic, the simply wrong, and the super-sophisticated. If the insights of the wise and the achievements of the social sciences are to be of much help they must guide us through the complexity of local social reality.

At this point, a brief comment is appropriate on the relationship between the ideas presented here, that the world we study is a world affected by our actions, and two core Marxist doctrines of an active epistemology. Marxism proposes that the best way to understand the world is to change it. Marxism also employs an holistic epistemology. Everything can, and does, change everything else, so that we would best study the world we occupy in terms of its historically changing totality. And we understand the present totality best by changing it.

If we suppose that to some degree everything does indeed affect everything else, it may still be that the best way to make progress in our study of the world is to concentrate on those parts of the world in which we can find relatively isolated phenomena and develop analytical understanding of their pure isolated form. We can use standard analytical techniques when we artificially isolate phenomena in controlled experimental conditions. This technique, of studying the world in small and manageable parts in a well-controlled immediate reality, has been highly successful. If the original holistic ontological hypothesis were true, we would expect that our analytical understanding of phenomena-in-isolation would have to be followed by a synthetic phase of dealing with phenomena-in-interconnection. We would also expect that among the phenomena we were making least progress in understanding would be those which were not amenable to the initial phase of isolation and analytical reduction. I argue in the next chapter that, even in an interconnected world, we can understand natural phenomena best by building our synthetic knowledge of the interconnections of phenomena on the prior successes of isolation and analysis. Marxism appears to deny that such a strategy, which has worked so well in natural science, will also work well in acquiring understanding of the special complexities of everyday life. It seems to me, however, that if we act in a immediate reality which maximises the interconnections of phenomena then we maximise the chance that things happen for reasons which diverge from our understanding of why they should happen. For example, if we build a social world in which solidarity within groups is maximised at the expense of polarisation between groups, then immediate reality will have both excessively congenial and uncongenial features.

Reality for natural science is, in many ways, simpler than social reality. One of the ways in which we have managed to keep science simple is by making the immediate reality of the laboratory experiment match our conceptual structuring of possibility. (Experiments would rarely work if uncooperative rival scientists competed in the same laboratory.) The rhetoric of experimental control I am offering is one of designing a local reality in which we can learn effectively about the limits of our ideas by putting our concepts to work in a world of experiment made representative of wider natural processes. This notion of putting our powers to the test has obvious moral limits if applied in the social reality of everyday life. Words like 'power' and 'control' have an obvious political significance. Seeking knowledge of everyday reality in the manner of natural science has a moral dimension.

5.4 From subjective certainty to scientific rigour

Traditional philosophy has sought to build knowledge on firm foundations, constructed with the critical tools of philosophical analysis. The idea of a foundation for knowledge in immediate consciousness, so important to philosophy since Descartes, has been undermined somewhat by Wittgenstein's argument against the possibility of a private language (Wittgenstein, 1953). And yet we continue to try to make a sharp separation between private thoughts and external reality. In the more holistic view offered here of knowledge constructed by active engagement with the world, it is most plausible to begin with a model of thought in which it is intimately coupled to the external world.

In this section I will apply the ideas already introduced in this chapter to a rather different way of thinking about the construction of knowledge. I will not seek comprehensiveness, but rather offer a new ideal type of thought process, which I will call *considered thought*. The essence of thought of this type is that we keep interacting with some part of the world in a similar way, the associated thoughts being enriched by feedback from the results of previous interactions and by variation in the context and details of each interaction. Introverts go through more loops of thought than extroverts for each cycle of action. Considered thought can give us a strong sense of certainty in our judgements about the world. When we draw upon every resource available to expose and eliminate error, we are able to move from subjective certainty to scientific rigour. I am suggesting that there is no purely rational procedure for the final construction of empirical knowledge; scientific rationality should be regarded as a limiting condition in the process of the formation of (individual and collective) considered thought in which errors have been searched for, identified and eliminated.

We human beings are, on the whole, less clever than some of the cognitive systems we create. The limits to human cognitive capacity are most obvious when we look for the first time at new situations. Our immediate awareness can easily be overwhelmed.

When we assimilate sensory information for the first time, our impressions are subject to error and illusion. To correct for these in disconcertingly new situations we are extremely susceptible to the suggestions of others. Your first exposure to a new art form might have produced such a response. As another example, many psychological experiments on human subjects are concerned with the immediate unconsidered responses of people in strange experimental situations. In these circumstances people do not behave as wisely as they do in the stabilised responses they produce to more familiar circumstances.

We have a very limited capacity to process completely new impressions. Our immediate thoughts must be contained within a short-term memory that psychologists tell us can only hold seven (or one or two more or less) items. (In less strange situations, these items may, however, refer to material we have already stored in our far more capacious long-term memory.) Our natural and spontaneous responses to novelty, then, tend to be shallow and limited, unless we can find appropriate ways of drawing upon past experience.

Human ingenuity has built up an immense range of stratagems for constructing knowledge in better ways than by merely relying on the immediate impressions of the moment. We are able to connect each new interaction with past experience (as animals also do) and with our reflections upon past experience (as animals cannot do). We are able to try out variations of the thought processes that accompany the experience and to try modifications of the actions involved. We are able to draw upon additional fragments of context that are potentially relevant to the matter experienced and explore any connection. We are able to reflect upon the relevant suggestions of other people rather than merely accepting or rejecting them as they stand. By doing many or all of these things, we can more readily discover illusions of immediate perception and personal errors of thought and action. In an environment which changes in ways we have learned to understand, we are able to anticipate and adapt to the change. In a stable environment, considered actions eventually become habits, no longer requiring very much accompanying thought.

Many of our strongest judgements arise when positive feedback loops occur in the flow of information in considered thought. We sometimes work ourselves up into an emotional state (perhaps of anxiety or fear) by looking repeatedly at the situation in terms of an emerging emotional attitude. (In the case of anxiety, the magnification effect can be self-sustaining even when the original stimulus goes away, as unanticipated physiological effects of anxiety themselves often make the individual more anxious.) Such positive feedback loops may involve other people as when mutual attraction grows into love or reciprocated anger grows into rage.

Our strongest sense of subjective (or intersubjective) certainty about reality comes, not from fresh thoughts or novel experiences, but from considered thought about repeated experience. The feeling of confidence builds up into a sense that we understand what is going on in the context of successful action. For example, we judge that illusions of particular perspectives and momentary mistakes have been cumulatively exposed and eliminated. However, such subjective certainty can easily be in error. Mere repetition of thought and of action may not be enough. We may, for example, slip into unthinking habits before we have given the cumulation of critical reflection a chance.

When considered thought leads to a sense of certainty, it is often presented as knowledge, but does not always meet the standards of knowledge. When positive feedback loops amplify individual idiosyncrasies or collective cultural enthusiasms, bizarre and intricate elaborations of thought may be constructed. And yet, if the thought is consistent by its own standards, the result may be quite acceptable as culture. Our pluralist culture often provides us with competing systems of thought that meet high internal standards of coherence but are inconsistent with one another. To abandon even consistency as a standard as postmodernism professes to do, is to yield too much to scepticism.

Can considered thought produce empirical truth? It will not be easy. Four problems suggest themselves.

1. The feedback loops of considered thought can cumulate error, so we must find ways to eliminate error even more effectively. One difficulty is that human language allows us to build up lines of thought in the absence of effective external checks. Our language can lead us far from the point of origin of a thought. Daydreamers know that they can build up their fantasies in an increasingly satisfying escape into unreality. Literature can spur the imagination of those who seek such an escape. When a powerful individual has acolytes who cater to all his whims, the primary empirical check on his thought may be a highly congenial immediate reality, so agreeably shaped by the mere expression of his thoughts that his control of his immediate environment can be a factor in his losing touch with less congenial wider reality.

2. In making our considered thoughts mutually consistent, we also tie them to a more elaborate frame of prior commitments and will find it more difficult to revise them. In adding to the range of ways we can rationally defend our beliefs we are constructing more reasons why we need not abandon them. It is well known that the larger the investment in our ideas, the more reluctant we are to give them up. Psychotherapists, for example, know that patients who pay large fees find it more difficult to come to the judgement that the therapy was a mistake, as this judgement would imply that the money spent on fees was wasted.

3. Aspects of richly elaborated systems of considered belief become stereotyped. As our repeated thoughts become habitual, we use short-cuts of thought that have proved perfectly adequate in the past. In some social situations, it can be difficult to appreciate that such short-cuts of thought are themselves part of the problem. For example, in political confrontations, there are strong pressures to polarise thought into black-and-white alternatives, as an aspect of coercing people to become involved in political action. (As in the slogan, 'If you are not with us, you are against us.') The subtle greys of sophisticated epistemological judgements cannot readily be sustained in the political sphere, making science and politics uneasy companions. All

cultures are full of habitual chains of thought which people cannot bring themselves to question.

4. Another problem with considered thought is that some of the feedback loops are not fully under personal and social control. When our emotions are caught up in social responses, we are sometimes unable to see all the options. For example, in many cultures, the appropriate considered response to a hostile act is often a reaction that is even more hostile, until one side or the other is not prepared to live with the consequences of further escalation. When such responses become stereotyped by both sides, they may be mutually destructive. Wars start from miscalculations in reactions of this type.

We should not see problems such as these as reasons to reject considered thought as too unreliable to produce knowledge. The danger from positive feedback loops in stereotyped symbolic thought is not inevitable. With appropriate strategies to identify and eliminate errors, considered thought can become more trustworthy. Scientific knowledge can be seen as an attempt to institutionalise the conditions which enable this ideal most nearly to be achieved.

The key to such strategies is that no part of the cognitive process should be too effectively insulated from any other which might be the basis of appropriate criticism. Considered thought should engage reality. Experience of variants of an action can build up a sense that reality is adequately controlled and understood. Considered thought should engage the individual mind, producing a sense of certainty through the judgement that readily discoverable illusions and errors have been identified and eliminated. It should engage as many external checks as possible on the relationship between private thought and public discourse. All the levels of cognitive activity should have become involved.

An important source of revision of considered thought is to enable the feedback loop to range over reflexive features. As repeated thought becomes stereotyped and habitual, it is not naturally revised unless the thought is itself made the subject of conscious thought. By making the repeated pattern of thought the subject of further thought, we may learn how we can be misled and how possibilities of error may be reduced or eliminated. For example, when cycles of escalating thought and action are leading us to an undesired end, as in conflict, creative effort can be given to finding ways of breaking the cycle.

As social beings, we have also learned to expose all but the most private of our convictions to the scrutiny of other people. We take account of the feedback they give us. Social life is an especially important check on the vagaries of individual thought and action. Our culture is rich in resources to aid us in avoiding error. Science, for example, has institutionalised many methods of error minimisation its educational processes. As we grow up, we acquire such cultural resources and learn to construct our beliefs in accord with what we think others will find acceptable. For example, since our more rigorous forms of explicit rationality are expressed in the social instrument of language, formal reasoning can be made accessible to public scrutiny. Those who have been well taught have learned how to avoid mistakes of formal reasoning in the first place or to eliminate them by private reflection. However, it can be more difficult to anticipate how others will react if they are using a different form of explicit reasoning from our own to deal with the same practical situation, or if they have allowed different interests to shape their informal reasoning. In a well educated society, weaknesses in considered beliefs about reality are more likely to be due to failure to interact fully with others or to collective failures rather than to purely personal inadequacies.

5.5 Implications for older theories of scientific method

The ideas introduced in this chapter give a powerful basis for challenging some widely held philosophies of science.

Induction

The traditional mainstream empiricist view of scientific method is inductivism, the idea that we should systematically collect facts, classify them, and by inductive methods, arrive at the general relationships which hold among the known facts and which give a rational basis for the prediction of new facts. This view has problems, for it presupposes a uniformity in nature which allows us to extrapolate from the known to the unknown. But pre-human nature is not uniform and so far humans have failed to make it so. David Hume set up the problem of induction in the eighteenth century, arguing that there is no logical basis for arguing from the past to the future, from the pattern in known events to unknown new events. However, such objections have not discouraged inductivists. Perhaps, they reason, we can formulate the rationale of induction more cautiously. Surely the pattern in the past is not irrelevant to our expectations of the future? Hume suggested that our proclivity for making inductions is a matter of psychology rather than logic. Perhaps we can formulate a subjective logic of inductive probability? Can we justify induction on the grounds that if any way of anticipating the future will work, induction will work? (Reichenbach, 1938; Salmon 1961, 1964).

I wish to suggest that induction is not merely invalid, as Hume argued, but that it is also unhelpful except in very limited circumstances. (For this reason, I am more attracted to Karl Popper's alternative to induction, to be introduced later in this section.) Unless we have built our knowledge out of passive observations in which our effect upon the world is minimal, inductive extrapolations require us to make generalisations within immediate reality or to move from immediate reality to the wider reality our actions have not affected. Induction can be expected to go wrong in such extrapolations unless great care is taken to make immediate reality model wider reality. In particular, if we extrapolate beyond a stable congenial immediate reality we may find that our inductions fail. A proper social basis for induction might reduce the problem. The congenial idiosyncrasies of one person's immediate reality need have no effect on the immediate reality of another, so that patterns in events which have been found to hold by many people are more to be trusted than those which have only been found by a single person. But there is no comparable way of checking the move from our socially shared immediate reality to the wider reality we do not collectively influence. In this wider extrapolation, induction is not to be trusted.

Consider, for a moment, the idea of human beings taking over a society run by machines. On present versions of the fantasy, a machine-run society would be an extreme version of an orderly bureaucratic society, with everything in its proper place and rules which maintain the regularities that machines can best cope with. It would have a immediate reality in which induction worked optimally. In the fantasy, humans would, on take-over, introduce spontaneity, originality, and creative escalation of competitive confrontation. Bureaucratic order would begin to break down. The immediate reality would no longer have optimal order.

Our present collective reality is not machine-like, but the natural production of spontaneous competitive beings. Living organisms seek to identify the order in their environment and use it to their competitive advantage. By acting this way, they change it. Order in organic systems is self-limiting. Induction does not work especially well in such an local reality.

Why, then, did the idea of induction have such a hold in centuries past? The attractiveness of induction in modern culture has to do with the attractiveness of an ideal of harmonious rational order. The philosophy of the Enlightenment sought to disseminate this vision more widely. To the extent that it is in our power, we try to make a reality of this kind. When we think about the divine construction of nature, we imagine that the natural world of an omnipotent, omniscient, and benevolent creator would surely be of this kind. Natural theology found evidence that the natural world is indeed so.

The Enlightenment vision of an orderly cosmos is an illusion. The world is as it is, and the regularity and harmony we find in it is to a significant extent the result of the kind of beings we are, the way we think and the way we act. In particular, most of the regularities we look for and the kind of order we seek to impose upon immediate reality are inspired by the structure of our language and conceptual systems. If our language was constructed differently, or if we thought otherwise, we would be inspired by a different sense of the order of nature.

Nelson Goodman (1965) offered 'a new problem of induction' which illustrates this issue. Consider the concepts 'blue' and 'green' and the generalisation 'all emeralds are green'. Suppose that we also have alternative concepts 'grue' and 'bleen'. Blue and green can be defined in terms of grue and bleen, or grue and bleen can be defined in terms of blue and green. For example, grue objects are green until midnight tonight and blue after that. Conversely, green objects are grue until midnight and bleen after that. Clearly, the same past evidence which gives support for 'all emeralds are green' gives support for 'all emeralds are grue'. But, after midnight, the inductive extrapolation from the greenness of emeralds will diverge from the inductive extrapolation of the grueness of emeralds. So what do we expect? That emeralds tomorrow will be green (i.e. they will have changed from grue to bleen) or grue (i.e. they will have changed from green to blue)? Our natural response is to insist that, all else remaining the same, emeralds will be green tomorrow. Goodman suggested that we prefer the terms 'blue' and 'green', which we have successfully used in the past, because these are well-entrenched concepts, while 'grue' and 'bleen' are strange to us. My use of this conclusion is to suggest that with a different well entrenched set of concepts, we would make different inductions. Induction is indeed relative to our language and to the concepts we have formulated within language.

Inductivism is often linked to a conception of science as the discovery of natural laws. In the present view, we must distinguish clearly between two conceptions of natural laws. There are passively observed laws, which report what naturally occurs. Many such 'descriptive laws' are known. They are, however, generally approximate, transient, and tentatively held. They are very often qualified by a *ceteris paribus* clause, which excludes complicating circumstances. They may not hold in domains affected by human action. For example, if evolution by natural selection were formulated as a natural law, it would fail in a context in which genetic engineers are designing and manufacturing new species (even though genetic engineers are themselves a product of evolution). If biologists need to produce a new laboratory animal which is incapable of pain or suffering and rather appreciates being cut about a bit, so that they can deflect certain lines of moral criticism of animal experimentation, they do not now have to work within the constraints of natural selection.

The other kind of natural law is that which always holds, in all immediate realities and in wider reality. Such an 'unrestricted natural law' provides a genuine constraint on the things we are able to do. This is rather like the old idea of a 'law of nature'. There are surprisingly few such laws known.[4] As human ingenuity grows, we can construct new immediate realities in which old unrestricted law statements fail. So such statements turn out to be nothing but reports on the current limits of human ingenuity. For example, once it appeared that the transmutation of one chemical element into another was impossible, and, in the nineteenth century, this would have stood as an unrestricted law of nature. But now transmutation of the elements is an everyday experimental achievement of the atomic physicist. The law can only stand as a limit on the power of those operations which we conventionally group together as 'chemical'.[5]

Induction is also linked to a notion that there are natural kinds in the world. These are entities which persist through experience and which we have found to behave in regular ways. The concept of natural kinds is a very large philosophical issue which I will only explore in terms of the claim that we can make reliable inductions about natural kinds. The natural kinds we generally identify have some features which naturally recur and others which we conventionally impose upon them. It is not always easy for us to distinguish what is natural and what is conventional about a natural kind. The most basic entities we identify in science, like atoms and electrons and quarks, are not things we identify in ordinary experience but are carefully crafted ideal constructs. They behave as natural kinds because we run our experiments in ways which force them to behave that way. It is very difficult to separate the real from the conventional with entities so highly abstracted from ordinary experience. At a more common-sense level, we surround ourselves with the kind of durable hardware which sustains its form and function. Inductions about such hardware work because we make them work. We have tables and chairs which we make capable of

---

[4] Reservations about building physics out of physical laws are also expressed by Cartwright (1983). Her book, *How the Laws of Physics Lie*, distinguishes between phenomenological and fundamental laws. The phenomenological laws describe particular states of affairs but do not explain them, and the fundamental laws explain but typically do not get the facts right – they *lie*.

[5] Even that would go if cold fusion* turned out to be possible after all.

sustaining a desired function; indeed, their manufacture optimises these features. We try to avoid materials which are unsuitable for such repeated use. We can only keep control of fluids like water through taking care about their source, subsequent treatment, and containment. Entities like fire are even harder to control perfectly. Fires can easily surprise us. Many other entities, like clouds, are still largely unmanageable.

The readily experienced types of entities which seem the most amenable to description as natural kinds are organic kinds, such as oak trees, dogs, and persons. But, as we learn more about the genetic and environmental basis for the forms such kinds take, we see that they too are, in part, idealised notions, inductions about which have a partly conventional character. Can an oak tree be grown 1,000 feet high? It has never happened in our experience. But why not, if we set ourselves the challenge, perhaps in a low-gravity environment? Can we be sure that dogs will never be able to conduct intelligent conversations with us? Well, the claim could be construed as a challenge to a genetic engineer, unless we prohibit him or her from trying. The conventional characteristics of the concept of 'person' are especially obvious. So much hangs on the attribution of this label that it is tied into the whole network of our concepts. At what stage in development should we describe as a person the developing system of an egg plus the sperm which fertilises it? Could a robot be a person if we worked hard enough at making ourselves more like robots and robots more like us? The entity, person, is as much a social construction as it is a natural kind.

In summary, then, inductivism does not appear to be a very good ideal account of the central rational inference process of science. It is rather worse than a half-truth. Even so, the philosophical point remains that science is impossible unless we use our experience of the past in coping with the future.

Pragmatism

One of the modern alternatives to inductivism as a theory of scientific method is pragmatism. This is the idea that we should always choose the theory which works, for that is the ultimate criterion of theory choice. This, too, has problems on the view offered here that places immediate reality between experience and wider reality. To say that we should prefer theories which work is very like saying that we should judge theories in terms of their status in immediate reality. But, in a congenial immediate reality, far too much works, and in an uncongenial immediate reality, too little works. In most of life we try to surround ourselves with congenial immediate reality. The standards of 'working' are too low unless we also put conditions on the kind of immediate reality in which the theory should work. The theory which works in the special kind of highly demanding immediate reality constructed by effective modern science is different from the social theory implemented by political leaders who only allow feedback in harmony with their own ideals and then say that their ideas are now proven. Sometimes, an idea works for a while, and then works no longer as circumstances change. How should distinctions between working in the short and long term be developed in pragmatism?

In adversarial situations, theories may no longer work because people are trying to prevent them from working. Is it too hard a pragmatic test of a theory that it should work even in the uncongenial immediate reality of people who do not want it to? I noted earlier that modern theories of economics do not work in many uncongenial immediate realities, in which people do their best to bring contrary values into action. Perhaps current economic theories could be made to work if only everybody would co-operate in trying to make them work – so producing a more congenial immediate reality.

My discussion shows that the pragmatic criterion of working or not working is quite unsuitable as a way to capture the rational essence of scientific method. Bad science might work in congenial circumstances and good science might fail to work in uncongenial circumstances. Although being able to work might be a useful working criterion, it needs to be supplemented with an account of the conditions under which it can be applied. Such an account must explain why theories sometimes work for the wrong reasons.

If some version of pragmatism were made the ultimate standard for science, then there would be no basis for its revision in terms of the criticisms argued here. Pragmatism could only become a working criterion if it were revised on non-pragmatic principles. A suitable higher principle to qualify pragmatism is realism.

Hypothetico-deduction

Another alternative to inductivism in modern theories of scientific method is the theory of hypothesis and deductive test, or hypothetico-deduction. One influential twentieth-century formulation of this view is Sir Karl Popper's falsificationism, according to which we should formulate bold conjectures, and then try to show them to be false by looking for counter instances to the predictions and explanations we deduce from them (in conjunction with background knowledge). This view will be discussed at length in Part II. It too is seriously affected by the idea that immediate reality is not always representative of wider reality. In particular, an uncongenial immediate reality may falsify a conjecture which is otherwise perfectly sound. For example, if a group of sceptics experimentally falsify another group's pet theory, the supporters of the theory may reasonably refuse to acknowledge the result, until they have satisfied themselves that it was not a consequence of some feature restricted to the immediate reality in which the sceptics carried out the test. In general, when we see that the observations with which we put our conjectures to the test are made in contexts which are meaningfully structured and reflexive, we find that hypothetico-deduction requires reformulation.

The conclusion I wish to draw in this section is that the concepts already introduced require the introduction of a modified theory of scientific method, adequate for the insights so far developed in this book. At the least, any such theory should require that in an extrapolation from immediate reality to general reality, every effort has been made to identify whether the pattern found in immediate reality is due to processes occurring as presently conceptualised or whether hidden processes are at work, shaping the local result. We can learn from experience what sorts of pitfalls are liable to occur in specific kinds of immediate reality, and find ways around them.

In the next chapter, a modified theory of scientific method is sketched out. It draws on philosophy and sociology of science, slanting the material in a new way to fit the insights already offered, and to facilitate its application to the issues of the rest of the book.

5.6 Summary

We do not merely make our thoughts correspond to reality, we also change reality to correspond to our thoughts.

Realism is most plausible when we take into account the effects of our actions, especially if, as in science, we strive for demanding tests of our conceptions of reality. We interact with a *local reality* that is partly of our own making. Within local reality, we are surest of *immediate reality*, the connection between our actions and our perceptions. As we try to build up a correspondence between our concepts and reality, the task will be more complex if immediate reality is *excessively congenial* (when our actions, guided by defective understanding, still produce the expected result), or *excessively uncongenial* (when our beliefs prevent our actions from being fulfilled). Knowledge claims constructed within excessively congenial immediate realities may be unreliable. Knowledge claims constructed within excessively uncongenial immediate realities may be self-defeating. Everyday reality cumulates complexity in a competitive society as rivals set about making reality correspond less closely to their opponents' beliefs.

On the view defended here, we should be sceptical about how well our concepts correspond to reality, though realism remains as a goal. We cannot always trust past experience. Induction and the descriptive laws it generates only hold in domains in which we make them hold. Unrestricted natural laws, which should hold in any domain, turn out to be statements of the current limits of human ingenuity: many have failed and few remain. The traditional idea of pragmatism is also unsatisfactory, for, in congenial immediate realities, too much works, and in uncongenial immediate realities too little works. The hypothetico-deductive approach of K. R. Popper also has to be modified, for hypotheses fail their observational tests too easily and unfairly in the unsympathetic hands of intellectual opponents.

6

A new account of the scientific process

6.1 Introduction

Traditional philosophical theories of scientific method concentrated on explicit systems of rationality and left out reference to the informal individual and social rationality which sustains them. The sociological turn in science studies instead made science into just another form of social life, the local complexities of which can be described and understood in the same terms as any other social practice. This chapter seeks to provide a theory of how science combines explicit rationality with individual and social informal rationality. The coherent fragments of explicit rationality which science has actually produced are provisional drafts of current philosophical ideals of knowledge. They are generated as the output of a nested set of cyclical activities. The cycles go on at every cognitive level, the results affecting subsequent repetitions. The knowledge content of science is continually being updated and revised by the interaction of the various loops of information flow. The theory moves goes beyond description of the epistemological practice of science into the construction of a ideal account that might also function prescriptively.

If we look at science as a whole, we may idealise the process in terms of the general form of a cycle which moves around the following stages:

1. Matching factual claims to reality.
2. Matching theories to factual claims.
3. Reconciling theories in networks of understanding.
4. Reworking reality in the image of our understanding.

Every cognitive level is (or should eventually be) involved in this cycle. The process is sustained within a complex social system and is driven through its changes by (socially acquired) individual motivation. The main rational incentives within the system are to represent reality and to make the pattern of understanding as coherent as possible. The changes that result simultaneously approximate to rational progress and to the evolutionary cumulation of socially selected variants.

I will try to show in the discussion which follows that this account escapes the worst consequences of social relativism (as at least some aspects of reality are caught up in the cycles of the scientific process), and that it is superior to traditional accounts of the scientific procedure. In Part II of the book, the theory will be used as the basis of discussion of the distinctions between science and non-science and between good science and bad science.

6.2 Matching factual claims to reality

At the beginning of the last chapter, I asked how we can build a correspondence between thoughts expressed in words, and things. In this section, I provide an answer: we induce nature to generate meaningful messages for us, in images or words, in immediate reality. This answer is offered as a development of the sociological account of the social construction and negotiation of new facts presented in chapter 3.

Even though science is locked in the artifice of language, it is possible to check its claims against nature. In the laboratory, we use our scientific understanding to produce effects which are recorded by our instruments as data, which we assimilate as facts and reconcile with our scientific expectations. The instrumental readings can be idealised as the messages we induce nature to write for us.[1]

One point to be emphasised is that the more directly we can make natural processes generate a desired message or image, the more we are persuaded that it corresponds to reality. Of course, the selection or contrivance of the situation which produces such a message requires great insight and practical skill concerning natural processes. But the high level of artificial control over the circumstances in which the message is produced only helps to demonstrate that the scientist must know what (s)he is doing, so that we can put more trust in the message (s)he claims nature has given.

---

[1] The term of sociological jargon, 'inscription' (as used, for example, by Latour & Woolgar, 1986), nicely indicates the nature of the raw material out of which scientists negotiate what thay come to call 'facts'. The word 'inscription' helps to remind us that such physical records do not come with a prior meaning unambiguously attached. We suppose that there is a meaning to be found, but it is we who attribute meaning to the inscription. It must be emphasised that, although inscriptions are the outcome of processes which the scientist makes happen in the laboratory, the physical records themselves are not contrived or constructed by the researchers but are the result of real processes in the world. The art of generating and interpreting such inscriptions is central to natural science.

In empiricist philosophies, it was once argued that nature can create partial copies of itself directly upon our senses. Although the secondary qualities (smell, colour, etc) were produced in our minds indirectly by the powers of bodies, the primary qualities (size, shape, state of motion or rest, number, and solidity)[2] were regarded as inherent in the nature of things and directly corresponding to the ideas they produce in our minds. However, all forms of sensation are subject to uncertainty and illusion, a problem which can be reduced, but not eliminated, by replacing direct sensation by more sensitive and reliable instrumentation. The scientist reads the instrument rather than the original phenomenon. The trend over the centuries has been to less direct forms of instrumentally mediated observation. The effect has been to make science less open to subjective sensory errors and more open to publicly checkable procedural errors which can be identified and eliminated in proper professional practice. We have moved away from allowing the world to register upon our senses to making the world tell us through our instruments what we want to know about it.

In the late twentieth century, we increasingly ask nature to produce pictures for us, rather than to write messages on instruments. Our society has become more televisual. In this culture, we readily appreciate the greater perceptual richness that can be packed into a meaningful visual image. A good way to introduce the manner in which science builds upon messages and naturally produced images comes from the history of perspective in art and the subsequent history of photography.

The discovery of perspective in art A well-known story in the history of Renaissance art tells how, in the fifteenth century, perspective was discovered as a method of enabling the representation of a three-dimensional scene in a two-dimensional image. Technical aids appear to have been important in the process. In the early fifteenth century, Filippo Brunellschi painted a panel which the viewer was to see by standing with his back to the original (architectural) scene, looking at a mirror through a peephole in the centre of the panel. The painting was framed in the mirror by the reflection of the original scene, and precisely replaced the part of the scene it represented. A few years after this, Alberti described how to construct a painting by regarding it as a window through which the subject is to be seen from a fixed viewing point. The image is in effect painted on the surface of the window. The technique could be made more mechanical by painting on the glass of an

---

[2] These happened to be the qualities which were both plausibly attributed to the underlying corpuscles and the easiest to represent mathematically.

actual window opening. Alberti may also have invented the camera obscura, in which an inverted image is produced in the back of the device by light from the scene coming through a pinhole. Very soon, Renaissance artists had formulated a number of rules of perspective to aid them in producing realistic paintings. The development of lenses enabled them to project images onto surfaces, where they could be recorded in paint, exploiting all sorts of tricks of perspective.

The success of drawings and paintings produced according to the rules of perspective is because they indicate precisely the positions, shapes, and sizes in which objects should be represented on a surface if they are to match the original scene when viewed from a particular point. The sense that reality has been captured by the perspectival image was especially strong when the method of production was guided by one of the technical devices. The optical contrivance was taking over part of the role of the artist in painting the picture. Although later artists learned that the perspectival rules should not be followed too slavishly to represent human experience in vision, realism had come to require a sense of perspective. It is quite easy to see why people were persuaded of the realism of such pictures. It was not from the reputation of the artist, though we are naturally guided by the judgement of those with artistic skill. Nor was it from the conformity to the rules of perspective (indeed, there are arguments that perspective theory is culturally variable, being nothing more than one of many methods of representation.[3] It was rather from the straightforward way in which each person could see for him or herself that the picture looked far more like the original scene than a picture which did not employ perspective. This was combined with the other advantages of art. The picture has a life of its own, independent of the object represented. It can, for example, survive the disappearance of the original scene. It can also be taken away for closer study elsewhere.

The development of photography in the nineteenth century took the same trend further. Photographs give so consistent and reliable a perspectival view that we are often prepared to accept them without feeling any need to check them against the original scene. The camera makes nature paint its own picture without an artist. In our culture, photographs capture an instant of the reality of a passing scene so that we can take it away with us.

A few sciences construct most of their facts out of photographs. Astronomers in particular record their most precious moments on film. Our rich knowledge of Martian (and more recently Venusian) topography is based on vast numbers of photographs taken from planetary orbiters. The

---

[3] Goodman (1976), a view attacked by Gombrich (1982).

power of this kind of science is further illustrated by the persuasive realism of satellite photographs of details of topography of our own planet.

The special quality of the photograph is that it presents its subject matter in a way that is immediately assimilable by the human eye and yet at the same time corresponds to the subject-matter in a persuasive way. A photograph can persuade us, where the report of an observer might not. Of course, we can only trust a photograph if we know that it is authentic. A photograph can deceive. Sir Arthur Conan Doyle was unable to see how the famous photograph of fairies at the bottom of a garden could have been faked. But subsequent confessions made it clear that it was.

Although the camera can be made to lie, within a suitable protocol for the production and presentation of photographs, it is an immensely powerful resource for persuading even the most sceptical that the world is as the photograph records it. That is why it is a useful example of how the messages and images produced in immediate reality function in science. Just as a photograph allows nature to produce its own picture, the instrument in immediate reality allows nature to write its own science.

The persuasive power of a photograph in the absence of knowledge of how it was produced is diminishing with the growth of computing technology of manipulating visual images. Similarly, we are less ready to accept indirectly produced instrumental representations of nature if there are intermediate stages of human intervention and interpretation in their production.

Genetic fingerprinting Consider an example from applied science in which the authority of nature's message has been given legal status. This is the major development in forensic science of genetic fingerprinting. The immense variability from individual to individual of certain readily cut out sections of DNA has been turned into a test that can (to a predetermined level of probability) identify an individual from whom cellular material (such as blood or semen) came. It can also indicate how closely two individuals are related genetically. In principle, the test is immensely persuasive if a well-designed and standardised procedure is carried out by competent individuals. Suppose an immigration service official doubts that someone now resident in India is indeed a son or daughter of a British citizen of Indian origin. The matter can be settled beyond reasonable doubt by genetic fingerprinting. Cellular material is prepared and processed to produce key DNA fragments than can be separated (by electrophoresis) and stained or made radioactive to produce a photograph not unlike a bar code. When enough DNA fragments match in the 'bar codes' of two individuals we are persuaded that they are related. And if someone is sceptical about a particular version of the test or of the way it was done, they can (in principle) arrange another, done a different way. The message in the DNA bar-code pattern is the result of human contrivance. But it is not purely a human contrivance. Nature is being made to write in bar-code form the message, 'yes, they share much of their genetic material,' or 'no they do not'. The sceptical immigration officer, who is inclined to think that the claimed family relationship is just a subterfuge to bypass immigration laws, finds that his scepticism appears unreasonable once nature has pronounced otherwise (at least until he finds some way of discrediting some aspect of the evidence or argument).

Not every case of genetic fingerprinting is as straight-forward as this. Suppose that a man is a suspect in a rape case. A match between enough of the bars in the test results of samples from the suspect and from the forensic material indicates that the same individual was involved. The design of the test in such a case involves a compromise between making most effective use of very small forensic samples and maximising the statistical certainty of the result. Although it is appropriate to challenge the result in a trial, particularly if the forensic sample is very small, if the testing agency can be shown to be unreliable, or if the statistical assumptions of a particular technique can be questioned (especially when the suspect and the actual rapist might be related or are from a gene pool that is not well randomised), it is still difficult to challenge effectively the general principle employed. If the forensic sample is large enough, any doubts about some aspect of a particular DNA fingerprinting procedure can be allayed by also using a variant procedure.

In the 1990s, this use of DNA testing provides interesting material for legal exploration of the limits to the authority of science. Seeing the match in the position of DNA fragments can persuade the technician involved that two samples came from a single individual. Although some steps in the process require human action and skilled judgement, the final result can be made to seem unambiguous. The technician either personally controls these processes, or is part of a co-operative endeavour with a shared sense of control. In the legal use of genetic fingerprinting, the result depends upon competent procedures linked to complex reasoning. Ultimately, however, the court is being asked to trust the representatives of the scientific community, who take responsibility for the tests. For the jury, nature is not speaking so directly in the test result. And, since technicians are quite capable of making mistakes, why should the jury trust the technicians? Eventually, we may expect that the best testing protocols will have been made nearly foolproof and will have proved themselves reliable to the satisfaction of all. But until then, what appears to the scientists as experimental proof is only a new kind of expert judgement to outsiders.

Our discussion so far has explained how we persuade immediate reality to produce meaningful pictures and messages for us. These are used as the basis of statements of empirical fact. Such a statement about the world is anchored in the claim that when we act on immediate reality in a certain way, we observe a particular effect and that this entitles us to assert the statement fact. In the more traditional cases of pure passive perception, such as seeing a white swan leading to the statement 'this swan is white', the connection between the observation and the statement is transparent. In more realistic examples from modern science, the observation is given significance by its context, and especially by background theory. The series of blackened bars in the photograph resulting from a DNA fingerprinting test carry the meaning they do because they can be supported by a persuasive account of how they were produced.

What is claimed as a statement of scientific fact should not be a mere report of personal experience, but an objective claim about reality that is independent of subjective factors. It is normally expected to transcend purely local conditions, to be the kind of public entity which anyone can, in principle, check for themselves. A claim about the general nature of my own dreams looks dubious as a scientific fact, unless there is some way in which other people can check the correspondence I claim to have found between my waking words and the sleeping reality. Therefore, an important part of the production of factual statements is to find ways of making immediate reality correspond to wider reality. Since many checks on a claimed scientific fact involve the checker constructing his or her own immediate reality in which the original claim can be replicated, the account of how the result was produced must be full enough to allow replication (normally with the aid of contextual factors that are part of a shared framework of both scientists). It must also provide some basis for the assumption that there is nothing misleading or distorting about the relationship of the local context in which the message from nature was produced and the general nature of reality. The discussion of the last chapter suggested that the special features of the circumstances of observation could produce reliability problems of this kind when immediate reality was either excessively congenial or uncongenial. It was argued that we learn through practical experience that certain kinds of local context are untrustworthy, and that others are subject to pitfalls which we must avoid. The result produced in immediate reality should only be considered representative of wider reality if the scientific account of what is going on is persuasive and any objections to the practical procedures followed have been anticipated and deflected.

This discussion of scientific facts exposes the complexity of their construction. The apparent transparency of the relation of simple observational facts to the reality they report is an illusion for scientific facts in general. It is a result of our tucking away the complicating factors in the framework of assumptions, where they are subsequently ignored. Acceptable scientific facts require construction and negotiation. Without successful negotiation, there is no scientific fact. This account is intended as a successor to the traditional empiricist view that we gain knowledge of reality by the directness of the process of sensation. It has been influenced in part by recent work of microsociologists of knowledge, discussed in chapter 3.

I hope, then, that the reader has been persuaded that scientific observation builds, not on direct observation of reality, but on our skill in making nature leave its own record – on our senses, on instrument dials, or in a photographic image.

6.3 Matching theories to factual claims

In the general cycle of scientific activity, the claims of scientific fact may be assimilated in quite different ways. The claim may be rejected, requiring the proposer to go back and rebuild his case, if he can. The claim may be tolerated or ignored, especially if it is unsurprising. It may immediately be welcomed if the factual support it provides is needed. Or it may be taken very seriously if it is judged well founded but challenges existing expectation. If it is linked to proposed adjustments in existing theory which promise the generation of more new facts, it may be welcomed as opening up further research. The network of understanding is then adjusted to allow its assimilation. Normally, the adjustment of the rest of the system is made as small as the practitioners find creatively possible. However, if a paradigm-sharing community is already in crisis, the assimilation of further new facts may force a more radical reworking of existing knowledge. Old facts come to be looked at in a new way as they are integrated into a new theoretical synthesis. This is the process Kuhn (1970a) described in his account of scientific revolutions.

This part of the process of science was made the key to understanding science in Popperian methodology. Popper suggested that we can learn most effectively how to cope in the world by formulating bold conjectures and trying (and sometimes succeeding) to show them to be wrong. I wish to accept that the empirical aspect of science is most significant when it forces us to change our understanding and expectation. If we proceed in such a way that we only assimilate that which does not require the revision of our understanding, then there is little point to interrogating nature at all.

The search for falsification is an oversimplified view of this aspect of science. Scientists are not primarily motivated to falsify theories by facts. (The most common circumstance in which facts *are* appealed to as falsifiers is when there is a need to demonstrate the superiority of a preferred theory by discrediting a rival. We will return to this topic in a discussion of incommensurability in the next section.) The process of science is both creative and critical. Scientists are motivated (individual motives are encouraged further by the institutionalised reward system of science) by the desire to predict, discover, and explain new and surprising facts. The fact should say something unexpected, and the explanation should link it to existing knowledge, demonstrating in what respect its unexpectedness is surprising and significant.

In more conservative traditions than science, the primary need for facts is to justify existing commitments. Although scientists, too, tend to favour evidence that supports prior values, they do so for non-scientific reasons; the institutionalised reward system of science gives its credit for making public something that has never been done before, the discovery of which can be treated as a kind of intellectual property.

Because the production of new facts is a central motivation of science, those who claim to have produced them are tempted to be too bold in their claims, and those who wish to use such facts must display critical caution. Popper emphasises the critical impulse in science. However, the critical impulse is at its most rigorous when applied to new claims of novel fact rather than when applied to existing theory. (For example, the concept of a *fraudulent theory* hardly makes sense, while fraudulent claims of new facts abound.) New explanatory conjectures are also likely to be criticised, but it is not expected that brand new ideas should be perfect when first formulated and criticism may be constructive as well as negative. Longer-established theories only occasionally come in for criticism. The critical impulse in science is therefore un-Popperian in its main effect – leading to suspicion over potentially creditworthy claims to fact, rather than to an enthusiasm to reject cherished ideas when new facts come into conflict with them.

Many factual claims in science remain unused and unrejected. Although their publication puts them on the record, no one finds the need to refer to them, so that they appear to be tolerated or ignored. This can happen because they are interpreted as supporting established ideas. Much of scientific activity yields routine unsurprising results, corresponding to the expected pattern. It may be referred to in literature surveys and the like, but it does not lead to further research or to practical applications. Such results are not completely pointless, because they give greater confidence in the relevant aspects of the network of understanding. Records of specific construals of messages from nature can be returned to if later work gives them an added importance. The lack of close checks on routine work which apparently yields no surprises means that some of it can be of low quality, or even spurious. The atmosphere of trust in science can easily be exploited by the lazy and the dishonest (e.g. Broad & Wade, 1982, chapter 9). Because of this, the messages nature is seen to be giving at any given time often appear to be clearer and more unambiguous in the scientific literature than they do at a later time, when conventional understanding has changed.

Another reason for tolerating or ignoring factual claims is because they are too enigmatic. Perhaps no accepted way has been found to show their surprising significance, or perhaps they will later be regarded as incomplete fragments of significant findings, waiting to be put together in a suitably prepared mind. Most often, their audience suspects that they contain obscure errors that no one judges to be worth the effort of demonstrating. Such claims remain ignored enigmas. Mendel's mid-nineteenth-century work on the genetics of peas appears to have been ignored for this reason until its famous rediscovery at the turn of the twentieth century, in a new context in which it no longer appeared obscure.

Other scientific results are immediately welcomed, easily surviving initial critical scrutiny. If they support new and controversial ideas or if they support longer-established ideas under challenge, it is immediately clear how they are to be assimilated. In the former case, those who produced them may receive scientific reward, in the latter case, there is less chance of institutional honour, but they are welcomed, all the same.

The most interesting facts from this quasi-Popperian point of view are those which are successfully negotiated and are yet in conflict with the prevailing network of understanding. Their assimilation forces further changes, which drives the process of science into the next step of the cycle. In rigorous quantitative science, they might be contrary to a carefully worked out prediction; in less formal exploratory science, they might hint at unexpected structure in nature. In either case, the challenge they offer is that a revision or elaboration of current understanding is required, success at which offers new scientific rewards.

One interesting finding of microsociologists of science is that not everybody is equally exposed to challenging facts that are contrary to previous expectation. It is the immediate group around those who produce the facts who have the first chance to assimilate them. Very often such a group is inclined to adjust understanding to fit the local regime of facts, and to pay less attention to issues elsewhere in science. The cumulative effect of this in a fact-driven science (but not in a science driven by central theoretical dogmas) is that localised forms of specialist understanding diverge from one another.

The consequences of new facts that force changes in the network of scientific understanding are not felt immediately, but take time to ripple through the entire network, and are especially slow to cross the boundaries between specialist communities. Their assimilation can be slowed by the kind of difficulty that Kuhn (1970a, 1970b) and Feyerabend (1975) refer to as 'incommensurability'. These issues are covered in the discussion of the next stage in the cycle of science.

6.4 Reconciling theories in networks of understanding

Cognitive change in scientific networks of understanding is driven by two opposing pressures, one towards diversity, the other towards integration. When novel facts open up new patterns of research, the scientists involved generate appropriate collective insights to weave the material together. Local research traditions on the frontiers of knowledge tend, therefore, to take distinctive forms by growing into their research material. In a social world of narrow specialisms, reality is found to be ever more complex. But, simultaneously, the urge to integrate and unify science is pulling scientific understanding back together. The interplay between these two pressures is complicated by the way that they work simultaneously at different levels of cognitive activity. An individual in a specialist branch of inquiry might, by unifying the new facts of that field, be making it harder to unify the specialist field with neighbouring fields.

The unification of knowledge Under these circumstances, there is no simple plan of predictable rational change in science. Our account of the scientific process would become very complex if it tried to work through all the possibilities. Rather than report the complex history of how interactions between the opposing pressures have occurred in the past, the discussion which follows focusses upon one rational ideal, the unification of knowledge. It is accepted in this discussion that the ideal is sometimes losing to the opposing process of the cumulative discovery of complexity, and also that the ideal can be producing opposing effects (at least in the short term) when operating at several levels at once.

The ideal of the unification of knowledge works at three levels. There is the psychological impulse to reduce internal conflict of beliefs. Secondly, there is the inspirational ideal of completely explicit rationality, leading to the construction of coherent formalised theory in a clear mathematical relationship to idealised sets of data. Thirdly, there is the social impulse towards consensus within and between like-minded groups in how they relate their explicit rationality to the world. These impulses achieve their most successful expression in the unified theories of high-level sciences, particularly physics. They aspire towards expressing the entire content of science in a formula which can be printed on a T-shirt.

In the tradition of thinking of science as organised knowledge within a single rational mind, it is clearly essential that knowledge should be coherent. Within idealist philosophies, the dialectical process involves a thesis leading to an opposing antithesis, the resulting tension being resolved in a synthesis, the third stage of the dialectical triad. In realist empiricist philosophies, it is required that all true statements about the world be consistent. So important is the coherence requirement for truth that there is a corresponding coherence theory of knowledge.

It is quite easy to defend inconsistent ideas, by regarding each as a complementary helpful approximation to the ultimate truth, each to be used in a different context. This is especially appropriate when seeking optimal answers that draw upon more than one set of discourses. It is less appropriate within a single system of explicit rationality. P. K. Feyerabend (e.g. 1975) argued vigorously in favour of the proliferation of competing alternatives in science, including contrary alternatives. It would appear, then, that there is nothing irrational about a view of science which emphasises how incomplete and imperfect it is and sees harmonisation and unification as a remote rather than an immediate cognitive goal. (It may not be an achievable goal at all.) In the practice of science, however, individuals and groups do appear to strive to integrate their understanding. Each individual seeks to reduce what social psychologists have called 'cognitive dissonance' (e.g. Festinger, 1957). In fully interacting groups of like-minded individuals with similar interests (including specialist scientific communities), the effect of sustained communication is to reduce disagreements. Such groups tend to reduce interpersonal cognitive dissonance and reach consensus.

In comparison with other intellectual enterprises, the scientific community has the reputation of displaying a high level of cognitive consensus, and the agreed knowledge of scientists appears highly unified. For this reason, those who stand outside the scientific consensus appear as extremely unorthodox, comparable to the heretical in religion. And yet, the mechanisms of consensus formation within science do not appear especially authoritarian. How is such a high level of agreement achieved?

For a single individual, the situation does not appear to be very different from non-scientific systems. The cultural pressures inducing an individual to reduce cognitive dissonance, are comparable in science and religion, for example. If anything, the high level of faith and commitment called for in religion puts greater pressure on the individual to eliminate inconsistencies and incoherences in religious thought, while the scientist is encouraged to keep an open mind and to delay judgement on unclear topics, to be honest about the extent of his or her own ignorance. The difference between the high level of consensus in science and the diversity of religious views in open, non-authoritarian society would seem to be due to contrasting social rather than psychological mechanisms. Given that individuals normally seek to unify their thought in intellectual contexts, our key problem in this section has become the sociological question, 'How is it that scientists so often come to agree?'

Following the distinction previously made between the social mechanisms that operate within small communities of like-minded scientists and those which operate between such communities, I will discuss consensus formation within expert communities first, and then go on to consider harmonisation of scientific understanding between such communities and between science and the rest of society.

Specialist communities of scientists construct coherent rational representations of their subject-matter. They develop explicit and rigorous language in which problems can be formulated, information reported, and arguments constructed. The practitioners' collective practice and informal discussion helps them to extend their explicit rational representation of the relevant parts of the world. Individual scientists may have diverse skills and purposes which would lead them to divergent judgements, but it is contrary to the collective interest of the specialism to sacrifice thoroughness for expediency, or rigour for quick-and-easy answers. There is no value in remaining ignorant of relevant issues, or in deferring to the authority of others. The institutionalisation of explicit rationality, then, drives scientists to seek a coherent consensual representation.

One way to identify the social mechanisms of consensus formation is to look at situations in which they have gone wrong. The normal pattern of trust, openness, and full communication can be contrasted to pathological states in which (a) lack of trust allows an individual to see others as conspiring against him or her, (b) lack of openness is converted into secretiveness, (c) the motive to share through communication falls away into non-communication, and (d) an external authority or pressure imposes its judgement on science. These pathological states are not in themselves irrational, but work against the institutionalised mechanisms of social rationality upon which orthodox science depends.

Science should be seen, then, as having institutionalised a process in which scientific findings are openly communicated and discussed in a system of mutual trust with the common aim of producing a coherent rational account of the world. This is an idealised view; it is related to the ideal view of science that R. K. Merton (1973) expressed in terms of his account of the norms of science. In practice, scientists are always a bit suspicious and secretive, with loyalties divided between the pressures for confidentiality in their employment and the urge to communicate the minimum required for professional recognition in science. Nevertheless, it seems that the ideal does work. The institutional pressures on scientists encourage them to be rather more trusting, open and communicative than most social groups, to an extent that facilitates the resolution of most disagreements, at least as they occur within small groups of like-minded experts.

It appears to be inherent in the human condition that when small groups of people with common aims and interests talk over an issue in a context where escalation into violent conflict can be resisted, they tend to come to commensurable views of it. That is to say, they agree about how it is to be described, and come to a common understanding of the basis of any residual disagreements. Normally, the people who come together in natural science do have sufficient common interests to come to a common understanding of this kind. They may not immediately agree, but the institutional context is one in which they can agree which matters might be resolved by further empirical research (or by development of the appropriate conceptual tools) and which matters are not likely to be settled by science and so can be regarded as non-scientific. If they behave in this way, they are able to move from mutual understanding to scientific consensus.

In our society, there is an important element of competition for status and economic reward in virtually every activity, and science is no exception. On judgements involving such matters, scientists are as liable to disagree as interminably as anyone else. However, in science, personal status and objective truth are usually distinguished. Personal status is tied to the respect one is given in one's group for such contributions to the growth of knowledge as the demonstration of new facts or explanatory insights. Scientists can quite often afford to agree on what constitutes scientific knowledge without having to put their reputations on the line.

The claims of the last two paragraphs were limited to small groups of like-minded individuals in full communication. The process is far more complex in more highly structured and highly differentiated groups. In larger groups of natural scientists, misunderstandings from imperfect communication and conflicts of interest are likely to affect the cognitive claims made.

The conduct and resolution of longer lasting disagreements in natural science has been a topic of continuing interest to sociologists of scientific knowledge.[4] In the heat of protracted scientific debate, the participants' perceptions that all is not well with their science often encourages them to look for reasons why. Under these circumstances, scientists often point to social factors caught up in the process of knowledge production. When social factors come to the surface like this, the task of the sociologist of knowledge is very much easier.

To some extent, the reflexivity principle can be a hindrance in such studies. Social scientists more often find it difficult to agree on cognitive issues than natural scientists. One of the factors at work is that social science is not so highly insulated from the socially divisive questions of everyday life. Social scientists inevitably differ in the political-economic-social-cultural values they bring to their intellectual practices. Only the smallest self-selecting groups in the social sciences can readily reach consensus. It is easy to see comparable factors such as conflicts of interest at work in controversies in natural science. However, in natural science, they are not part of the normal state of the activity, and when they are pointed to by the participants in a controversy, it is because the scientists, seeing that something is wrong, look for the cause.

I will not try to survey the wide range of issues in the literature on scientific controversy. I will focus upon two of the most discussed causes of failure to agree, and try to show why, in spite of them, conflicts normally disappear from science. These are incommensurability and conflict of interest.

Incommensurability. The notion of incommensurability was introduced by Kuhn (1970a, originally 1962) and Feyerabend (e.g. 1975) as a way of pointing out how, in cohesive holistic scientific approaches (as they claimed paradigms and theories to be) there is no neutral observational ground, no set of unbiased facts which can provide a comparative measure of each theory. The notion came to stand for a fundamental problem of translatability. For if, as Quine (e.g. 1960) had argued, there is an essential arbitrariness about how

---

[4] I will refer in particular to an example that illustrates my points especially well – Dolby (1976). But there is a large literature on scientific controversy. Collins (1981), is representative of the study of controversy in sociology of knowledge; Engelhardt & Caplin (1987) is representative of the American study of public controversies in which science plays a part.

observational concepts attach to objects in the world, and if as Hanson (1958) had argued, every observation is theory-laden, a change of theory also changes the content of observations. Of course, the proponents of the concept of incommensurability were not saying that an observation statement like 'the pointer on this meter points to the number 2 on the dial' identifies a different observable state of the meter in the two theories. They were saying that the *significance* of such an observation could be different. That is enough to set up the problem of incommensurability, for each theory only gives significance to an observation when it is embedded in the relevant context. In particular, when the Popperian process of testing one theory is conducted with the aid of a second theory to generate the falsifying facts, there is a danger that the advocates of the two theories will not agree on what each observation signifies, and will descend into mutual incomprehension.

Kuhn used the idea of incommensurability to explain the fundamental nature of the reworking of ideas which occurs in a scientific revolution. He argued that, in a revolutionary succession, it is as if our perceptions have to undergo a total re-organisation, a gestalt switch as in moving between seeing the old woman and the young woman in the figure which follows. Those locked into seeing the old woman would fail to understand those describing their perceptions in terms of the young woman. In a revolutionary controversy, people talk past one another in a failure of communication.

Such a failure of effective communication is a feature of the initial stages of any scientific controversy which involves different paradigms. In discussions in philosophy of social science occurring when Kuhn's book was having its maximum impact, attempts were made to develop this point into a major philosophical issue (with a different, Wittgensteinian, ancestry). Perhaps we can never fully understand those whose language is anchored in a different form of life from our own (e.g. Winch, 1964). But, in natural science, at least, this discussion exaggerates the problem. If we took the problem to its logical extreme, none of us could ever precisely understand the utterances of another, because each of us has had a distinct personal history. That is an unhelpful view. Unless we are autistic, we devote considerable effort to trying and succeeding in understanding one another, even those somewhat different from ourselves (such as the opposite sex). Science presents no special difficulty. If scientific controversies last, each side's advocates *do* come to understand the theoretical gestalt of the other side, even when they do not agree with it (Dolby, 1976). We are well able to learn to think within two (or more) systems of explicit rationality or frames of discourse. Perhaps the critics express the doctrines of the other side in a different way, seeing gaps in the argument as weaknesses rather than as a stimulus to further research. But, on the whole, incommensurability is at most a characteristic of the *initial* phase of confrontation between rival sets of scientific ideas. It is a phase which may last longer in the absence of effective communication, but is eventually worked through.

Within closely knit scientific communities, then, any incommensurability is transient. In controversies between distinct groups of scientists, perhaps based in different countries, or with primary allegiances to different specialist topics, the problems of effective communication can make the phase of incommensurability last longer, perhaps until the protagonists are displaced by their successors in the next intellectual generation with differently aligned groups.

There are other mechanisms tending to resolve disagreements in science due to incommensurability, which also apply to the second source of disagreement I wish to discuss, conflict of interest.

Controversies and conflicts of interest One especially clear and simple theory in the sociology of scientific knowledge since the 1970s was the doctrine that what is held as knowledge was shaped by the interests of the practitioners (e.g. Barnes, 1977). The impulse towards integrating individual and collective thought is as much about interests as it is about beliefs. The added complexity of social interests is that they are often in conflict, with no clear means of resolution. Scientific controversies can be seen in terms of such conflicts of interests (e.g. MacKenzie & Barnes, 1979). This use of the term 'interest' descends from a Marxist view of society, according to which we tend to act in ways which defend our class interests, whether we acknowledge it or not. It is easy to argue that some disputes in social science show the effect of such interests. For example, as Marx pointed out, Malthus's lack of sympathy for the children of the poor in his arguments on population exactly fits the interest of the classes with whom he identified (e.g. Meek, 1953). Studies of IQ have often produced results favourable to the interests of the white members of the professional class who carried them out. Interest theorists in the sociology of scientific knowledge have searched for analogous external factors in natural science, such as the commitment of believers to their religious faith in their science or the interest of male scientists in demonstrating male superiority.

The evidence for interest theory tends to occur as a correlation between the interests attributed to each individual and the side taken in a controversial scientific issue. Occasionally, it is found that an interest played a conscious motivating role in the creation of the scientific doctrine with which it is associated. It is also possible for such interests to be drawn in to escalating scientific conflicts, not because they played a role in creating the related viewpoint, but because someone later saw an advantage in constructing a linkage to a wider interest.

One problem with interest theories of scientific knowledge is that whatever the evidence for the role of interests in the creation of scientific ideas, most practitioners would regard harmony with external interests as irrelevant to the properly scientific status of knowledge; they are pleased to identify the effect of an interest in the ideas of an intellectual opponent whom they seek to discredit, but cannot afford to acknowledge comparable interests in their own work while claiming it to be scientific. I discuss below mechanisms by which conflicts of interest, once identified by the protagonists, come to be excluded from properly scientific controversies. Science that seeks to legitimate itself by appeal to external interests or authorities is seen as pathological.

Interest theory has difficulty in explaining why it is that controversies appear to be relatively rare in science and why so many get resolved. Indeed, the interest of interest theorists has been in showing that controversies are more widespread in science than its public relations would like us to think, and that more controversies are suppressed than are resolved. In the discussion which follows, I will accept that the more controversies you look for in natural science, the more you find. And after any given controversy it is usually quite easy to find disgruntled individuals who never concede that the argument is over; they have merely judged that it is fruitless to try to continue.[5] It seems to me that the distinctive nature of science does not lie in the absence of controversies, but in the effectiveness with which they are damped down and disappear.

I wish to discuss three interconnected mechanisms in the conduct of scientific controversies.

1. The strategies employed to win scientific controversies include localising the conflict to a specific matter of fact, or generalising it by linking it to wider issues.

2. However, if a conflict includes issues which are identifiably non-scientific, then one widely used strategy is to label it as non-scientific and refuse to deal with it as science.

3. Science is sustained by spreading to new generations and new geographical centres. Those who were not initially involved and have not been caught up are likely to go on with the science in a way that omits reference to issues linked to the controversy, so that the controversy dies with the declining importance of its protagonists.

It can be productive to try to understand science in terms of the options available to, and taken up by, individuals who are motivated in some fairly simple way. This kind of modelling of scientific activity has been taken up by Latour (1987). In such terms, we can look at scientific controversies as attempts by protagonists, not simply to win, or even to win with honour, but also to gain credibility for their own contributions to science. Poisoning one's opponents, or persuading the authorities to lock them up as a threat to society, although it might win the immediate argument, would have an adverse effect on the credibility of the scientist who did it. (Training many successful students who retain loyalty to their teacher might be a more effective non-rational way to win in a scientific controversy). The neatest way to win by playing the game fairly is to argue the issues through with one's opponents to the point where incommensurability has been minimised, and both sides are prepared to let the resolution of issues depend on specific factual findings. If this is done, then the issue may actually be resolved, for a scientist who goes back on such a commitment risks losing credibility. However, a scientist might wish to win an argument even at the expense of his scientific credibility, perhaps because it is important for non-scientific reasons, as when it advances an external interest. Under such circumstances other strategies may be employed. One technique is to try to draw in unin-

---

[5] For example, E. H. Armstrong, as discussed in Dolby (1976).

volved people to one's side. The most straightforward way of doing that is to emphasise links between the controversial issue and other issues on which the bystanders have an interest, and to do it in such a way that they see it as worth their while involving themselves. Such building up of allegiances in conflict is an important political strategy, and it certainly sometimes occurs in science. But, when used in science, it has the disadvantage that if the new, widened conflict clearly involves non-scientific issues, more scientists will regard the matter as non-scientific and avoid getting involved. The scientific credibility of the participants in the escalated controversy will not go up, and may go down.

The strategy of labelling as non-scientific a controversy that is not readily resolved, is suitably illustrated by cases in which the relation between neighbouring disciplines is soured by an ongoing programme of the reduction of one to the other. For many centuries, for example, debate on the extent to which biology could be reduced to physics and chemistry continued, echoing the seventeenth-century disagreements between Aristotelian vitalism and Cartesian mechanism. It was an issue that was close to the professional interest of rival forms of training for biology – should one learn physics and chemistry in order to bring about the reduction, or learn only biology on the grounds that biological phenomena are distinctively different from those of physical processes? This conflict of professional interests was not readily resolved, and the matter was excluded from science as philosophical (the vitalism question). It is interesting to note that the currently dominant form of biology claims to have found how to make the reduction. Even the deepest mystery of all, the genetic control of embryological development is now losing its secrets to biological chemistry. The matter is no longer philosophical but again part of the programme of testable science. The training of most biologists has adjusted to the new situation.

In general, no matter what the cause of a sustained conflict, those who are not directly involved will tend to see it as less central to science, and will prefer to work on less problematic matters. This process can look like controversy suppression, for the participants in the controversy soon find that they are no longer being listened to. In a rapidly growing science, the effect is very powerful, for new people who do not get drawn into the old arguments quickly outnumber and displace the old. If a field is doubling every twelve to fifteen years, as was the traditional pattern (Price, 1963) until recently, then even the most heated unresolved controversy disappears from prominence within a decade or so, finally to die out with the marginalised but unrepentant participants.

As scientific controversies die down, polemically inclined individuals who refuse to give up find that they have been marginalised, and that disciplinary boundaries have been reconstructed to exclude the once-sensitive issues. By then, it is very difficult for anyone to recover interest in the controversial question.

In this discussion I have tried to show that even though the process is not as purely rational as its philosophical ideals, scientific disagreements do tend to get resolved within scientific communities and moderated or suppressed between communities. If you listen carefully, you can find all sorts of suppressed discords in the apparently harmonious music of science, but they do not appear to matter in the normal process.

Finally, I wish to discuss the manner and extent to which knowledge at the levels up to institutionalised science may be reconciled with human understanding considered more broadly.

My discussion so far in this section implies that science is successful in producing an unusually high level of consensus at the lower levels of cognitive activity in science. What about the higher, non-scientific level? In the Enlightenment goal of a universal conception of our collective interests, it would seem that there would be no great difficulty in scientific activity leading to common knowledge. Perhaps in a closed or totalitarian society that vision is still realisable, but in the absence of an imposed set of collective interests, our society fills with competing standards and conflicting interests. The mechanisms by which scientists localise properly scientific disputes do not apply to issues which legitimately involve non-scientific interests. It seems that, in an open society, outside science we drift apart too easily and fail to achieve universal intellectual consensus.

There has been a great deal of scholarly study of public disputes in which science plays some part.[6] The disputes typically involve conflicts of interest. They may be complicated by initial misunderstandings comparable to those previously discussed in terms of incommensurability. Within them, conflict quickly acquires an institutional setting. For example, issues of nuclear reactor safety may be discussed at a quasi-judicial inquiry over the proposed construction of a new reactor. Similarly, reference was made earlier in the chapter to the use of DNA fingerprinting, where many issues have been debated in the context of criminal trials.

In such public disputes, the rules of conduct which emerge are rarely those of science itself. It is not possible, for example, to make an issue like nuclear

---

[6] D. Nelkin has been a prolific author on public controversies. See for example her (1992).

reactor safety depend entirely on matters of proven scientific fact. Public fears about the irresponsibility of others is a legitimate part of the equation.

In wider society, science is often drawn into political, legal, economic, technological, and medical issues, to be handled in diverse ways. We have not one form of definitive expertise, but many. Similarly, whenever we try to define the general form of life within which our knowledge is constructed, something can be found outside it which may challenge our assumptions and conclusions. We have not one domain of collective discourse, but many.

Perhaps, in a capitalist society like ours, there is a final level at which all conflicts among competing ideas is to be reconciled: the market place of ideas. Any cluster of claims to knowledge can be viewed in terms of its value to potential users. Knowledge in the market-place can pose risks and promise benefits. Orthodox medicine competes with its radical alternatives. Knowledge of a drug which might possibly cure a specific illness has a market-value (even though it might in the end turn out to be ineffective and dangerous). Knowledge of a new technique for the compact physical storage of information, or of a way of controlling the expression of genes each has its market-value. In such a market-place, conflicting knowledge claims may be worth pursuing simultaneously, each having value in a suitable cultural niche. For example, astronomy seeks one market and astrology another. The shared level of common discourse is not universal rationality but a language of the relative value of alternative means to alternative ends.

The vision of a market-place of competing ideas in a heterogeneous society is compatible with the view that, at the level of wider society, science must adapt to the prevailing constraints. If we are moving into a stage in which knowledge is to be assessed in terms of its economic value to the market, then such an account will become pervasive. In the long term, however, the market must adjust to stay within the boundaries of its own external constraints, among which is the nature of reality. Economic bubbles in the demand for ideas will eventually collapse if they are illusory.

In summary, I see the normal process of the harmonisation within networks of scientific understanding as a matter that long ago went beyond what can be achieved by the rational co-ordination of the thought of a single individual. The old Enlightenment dream of the unification of knowledge might seem to be just around the corner in sciences which are driven by unifying theories. But hidden away from the inspirational glow of such theories there are many dark corners of specialist science in which knowledge continues to become more highly differentiated. Nevertheless, the mechanisms for the resolution and suppression of dissent on properly scientific matters are so powerful that a working harmony is sustained. The key sign of this is that the transmission of scientific knowledge from generation to generation and from specialism to specialism continues in a way that tends to leave dissent behind. Those who take up science do so in order to produce more knowledge themselves, and find it appropriate to build each new contribution on older work that enhances rather than reduces their credibility. That is the normal process in orthodox science. This account does not, however, assure the maintenance of unified knowledge beyond the limits of the institutionalised scientific orthodoxy.

6.5 Reworking reality in the image of our understanding

The final stage in the cycle of the construction of scientific knowledge is the use of existing networks of understanding to construct new forms of action and control. Science is used to construct new ways of forcing nature to tell us its secrets and to transform the world in accord with the interests of its users.

In this section, discussion will begin with the use of existing knowledge. The relative importance of formal and informal rationality in the performance of actions will be assessed in terms of the extent to which formal scientific rationality can be translated into machine-controlled action. Finally, there will be a brief discussion of the limits of pragmatism as an account of this phase of science.

The use of existing knowledge The network of scientific ideas does not function identically as a resource for all people. As science has grown in scale over the centuries, the sum of available knowledge and know-how has increased and the accessibility of knowledge from specific intellectual locations has become more of a problem. Each individual and group has full acquaintance with a smaller proportion of the total resources available. Even though each group seeks to assemble knowledge relevant to its tasks, the specialist is an outsider to all of science except his own specialism. It is not insider knowledge but outsider knowledge which is the predominant shared resource of the scientific community as a whole. The network of scientific understanding has to be understood in terms of the institutional complexities of the social system that sustains information flow. The techniques by which the flow of information through science is facilitated in a usable form will become increasingly important in the sociology of scientific knowledge.

The issue is complicated by the elaborate rituals of academic and professional life associated with the public acknowledgement of past knowledge put to present use. The academic system introduces students to existing knowledge, making sure that they can use it competently in standard exercises.

When science was professionalised and assimilated into the academic system in the nineteenth century, the traditional concern of men of science to recognise the priority of their predecessors was a central feature. Although the ability to use the knowledge of the field in a professionally competent way is not strictly necessary to practice curiosity-driven research (after all, science began as an amateur activity and amateur science can still contribute to some branches of knowledge), it is a normal requirement for employment as a scientist and for those who wish to gain research grants.

Orthodox science has always been elitist. From the outside it has looked like a mutual admiration society which has sought to ally itself to the elite of wider society. Although the status of accomplished scientists can be judged from their past accomplishments, there is more of a problem about those rising individuals who have not yet accomplished much. An important guide in assessing the potential of such people are the indications of competence displayed in scientific training – that is, they are based on how such people make use of the shared body of scientific knowledge of the discipline.

In such a social system, when there is competition for scarce resources such as research funds, the pressures are high on scientists to display skill in their use of existing technical knowledge. Ignorance of practical precautions, incompetent use of abstract techniques of conceptual manipulation, misuse of technical language, and failure to refer to important earlier work, must all be avoided. Many scientists judge it prudent to exaggerate their displays of competence. They use the most complex equipment they can gain access to, they use the most difficult technical arguments, they pack their communications with unnecessary jargon, and they cite far more predecessors than they need (being keenly aware of who should *not* be cited as well as who should).

Such diligence in the use of existing knowledge is a result of the way the institutional pressures work. In my authorial role of outsider, I must be sensitive to what is going on. Even though it is inevitable in writing the last paragraphs that as their author, I should feel reflexive anxiety about how my own work measures up to such standards, the outsider role generates suspicion that much of this ritual is unhelpful. A little less technical complexity, a little less jargon, and a little less concern with honouring intellectual ancestors would do science no harm. Perhaps such a view is controversial. Perhaps the efficient conduct of science *requires* that conspicuous displays of competent use of existing knowledge always be made. I leave it to the reader to decide.

Misuse of existing knowledge is a serious problem for science and for other forms of intellectual practice in which originality is highly valued. Plagiarism, the unacknowledged use of the work of others as one's own, can be seen as a serious academic failing.

The role of formal and informal rationality in the performance of actions The manner in which scientific knowledge is used as a resource to guide actions is closely analogous to the use of knowledge in non-scientific intellectual enterprises. The special knowledge of witch-doctors in primitive societies, or of teachers in modern schools, or of any of the technologically oriented variants of science can be described similarly. Explicit formal knowledge is combined with implicit informal 'know-how' to produce practical effects which are difficult or impossible for less informed and less skilled outsiders. Informal technical practice is, however, often far less demanding of its explanatory systems than the research routines of science. Outside science, we limit the practical challenge to our ideas to ensuring that they work well enough for current practical purposes. Within science, we construct explanations which also work in the especially demanding milieu of making nature write messages which can be given publicly shared meanings and which we can be persuaded are representative of nature more generally.

In chapter 10, the possibility of science being done entirely by computers will be discussed. The present scientific use of computers is leading us in that direction. It is central to late twentieth-century science to find ways of implementing chains of formal reasoning in machine processes. Computers can operate scientific instruments, recording and mathematically processing the results. Increasingly, they can also carry out control operations, particularly in experiments of standard types that are frequently repeated. Computers do not yet, however, perform the equivalent of informal rationality. They operate effectively only within fully articulated contexts constructed and sustained by humans. Within such frames, their power and limits offers a revealing insight into the current boundary between formal and informal rationality.

Pragmatism One of the most striking achievements of science is the way we have used it to transform our world. It has steadily enlarged our repertoire of effective actions. It is in the context of the importance of practical actions that the philosophy of pragmatism emerged. One of the arguments developed in the last two chapters is that making our ideas work is not in itself a sufficient criterion for scientific knowledge. The slogan, 'it works reliably so it must be true' can be very misleading, especially in congenial immediate realities. In everyday life, we often reach practical results by constructing loops of the 'try-it-and-see' kind. That is, we form some provisional guess about a possible successful action and try acting in terms of that guess. If the outcome is successful, then we treat the basis of our guess as true (at least until it no longer works). Those who begin with strange initial speculations may easily find them working (especially in a congenial immediate reality), and so build up bizarre superstitions.

My discussion leads to the view that we should clearly distinguish between the frameworks of action-guiding ideas constructed by the processes I have been describing in this chapter, and action-guiding ideas generated randomly or by non-scientific sources. Pragmatism, with its emphasis on the success of action, fails to take adequate account of the need for constraints on ideas in addition to the requirement that they are seen to work.

I have argued that the way we see things in science is shaped by innate factors, by culture, by context, and by theory, and have developed the argument that simply observing phenomena is perhaps not as important as actively engaging nature by making it write messages or produce images with our instruments to which we persuade others to attribute a specific meaning. The final test of our competence is that we can make our ideas work well in practice, in comparison with more traditional bases for action.

I have now given some idea of the general nature of the procedures for the creation of scientific knowledge. In Part II, I will go on to apply this view to questions of the distinctive nature of science.

6.6 Summary

This chapter offers an account of the cognitive process of science suited to the version of realism already offered. There is no way of moving directly from foundational certainties to general forms of scientific knowledge. Instead, the process involves repeated cycles of extension and revision of present knowledge, with the curious feature that we are continually building upon past achievements which may then be discarded. The cognitive process is most effective when it combines openness to criticism with effective methods to resolve conflict.

There are four phases in the overall iterative cycle.

(1) We seek to bring our factual claims into correspondence with reality, often by making nature record a meaningful image of itself in our perceptions or upon our recording equipment

(2) We harmonise our theories with the facts as well as we can.

(3) We reconcile our theories within wider frames of discourse. Different reconciliation processes occur at each cognitive level. Science has powerful

mechanisms for the resolution of any conflicts that arise, especially by loca-lising the conflict to specific matters of fact and by excluding as non-scientific any issues which cannot be resolved in this way. However, these special mechanisms do not help beyond the limits of institutionalised scientific orthodoxy.

(4) We use our present understanding in acting further upon reality. It is crucial that science be active rather than merely contemplative, particularly in experimental method and in the technological uses of science that justify its social support. All systems of belief work to the standards of those who use them, but science is able to surprise us by extending the limits of what we can make work.

Part II

Does science have distinctive qualities?

7

What, if anything, is distinctive about science?

7.1 Introduction to the distinctiveness issue

'Science has transformed modern society. Whether one approves or disapproves of the way society exploits the power it provides, science appears to be distinctive in its revolutionary force.'

Are these platitudes correct? *Is* science distinctive? Is it *science* that is distinctive, or the technology to which it is so closely linked, or the economic systems in which it is embedded? Perhaps it is not the whole of science that is distinctive, but some limited part? Perhaps there is an error in the assumption that science is the kind of stable thing which could be distinctive? Perhaps science is a highly variable social construction which is renegotiated differently on each occasion?

I will try to show in this chapter that the question of what, if anything, is distinctive about science, is important and worth probing deeply. Our more casual answers are not good enough, because they are shaped by unthinking assumptions. We will come up with different answers if we concentrate on science as established knowledge, or on science as a process for producing new knowledge. We will see our problem quite differently if we idealise to the explicit rationality of the philosophical tradition or if we look at science in relation to one of the specific contexts in which it has a practical significance.

The practical significance of the question of the distinctiveness of science is an issue of science policy. Its importance is easily demonstrated in an age in which more economic and manpower resources are continually requested for science on the grounds that the investment makes good economic sense. Science is claimed not merely to be a form of culture but also to be a worthwhile overhead on technological development. Science became a major area of interest for modern governments in the mid-twentieth century. It had recently demonstrated its value by the key role it had played in the conduct of the two world wars (where science-based weaponry and defences made a major difference). It was also the time when there was a campaign (based on the problems of science in Nazi Germany and the Soviet Union) that science was at its most effective when it was interfered with least (e.g. Merton, 1973, chapters 12 and 13, originally 1938 and 1942). In this context, science policy emerged as a set of practical questions about how to identify those scientific projects which would best repay investment and how to minimise the level of spending within them without loss of efficiency.

Although the subject grew into a form of planning science, it was no more successful than other planning sciences. Much of the expertise was provided by scientists. However, this immediately raised the problem of the extent to which we can take at face value the pronouncements of the various authorities. It is always possible to argue that those who claim science to be distinctive are merely seeking to legitimate their prior interest in promoting the importance of science. Scientists, for example, are likely to be prejudiced in favour of science because they cannot afford not to be. Indeed, there is a paradox in this. If we say that those who are best informed about something are always most competent to judge it, we run up against the difficulty that the only people who are likely to have made such a commitment to its close study are those who are, or have become, biased in its favour. Conversely, those who claim that science is *not* distinctive may be under the pressure of a prior commitment motivating them to dethrone science from its privileged status. The greater the economic level of scientific activity, the more plausible it is to attribute ideological motives to the arguments of our opponents (and for them to attribute ideological motives to us).

The best that could be done in the 1960s and 1970s on the question of how to optimise government spending on science was to divide the question into the political issue of how much money should be spent on science as a whole and to develop peer review procedures within science for the question of how the money should be divided up.

Since the 1970s, the constraints on spending has increasingly been framed in terms of commercial calculations of immediate economic benefit. The older claims, that science produces its greatest economic benefits in unexpected ways at unexpected times, were sacrificed to short term considerations. Science policy issues, then, demonstrate the practical importance of the question of the distinctiveness of science, but do not provide an answer.

Some approaches to science presuppose that it *is* distinctive and look for distinguishing features. Other approaches make no such assumption. If an account of science finds no respect in which science is any different from vaguely similar enterprises, we may either accept that this is the case or suspect that the account is inadequate.

I will begin by looking at the philosophical search for an adequate account of the distinctiveness of science. I will then consider the opposing view, offered in the sociology of scientific knowledge, that science should be studied as just another social activity, and that the contrast between science and non-science is simply the result of the construction and maintenance of social boundaries. When these two approaches are combined, the problem becomes one of deciding whether the reasons that scientists, philosophers, and others offer for the distinctiveness of science are mere social devices, or if they have some kind of general validity which can go beyond the specific domains of their primary use, in which they are obviously linked to the interests of the people involved. I will argue that both views contain important insights but are inadequate as they stand. The implications of the synthetic conclusions drawn are applied to three specific questions: the relation of science to common sense, to quasi-scientific enthusiasms of popular culture, and to the intellectual activities of university culture. The chapter ends by summarising some of the distinctive features of science.

7.2 Philosophical answers

The dominant philosophical images of science are expressed in terms of its use of explicit rationality. For hundreds of years, philosophers have been impressed by the success of science at establishing a secure rational basis for knowledge, and have sought, with the aid of sympathetic men of science, to construct theories of knowledge which can account for this success and overcome the philosophical doubts. However, their theories of scientific knowledge have never completely overcome scepticism. Indeed, new sceptical arguments have been produced to meet each proposals of a new theory of scientific knowledge. Some philosophers have retreated into subjective forms of idealism, abandoning the respect of their predecessors for science as a paradigmatic form of knowledge. Others have taken the view that the sceptical arguments lead to unsustainably shallow viewpoints. For example, even if you cannot personally find a complete answer to Descartes' sceptical worry that we might be dreaming and life could therefore be an illusion, you might well conclude that yes it might be, but as long as there is no richer or more coherent form of experience around, perhaps we had better get on with it, whether it is a dream or real life.

The philosophical quest for a firm philosophical foundation for knowledge in general, and for scientific knowledge in particular, has led to the construction of the ideal of explicit rationality. To the extent that science is contained in explicitly formulated objective facts and laws, which are open to any form of critical scrutiny, then we can make science as rigorous as is possible. The logical positivists constructed such an ideal. The failure to achieve the goal of perfect rigour for scientific knowledge did not end the philosophical quest. Perhaps, as Popper argued, past and present statements of scientific knowledge will be discovered to contain errors, but a proportion of such errors can be cumulatively eliminated.

This book has argued that although the build up of explicit rationally based descriptive and explanatory systems of scientific discourse has been a mark of the success of Western science since its beginnings, these systems have never been self-contained, but have gained their significance and power from the informal individual and social rationality with which they have been combined. Any answer to the question of this chapter must recognise the essential incompleteness of explicit scientific rationality. Science may be the business of constructing and improving such a form of rationality, but until the task is completed, the various provisional fragments of explicit scientific reasoning must be regarded as uncertain knowledge.

It is in the context of this general view that I wish to persuade the reader that there is no single satisfactory philosophical answer to the question of how to demarcate science from non-science. We are, for example, likely to come to a different judgement depending on our choice of the type and level of analysis. If we are considering the rational form of completed science, as the philosophical tradition has traditionally done, our preferences may vary along the continuum between logical atomism and philosophical holism. Some philosophers have concentrated on the distinctive nature of scientific concepts (as in operationism), others on scientific statements (as in logical positivism), others on scientific theories (as Popper did), others on the way theories develop over an extended period (as Lakatos did). Those whose preferences lie nearer the holistic end of the continuum, look at science as a whole, seeing it as a coherent and interconnected network of understanding. An even more extreme view would be to say that it is not our science which is distinctive but the knowledge-producing activity of the whole society within which modern science is embedded.

Philosophical literature does not offer us just one philosophical principle of demarcation, but many. Each can be used as the basis of criticism of the others and there is no definitive way of deciding between them. For example, here is a list of philosophies of science since the early nineteenth century from which explicit or implicit prescriptive criteria may be extracted (from Dolby, 1987).

1. *Sciences are limited to the subject areas included in authoritative classifications of scientific knowledge.* A co-ordinated scheme for knowledge was constructed within which a place was given to approved forms of knowledge while others were excluded. The famous classification of Auguste Comte, for example, excluded psychology and political economy, but allowed phrenology* as a branch of physiology.[1]

2. *Empirical science is limited to knowledge produced by acceptable inductive procedures.* Sciences were required by inductivist philosophers including J. S. Mill (1875, originally 1843), to be built up from a trustworthy and systematic observational base. When a sufficiently wide repertoire of observations has been collected, they should be classified, to organise the data systematically. The canons of induction should then be applied to establish the general laws consistent with the observations. Every effort should be made to keep the process free from the distortion of bias, as when an enthusiast collects only the evidence favouring a pet hypothesis. Any attempt to produce science which departs from the proper inductive pattern should be regarded with suspicion. For example, the leading philosophers of the inductive method pronounced in this way on the scientific status of Charles Darwin's presentation of the theory of evolution by natural selection. (For historical accounts of the methodological discussions of Darwin, see, for example, Ellegård, 1957; Feibleman, 1959.)

3. *Acceptable science is a matter of convention.* According to conventionalists like H. Poincaré (1952, originally 1902; 1958, originally 1905; 1914, originally 1908), the demarcation of science should be left to agreement in the relevant community of scientists, who draw upon such considerations as simplicity, elegance, and coherence. For example, the rejection of the Velikovsky affair* as pseudo-science by Polanyi (1969) was based on a conventionalist appeal to the judgement of scientists. Polanyi invoked the tacit knowledge of the relevant scientific experts.

4. *Acceptable science must employ only operationally defined concepts.* In P. W. Bridgman's system of operationism (e.g. Bridgman, 1927, 1950, 1959), every legitimate *concept* in science should be tied to the unique and unambiguous set of operations by which it was to be observed or measured. Theories based on non-operational concepts should be rejected from science. For example, behaviourist psychologists (e.g. Stevens, 1935; Skinner 1945) invoked operationist criteria to reject the idea of a psychology based on introspective concepts.

---

[1] Comte (1969). For the argument rejecting psychology in favour of phrenology, see vol. 3 pp. 761-845, and rejecting political economy in favour of social physics, see vol. 4, pp. 264-286. See also H. Martineau's English summary in Comte (1974), pp. 380-398 and 446-50.

5. *Logical positivism: science is limited to what is derivable from true protocol statements.* According to logical positivists (e.g. Ayer, 1959), science should be capable of rational reconstruction in which the truth of all *statements* could be established simply by showing their logical relationship to the basic (protocol) statements of sensory experience. All other sentences should be rejected as meaningless. In particular, theological statements about the attributes of a transcendent deity were, they said, meaningless.

6. *Popper: acceptable scientific hypotheses must be falsifiable.* According to Popper (e.g. Popper, 1959, 1974, 1979) all scientific *theories* should be capable of being falsified by empirical test, and must be exposed to such tests. Any theory which was unfalsifiable was metaphysical, rather than scientific, and the practice of refusing to risk the falsification of a theory claimed to be scientific was actually pseudo-scientific. On these grounds, Popper denied that astrology, Marxism, and psychoanalysis were sciences.

7. *Lakatos: acceptable science is the outcome of progressive research programmes.* Lakatos (1970) modified Popper's criterion to consider sequences of scientific theories, which he called *scientific research programmes*. Because good scientists often seem to be justified in protecting their theories from falsification, Lakatos denied the possibility of instant scientific rationality and suggested that we should judge what is properly scientific by distinguishing progressive research programmes which successfully predict new discoveries from degenerative research programmes which are merely adjusted to make sense of discoveries discovered elsewhere. The progressive research programme is scientific because it manages to increase its testable content over the long term. The degenerative research programme can, in long-term retrospect, be judged pseudo-scientific. Marxism, for example, has never predicted a stunning novel fact, and has some famous unsuccessful predictions (Lakatos, 1981, p. 119).

8. *Feyerabend: 'anything goes'.* An extreme view was taken by Feyerabend (e.g. Feyerabend, 1975, 1978) of rejecting the exclusive application of any one such criterion, preferring such slogans as that in scientific method, 'anything goes.' On Feyerabend's view, there is no pseudo-science, for every alternative has its place.

This list is not intended to be exhaustive. For example, the philosophical doctrine of pragmatism implicitly provides the criterion that if ideas have greater heuristic value then they are more scientific. Marxist philosophy offered a variant theory of science in terms of dialectical materialism (e.g. Engels, 1946). Marxists refuse to abstract science from its location in society. They have offered the demarcation criterion that if the framework of assumption within which an intellectual enterprise is constructed incorporates ideologically objectionable values, then the enterprise is pseudo-scientific (Rose and Rose, 1976).

None of these criteria ever gained universal acceptance. If just one criterion had filled the thoughts of all, it might have appeared satisfactory. But supporters of each viewpoint have found it easy to generate criticisms of their rivals, and such mutual criticisms were never satisfactorily resolved. The discussion goes on (e.g. Charlesworth, 1982), even though it has been argued (Laudan, 1982) that it is time it ended.

7.3 Sociology of knowledge answers

The problem of finding a rational principle of demarcation took on a different appearance in the wake of Kuhn (1970, originally 1962). Kuhn emphasised the inadequacy of the available rational techniques for characterising science, and sought to supplement them with psychological and social description. (I am following this approach in the present book by adding informal rationality to explicit rationality in science.) Kuhn believed that he could characterise the distinctive qualities of science in terms of its character as a 'puzzle-solving tradition.' (Kuhn, 1970b) However, this criterion appeared to some commentators to apply equally well to non-scientific practices such as art and theology. Kuhn was not so sure (Kuhn, 1977, chapter 14). The new forms of science studies which grew up after Kuhn's work, especially as practised by those with no background of commitment to the values of science, tend to represent science as being just like any other form of culture. Kuhn himself was not happy with this use of his work.

Perhaps, for example, science is simply the knowledge system of modern industrial economic life. Such a view, however, might itself be an artefact of the currently fashionable kind of socio-psychological reduction of science.

The social study of science has returned to the question of the relationship between orthodoxy and unorthodoxy in science (e.g. Mauskopf, 1990; Ravetz, 1990). Now the problem is seen as one of boundary construction and maintenance (e.g. Dolby, 1982). When it is in the interest of scientists, and perhaps also non-scientists, to construct or to maintain some sort of boundary between their separate activities they will do this, even though there may be other pressures that simultaneously break the boundary down. Recent work includes Jasonoff (1987), and Starr and Griesemer (1989), which Wynne (1992, p. 297) sees as making the point 'that the boundaries of the scientific and social are social conventions, predefining relative

authority in ways which may be inappropriate, and which are open to rene-
gotiation'.

It is easy to see why scientists and others might wish to have such boundaries. A self-defining group very often constructs boundaries around itself,
either informally by the display of various marks of membership (e.g. accent,
the use of special jargon, or dress), or formally (as some professions use
educational qualifications and certified membership). In this way, they can
keep out outsiders who might dilute or undermine the shared values of the
group, and they can maintain standards and solidarity within the group. The
situation is similar in science. If an individual is recognised as a relevant
scientific authority, his or her judgement on cognitive matters counts for
more. Work which is regarded as scientific acquires some of the prestige of
the best of science. It demands to be taken seriously. Since society needs such
authorities, it encourages and recognises the boundaries produced.

The easy answer to the problem of this chapter that comes from this form
of sociological analysis is that there is nothing *fundamentally* distinctive
about science. Science is what society allows scientists to say it is. Scientific
rationality may emerge as a part of the process of maintaining boundaries
between science and non-science, along with apparently incidental devices
such as specific training in experimental and conceptual skills and the copious
use of technical jargon which only insiders can understand. In the same way,
theology is the intellectual domain carved out by theologians with the consent of others, and jurisprudence is the domain of those who study the law.

Science is less a system for the exercise of power than some other comparable professional practices, though some analysts seek to describe it in terms
of power relationships. Science is more like a mutual admiration society, in
which individuals present their findings to their colleagues in the hope that
they will be appreciated and that their work will cumulate credibility. Success
in science does not involve denying opportunity to others as much as other
social games in which the players compete directly for power.

The professed ideals of science make it extremely open. There are many
respects in which scientists do not wish to cut themselves off from ordinary
life. Because of its emphasis on explicit rationality, science is presented as
readily teachable to others. In an open society, there should be no barrier to
anyone learning science (though feminists, for example, say that in practice, it
is harder for women to enter and get on in science than it is for men). In
principle, then, anyone can learn science and apply their learning to their own
research.

Perhaps, then, the claimed distinctiveness of science lies less in the boundaries which people construct at the *periphery* of science than in the way the

central processes of science are claimed to be distinctively different from other activities. It is not simply that there is good science and there is non science. The two shade into one another through large grey areas of less good science and near-science. Boundaries can readily be constructed along only some of the dimensions involved.

In the next section, we will explore this idea, by looking at the distinctiveness of a number of aspects of science and the manner in which they are fitted together.

7.4 Which aspects of scientific activity are distinctive?

This section considers a number of standard distinctions in the nature of science, and links them to the view of science developed in Part I of this book. We will see why it is that from some viewpoints science looks less distinctive than it does from others.

There are two relevant dictionary meanings of science:

(a) Organised knowledge, often on a particular subject, as in, say, the science of physics.
(b) The pursuit of such knowledge.

Our use of the word 'science' rather than such precursor terms as 'natural philosophy' reminds us that the central characteristic of science is, simply, that it is knowledge. However, in the context of the philosophical ideal of objective knowledge, it has seemed that the key to science is that it is a process for the production of new knowledge. It is a secondary question of whether or how it is put to use. Although the knowledge-creating process sometimes has a use in mind, the knowledge produced is not locked into any particular eventual use.

We should, then, distinguish science as process and as product. The issue of the distinctiveness of science will take different forms for each.

The distinctiveness of science as organised knowledge

In this sense, science is widely thought of as displaying the following important features (though not all would agree):

(a) It is based on experience (though the term 'science' is often also applied to formal sciences and to mathematics in particular). This is clearly not unique to science. All practical life has a place for experience.

(b) It is rationally co-ordinated, very often by the rigorous use of mathematical language, but always by the application of logic. If superstitious thinking cannot survive logical scrutiny, then it is not science. This is a general characteristic of intellectual life in Western culture, and of comparable cultural enterprises in other cultures. However, science is at an extreme in its use of formal rationality.

(c) It is explicit, rather than implicit or non-verbal. For logic cannot so readily be applied to non-explicit thinking. If craft knowledge is intuitive rather than verbal, then it is not science. To the extent that science is anchored in the tacit craft knowledge of research expertise (Polanyi, 1958), the aim of the scientist is to make such tacit knowledge explicit.

(d) It is publicly shared. In principle, scientific knowledge is accessible and open to criticism to all who are competent to evaluate it. It is knowledge because it survives such criticism. The belief systems of secretive mystical traditions are therefore not science. Trade secrets and secrets of military science remain marginal as science (though they may be acceptable as technological knowledge) until they have undergone such public scrutiny. This kind of openness was most marked in the growth of science in educational settings, and is also a characteristic of other forms of knowledge which claim to be open to all who are willing to learn.

(e) The status of scientific knowledge has no reference to its source, but simply to the rigour of the supporting evidence and argument. Scientific knowledge is objective because its rational basis does not presuppose any particular set of values, it is accessible to all. On this ideal view, it is not the peculiar nature of scientists who make science distinctive but the way the knowledge is proved by methods that anyone can check out for themselves. In contrast, theological knowledge draws much of its strength from reasoning closely linked to revelation. That is, the source of the knowledge is a key factor in its acceptance. If some theological doctrine (the Catholic rejection of artificial contraception, for example) is objected to on rational grounds, the objection can be resisted, by remarks such as that it is not to be expected that mere humans can fully understand the mind of God; for humans it is a matter of faith.

This discussion has looked at the question of whether scientific knowledge is distinctive as a form of knowledge. The issue it addresses is at its sharpest if science is a fully explicit objective form of knowledge. It is less sharp if the argument of Part I of this book is accepted, that the explicit structures of scientific knowledge are always relative to informal frameworks. Even the most rigorous parts of scientific knowledge carry some of the uncertainty of the manifold assumptions on which they build. Informal rationality is a key component of scientific knowledge.

The distinctiveness of scientific knowledge in use

One point which should be made against the sociological view of scientific knowledge being shaped to fit specific contexts is that scientific knowledge is constructed so as to be transferable to other contexts. It can be understood and used largely independently of the manner of its origin by those capable of understanding the explicit reasoning involved, especially if they can reconstruct a new basis for appeal to it in informal rationality. Scientific discoveries spread around the world, from laboratory to laboratory, from discipline to discipline, and from pure science to engineering. The various social bases of science facilitate the creation of cognitive products which can move away from their point of origin. Just as much of nineteenth-century science was constructed in ways suited to being taught to new students, so much of present-day science is constructed in a context preparatory to its practical use in the development of technological products. These same modern contexts of technological development also use common sense, craft knowledge, and the tradition of technological invention. Scientific knowledge in the context of its use is therefore diluted by knowledge from other sources.

The extent to which present-day scientific knowledge is systematically diluted by technology has meant that those (Marxists, for example) who seek to define science in terms of knowledge created in response to practical need tend to blur science and technology together. On such an approach, science differs from technology only in that it is more obviously contaminated by the idealist ideology of bourgeois class interests. Such an answer collapses questions about science into the question, 'what is distinctive about technological knowledge?' It is undoubtedly the case that science and technology are closely related, even though the knowledge of each is constructed in terms of different aims. Science is concerned with the advancement of knowledge, technology with the pursuit of effectiveness in producing objects. In spite of this difference of aim, science and technology interact very closely. Science is often stimulated by technological problems; it draws upon the technology of instrumentation, and its findings are regularly applied in technology. Novel scientific discoveries can even lead to whole new technologies. Many modern institutionalised forms of inquiry fuse the knowledge-advancing interest of science with the goal of greater efficiency of a particular technology, so that there is no institutionally drawn line between the two. Science and technology have long blurred into one another in medicine. There is no simple sustainable distinction between medical practice and medical science, although the contrast is clear at the extremes. As forms of knowledge, a key distinction between science and technology is that technology is best evaluated in terms of the pragmatic criterion, 'does it work?' while scientific knowledge is evaluated by the criterion, 'is it true?' (This distinction would, however, be rejected by pragmatists.)

Science as the production of new knowledge

As a *process*, science is normally thought of as having the following features.

(a) It follows a rational methodology. Methodology sometimes covers the process of generation of new insights (or this may be attributed to creative genius), and it always covers the process by which the insights are established as knowledge. Since the seventeenth century, competing accounts have been given of scientific methodology, which vary in their emphasis on precise description and measurement in a suitable mathematical language, systematic observation, and carefully constructed experimentation. The primary claim for methodology is that it reliably yields knowledge as its outcome or at least that it minimises the likelihood of error. Intellectual practices that are based on uncritical acceptance of a tradition (such as, say, popular astrology*) are regarded as non-scientific and as more likely to be diverted into error by facing new challenges in a changing world.

(b) It is sustained within a distinctive institutional framework, the primary purposes of which are to generate new knowledge and to communicate to others that which survives exposure to criticism. Within the higher-education system, the institutional framework of science shares the wider institutional purpose of training new generations in the subject, and more specifically in how to do scientific research. Much of modern science is so esoteric that it can only flourish within this institutional framework. The explosive growth of such institutional frameworks over the last two centuries has been a significant factor in the increasing differentiation and fragmentation of modern culture. Intellectual enterprises carried on outside the institutional framework of orthodox science, are often readily dismissed as unscientific.

(c) Scientific activity leads its practitioners away from traditional forms of common sense. The activity of scientific investigation employs an ever-growing repertoire of technological aids, particularly as instrumentation. Scientific research originated from but is no longer anchored in immediate human experience. Rather it is constructed within special institutionalised frameworks of indirect instrumentally mediated theory-guided expectation. Intellectual enterprises which seek to justify themselves by appealing to intuitive common-sense judgements, rather than to the special practices of a disciplinary specialism are readily dismissed as unscientific. In the wake of unresolved scientific disputes in its early years, the main supporters of phrenology* sought such a popular justification in the nineteenth century.

A standard distinction of philosophies of scientific method is between the discovery and the justification of (new) knowledge. Where early accounts of scientific method, such as the inductive method, presented themselves as ways of finding new conclusions as well as for justifying them as true, twentieth-century accounts have made a sharp distinction.

Discovery. Popper (1959, p. 31), for example, saw nothing distinctive in the discovery process of science. Science, he thought, imposes no requirements on how discoveries are made, only on how they are justified. Some of Popper's successors argued for a logic of scientific creativity. (The subject is reviewed in Brannigan, 1981.) It can plausibly be argued that in specific contexts, some methods of generating discoveries are more reasonable than others, for they are more likely to give results. However, such arguments about heuristics do not show that there are distinctive discovery methods in science; rather there are a range of methods many of which are shared with other comparable activities. This is true for both the psychological and the social creative processes.

The standard psychological literature on creativity grew out of the introspective remarks of creative people about the creative process (e.g. Hadamard, 1954). This can be boiled down to the slogan of the four stages: 'preparation, incubation, inspiration, and verification'. As conventionally illustrated, in the anecdote, Archimedes had a problem about measuring the density of a royal crown. It was easy to weigh it, but harder to work out the volume of such a complex shape. When taking a bath, he noticed the water was overflowing as he got in and suddenly realised that he could measure the volume of the crown in terms of the volume of water it displaced in this way. He later verified his insight.

Work on creativity in the 1960s focussed upon measuring and encouraging 'ideational fluency' – the free flow of new ideas. The psychological processes involved appear to be similar in all creative fields, science being distinctive only in that most of its activity favours 'convergent' thinking – solving standard puzzles – rather than 'divergent' thinking – the creation of radically new ideas (Kuhn, 1977, chapter 9).

Just as the psychology of scientific creativity is not distinctive, so the sociology of creativity appears to follow patterns widely observed outside science. Scientists stimulate one another by interacting around collectively constructed cognitive goals, and the normal process of scientific discovery involves extensive social interaction before the production of what is later regarded as a discovery (Brannigan, 1981). Particularly in team-based research, it may not be possible to sort out the separate roles of the individuals involved. In general, the social processes involved in creativity in academic science are comparable to those of academic life in general and those in commercial science are comparable to those of technological development in general.

Although scientific discovery does not have a distinctive rational basis, it is plausible to argue that the creative processes it employs are well tuned to its needs. The high level of competition for the personal and social rewards of scientific achievement interact with the high levels of competition for status and financial resources between rival scientific institutions to produce creative practices which are the best that can be done in the prevailing circumstances. As new creative resources emerge and as the circumstances change, the creative processes change correspondingly. Of course, the competitive pressures that drive change within such a system are also wasteful. There is bad science as well as good, and what was good can become bad simply by not adapting fast enough or by adapting in inappropriate ways. In all this, science reflects the more dynamic parts of the wider culture.

Justification of scientific knowledge The main historical sequence of discussions of scientific method was devoted to the manner of demonstrating scientific conclusions. Higher standards of objectivity and rigour, higher levels of certainty about what the world is really like, were claimed to be achieved by science than by any alternative method. However, as was claimed in 7.2, this whole approach has failed in terms of its original aim. Empirical science does not produce certainty, and anyway, there is no agreement on the optimal form of scientific method.

It is argued in this book that the procedure of science includes informal as well as explicit rationality. This argument does not eliminate the importance of justifying the conclusions reached. Indeed, it is a central feature of science to turn knowledge proofs into explicit public objects which all can check for themselves. However, in the absence of definitive standards of scientific proof, we cannot offer justifications that are independent of all framework assumptions. Sometimes it is easier to say what cannot be satisfactory than what must succeed. The distinctiveness of science has as much to do with the avoidance of known sources of error as with following procedures which have proved optimal so far. In chapter 9, I will discuss the pathologies of science, known pitfalls which science seeks to avoid.

On the view presented in this book, the justification of scientific knowledge is dependent on one's position in time, in social space and the level of cognitive activity involved, so that different questions about its distinctiveness arise. I will consider the time dimension first.

Can we see in a moment's glance that there is something distinctive about scientific observation and scientific reasoning? Lakatos (1970) ridiculed this possibility as 'instant rationality'. From moment to moment, scientific judgements are like any others. Should we wait, then, for the completion of the first exchanges of scientific views on an issue, a matter of days or months? It may take many years for reactions to original claims to settle into a clear consensual judgement. Is that the right timescale on which to look for the distinctiveness of science? Or is that too long a timescale, because by that time, the intellectual environment has begun to change, and, furthermore, what is being judged is no longer the same thing as was originally proposed? An even more extreme view might be that the whole of the history of science does not yet give us time to judge. Perhaps science has been a Great Mistake, which we will all be forced to acknowledge when the problems of science-based society have cumulated far enough.

Similar issues arise as we view the distinctiveness of scientific judgements across space. We have a choice in how widely we look through space. Very often our concern is to judge science in a locality – a particular experimental practice, a local laboratory, a national institutional practice, or the science of a much wider area, such as the under-developed world. Different questions arise about the distinctiveness of science at each regional level.

Further issues arise when we consider whether there is a most appropriate level of cognitive activity at which to target our enquiry (cf., chapter 2). Are the distinctive qualities of science to be found in how our perception engages reality, in how individuals reflect upon experience, in how groups with a shared purpose come to intersubjective agreement about the natural world, in how we institutionalise our social interactions in order to come to rational agreement, or in how everyone in society collectively judges such matters?

This book assumes that there is no simple correct choice among these alternatives. Ideally, we should consider the interaction of all the levels on all these scales. In practice we deal with limited domains. We may only be interested in a local form of science (a person, a laboratory, a discipline, or a country, perhaps), and need a judgement within a limited time. In the discussion of Lysenko in chapter 4, we saw that Stalin wanted advice about the best and most practically useful forms of science within a time-scale so short that it encouraged charlatanry. It is very often the case that practical decisions need urgent answers, which may be regretted subsequently. For impractical contemplation, the wider the region we can consider and the longer the timescale, the better the judgement, though we must recognise that the local connection of science and social context becomes even more noticeable when we look at things from far enough away to see them on the largest scale.

The view I have sought to establish is that science is most distinctive as a way of producing new knowledge, that it does this by constructing immediate realities in which there is a satisfactory correspondence between our ideas and the messages we make nature construct for us. Our representations of what is going on in such immediate realities are also designed to model wider reality, although this is one of the most uncertain aspects of science. The process by which we do all this involves action and experience (or instruments which supply the same information), individual rational thought, and social interaction in shared formal and informal discourses. The procedures do not arrive at certainty, rather they optimise the output under the simultaneous constraints of the immediate reality, the individuals who are involved, the local social form of scientific practice and the numerous pressures from the wider society.

This account may be clarified further by applying it to contrasts between science and similar non-scientific activities. When the positivists contrasted positive science with theology and metaphysics, the targets of their contrasts were formulated most unsympathetically – being identified primarily in terms of their failure to meet the demarcation criterion being proposed. In the present study, I will not concern myself with refining the positivist abstractions of theology and metaphysics. I will consider contrasts of science with ordinary common sense, with quasi-scientific enthusiasms in popular culture, with various forms of institutionalised intellectual practice and with social science. In the next chapter, I will discuss the contrasts of good science from bad science, along the continuum from science to pseudo-science.

7.5 Science in relation to common sense

In general parlance, common sense is the understanding that virtually all of us share and which can therefore be presupposed in specially developed ideas and arguments. The content of common sense is never made precise, and it is often said to be intuitive rather than explicit. We assume that common sense is built upon what each of us naturally construes from our similar experiences, especially as a result of being socialised into the same culture. We do not expect common sense to be quite the same in a radically different society with different forms of life. One use of the notion of common sense is that its fuzzy representations get to the intersubjectively important essentials of things, remaining stable across variations in personal and factional viewpoints.

It is often suggested that science is just an extension of common sense. Certainly its origins were close to one form of educated common sense of the seventeenth century, and every student of science comes to it from the common sense of modern school education. In the historical route, there was, however, a sharp (revolutionary) transition between older craft practice and science, in that the science involved a new kind of self-consciousness, one which gave craft understanding an explicit nature so that it became part of the task to think about it, critically examine it and creatively extend it. As a result the growth of knowledge accelerated very sharply. This transition is discussed further in 8.5.

One way to give substance to the suggestion that science is an extension of common sense is to represent both as resources for solving problems. On this view, the distinction between them is, it may be suggested, blurred. I have already argued that when we focus upon science as knowledge-in-use, there is no sharp distinction between science and other sources of useful knowledge, among which we may count common sense.

It can be helpful to see science as a form of problem-solving. However, there is an important difference between science and common sense in solving problems which can be used in the argument that science is *not* simply an extension of common sense.

We are continually faced with problems, and our survival as individuals and as a species depends on our being able to solve the most urgent of them. However, as a species, we naturally tend to be satisfied by the first solution we find for each problem so that we can get on with life (e.g. Giere, 1988, pp. 157-61). If our problems are urgent and unexpected, we may not have time to consider more possibilities. We do not normally pause from the hurry of the rat race to seek the solution which survives the most demanding tests. Nor do we naturally tend to try to establish the most general forms of solution. Survival depends merely on getting an adequate answer that is needed in the *present* context. Very often, such first solutions work in a congenial immediate reality for reasons other than those we think make them work. As a provocatively oversimplified example, all surviving rain dances work, because every culture which has used a rain dance has either had rain fall before it was too late, or the sustained drought has destroyed the culture and those who might have judged their rain dance to have failed have died. Droughts have generally been short enough for the human species (and its rain dances) to survive. Nevertheless, (most of) we Westerners are inclined to judge that the rain fell because of factors independent of the rain dance. Similarly, the perceived solutions to many of the problems we face in every-day life work only because of favourable contextual factors. If those factors do not change much, the solutions will continue to work. The common-sense view is that if a problem solution continues to work it must be true. But such an attitude is not adequate for science.

Science is not usually motivated *simply* by the search for adequate solu-tions to practical problems. S. Toulmin (1953) suggested that scientific knowledge is like a map. A map can be of help in solving geographical problems. It can, for example, help us to work out how to make many journeys, including ones which have never been made before. We can see science as a cognitive map which acts as a resource in helping us to solve problems. If the only method we employed to construct a map was to incor-porate guesses about the layout of the land which led to successful journeys, our map would be unreliable for new journeys. Many incompatible maps could arise from such a procedure. A good map is difficult to construct out of the experience of many attempted journeys. It must be built up by different and more systematic methods of surveying the terrain which are exposed to more challenging tests.

The common sense of the successful solutions to many problems does not constitute science, even though the cognitive map of a science may provide a source of solutions for common-sense problems. The difference lies in the way in which a cognitive map of science is constructed. The ideal problem solution provided by science is a general solution which can be proved by being derived rationally from independently established general principles. At the least, an authentically *scientific* problem solution must have been exposed systematically to the most demanding forms of testing available. It is not enough to employ the first solution that works. This difference between science and common sense is in accord with Popper's philosophy of science.

We see, then, that science differs from common sense in its standards for the construction of cognitive maps and the solution of practical problems. The contrast is analogous to that between advocacy of a medical remedy used to treat a patient who recovered, and scientific understanding of the causes and nature of the disease (from which many successful cures might eventually be produced). Finding a cure for a patient in the crisis of an illness and investigating general causes of the same illness require different strategies. Since investigation to scientific standards is often not convenient in practical affairs, scientific language also differs from that of ordinary life in making a great show of caution in drawing conclusions. The rhetoric of firm *belief*, of *faith* and of *commitment* to an idea is inappropriate in scientific problem solution. Instead, the scientist strives after ideal standards of objective demonstration, even when practical pressures seem to make it unnecessary.

7.6 Science in relation to popular culture

In a scientific age, any topic of popular culture that has to do with the natural world is likely to be dressed up in the garb of science. There is, of course, a great deal of popularisation of orthodox science. But there is also a pressure for many cultural products and practices (such as advertising) to claim a scientific basis. Since the public relations exercises of impoverished intellectual establishments must use advertising techniques in shaping their messages to the requirements of the communications media, it is not easy for the ordinary person in the public domain to tell the difference between respectable scientific orthodoxy and disreputable charlatanism. Even the mass media often find it in their interest to blur the distinction between good science and dubious self-publicity.

In this context, it is easy for quasi-scientific enthusiasms to flourish which have little or no support from the current scientific orthodoxy. Many alternative belief systems often avoid competing directly with science by not making any claim to be scientific. If tensions arise, they offer arguments for why we should all be suspicious of science. Some flourish in the university system, especially at the fringes of philosophy, social science, and literature.

A rich area for alternatives to science is among those who seek to conserve specific values and beliefs. Even extreme examples from popular culture provide illustrations. For example, variants of the idea that extraterrestrials might have visited the Earth in historical times (Däniken*), that they might have an interest in human affairs, and that they might still manifest themselves as flying saucer phenomena, have been popularised many times since the mid-twentieth century. Such ideas gain support, not because of the carefulness of the supporting observations, nor because of the rigour of the supporting reasoning (both of which are conspicuously lacking in this example), but because they tie in with prior beliefs and values.

An especially rich area of alternatives to science is among those who define themselves as 'seekers' and look for solutions to problems where orthodoxy has failed. If you see your problem as lying in yourself, you may look for medical or psychotherapeutic help, and, if the orthodoxy has nothing to offer, move on to the more unorthodox. If you see your problem as lying outside yourself, in society, you may look to politics to provide a means of correcting the problem. And, if orthodox political parties cannot help, you may hunt among more extreme political movements. If you see the problem as lying outside ordinary everyday experience, you may look to religion for your answers, and, if orthodox denominations have no answer, move to more extreme cults and sects.

Such alternative systems are rationally defensible in a science-based society, especially when it is appreciated that the problems of an individual are not always well served by science. For example, if you are dying of a fatal disease for which modern medicine has no answer, it is of no use to you to wait for progress in science-based medicine. If there is a faint chance that an alternative medical system can help, you may find it reasonable to try it, even if it requires commitment and dedication, and even if the rest of society point out that the practitioners may merely be exploiting such grasping at straws.

The rationality of popular alternatives to science is rather different from that of orthodox science. Where orthodox science strives to make its reasoning explicitly follow the canons of formal and mathematical inference at as many points as possible, and anchors novel claims in their links to the past science known to its scientific audience, popular science more often links its ideas to wider cultural values, supported if necessary by appeal to common sense. An individual is attracted to a popular idea which can be linked to his or her religious beliefs, political views, worries over health, self-understanding, or concerns for the future. There is very little need to link popular science to previous popular science. Those alternatives to science which have widespread appeal are especially likely to succeed. Technical reasoning can play little role in popular alternative science, simply because so few people have access to any specific form of technical training. Even mathematical reasoning puts off most people.

A key feature of the social basis of popular alternatives to science is that they be exposed to enough people who are inclined to take them further. The lower the percentage of those who become interested, the more important it is that exposure be widespread if the issue is not to disappear from view. The ideas which work best are those which are well tuned to the media of communication and which make minimal intellectual demands on their recipients.

One especially significant phenomenon of modern culture is its need for novelty. Our mass media continually feed us the excitement of the new. We have come to crave the news. The same trend in popular entertainment encourages the market pressures that turn novelties into fads and crazes. There is a honeymoon period for novel ideas, during which they are open questions, so that it is quite reasonable to *enjoy* finding out more about them. The psychic metal-bending powers of Uri Geller* became such a topical talking-point in the 1970s.

The scientific orthodoxy cannot give *considered* evaluations and rejections of even the silliest of brand new ideas, just because they are new. Later, as they lose their novelty and newsworthiness, the uncritical enjoyment of the many is turned into the more committed belief of the few. By then the content of yesterday's novelties gets spread less effectively by mass media, and if they turn out to be unacceptable to scientific orthodoxy, its opposition becomes better organised. However, a more committed minority may by then have found one another and set up their own communication network, in which fading popular ideas can be sustained.

The characteristic rise and decline of interest in novelty tends to be followed, then, by a long-lasting residue of committed interest from a small proportion of those who were caught up in the fad. The residue of interest in a once popular idea is sometimes concentrated in quasi-religious cults and sects at the fringes of modern society. Here, the rational standards are so different from scientific orthodoxy that such popular phenomena very often become labelled as pseudo-science.

In this process, the critical attitude of the scientific orthodoxy may play a minimal role, for its considered views have difficulty catching up with the popular alternatives they chase. It may even be more important that satire comes along with criticism of novelties past their sell-by date. When we are asked to laugh at the idea of tourist trips into the Bermuda Triangle, then it is harder to take the original idea seriously enough to become one of its converts.

*Astrology** The limiting case of the manner in which popular quasi-scientific interests can be sustained in spite of establishment disapproval is astrology, which flourishes in the popular culture of the West in something like a steady-state. The high percentage of people who know their own zodiacal sign and read horoscopes for fun are not discouraged by scientific criticisms, because if it is just fun then the scientific critics are behaving like killjoys. Just enough of such people are tempted to investigate the matter a little further to provide a market for more specialised astrological practices and publications. And sufficient of the serious users of astrology become even more committed to become practising astrologers themselves and sustain the whole enterprise. Astrology has grown into a system which, like orthodox science, is open to all to learn, with its own technical rationality, supplemented by its own forms of skilled subjective judgement. As with science, the explicit reasoning, by which the positions of the planets are calculated at the moment of an individual's birth, is inadequate by itself, and must be complemented by the less explicit process by which the horoscope is interpreted.

As the case of astrology makes clear, most of what passes as popular alternatives to science parallels the *use* of existing scientific knowledge rather than the *creation* of new knowledge. Astrologers are more like applied scientists than pure scientists in the way they relate to their system of learning.

Although the contrast is clear between the rationality of science and its popular alternatives, the two are not completely distinct. For they run in parallel. Because popular culture is less critical, many alternative ideas can flourish within it. Ideas can survive in the limbo between playful speculation and serious argument, waiting their time in the hope that science or some other orthodoxy may eventually be able to deal with them more productively. Many a scientist who has some idea unsuited for or rejected by the scientific establishment has turned to the popular domain in order to give his unorthodox views an airing, in the hope of finding a following. The transition has always been easy in astronomy, which has a strong amateur base. Percival Lowell's enthusiasm for life on Mars* and Fred Hoyle's enthusiasm for the idea that the dust clouds between the stars contain large quantities of living matter, are typical examples. And the popular domain is not entirely a dead-end, for there are routes back to scientific respectability. For example, evolution before Darwin* had come to survive only as popular unorthodoxy, and A. R. Wallace was able to move from this popular belief to becoming an independent co-discoverer of the theory of evolution by natural selection. After the orthodox rejection of Wegener's theory of continental drift* in the late 1920s, the idea only survived in the backwaters of culture as a kind of geopoetry, until it was brought back in the late 1950s to explain certain anomalous results in studies of the sea-floor.

7.7 Natural science in relation to academic practices

The institutional form taken by modern science was modelled in part on earlier non-scientific patterns for the social organisation of learning. In modern institutions such as universities, the feedback between science, mathematics, technology, the learned professions, social science, and the humanities makes them all superficially alike. The academic disciplines, into which these intellectual areas are divided, have a great deal in common, a result of their shared institutional purpose of combining teaching with the conservation and advancement of knowledge. For example, all seek their own kind of intellectual rigour, taking advantage of the obsessional nature of specialised study and the reduction of the sense of urgency in academic life to raise intellectual standards for the creation of knowledge.

The contrast between science and other intellectual practices has traditionally been seen in the manner of the former's distinctive use of explicit rationality, as it builds up systematic knowledge by rigorous chains of reasoning that begin with unassailable scientific facts. Even philosophy is thought often to fail to meet the rigorous standards of science (because it is so often concerned with the grandest and hardest questions). Only mathematics and logic as pure forms of inquiry are seen as more rigorous, and they differ from science in being minimally concerned with the empirical world.

This book has argued that explicit rationality is less central to science than is normally thought. Facts only become settled through negotiation in the context of uncertainty. The chains of explicit reasoning always depend on uncertain framework assumptions. It would appear, therefore, that, on the present analysis, science should be less distinctive from other intellectual practices in similar institutional settings than has been thought. For example, the image of science as exceptionally objective and neutral is undercut by arguments that even in science, account must be taken of the observer, although this is not done self-consciously. Similarly, the notion that science seeks and finds universal laws has been undercut by the argument that science is no longer organised conceptually by statements of law and that there are fewer satisfactory scientific laws than is generally supposed.

How does science relate to the older disciplines of the humanities? The main point of contrast would appear to be the subject-matter. The natural sciences are concerned with the external world, while the humanities are concerned with mankind and its cultural products. There are also differences of method, which have been the topic of interminable discussion. History, for example, is concerned with specific events more than with the general pattern of events. It is always concerned with human agency, and very often tries to give insight into the actions of historical figures. It generally presents its conclusions in the form of narrative. Natural science, in contrast, is not concerned to understand human agency, but rather uses prediction and control as ways of substantiating its claims to have understood phenomena. Philosophy is more diverse in its concerns and methods than natural science. But philosophical arguments often try to build on universal principles rather than the elaborate suppositional frameworks used in natural-science communities. Philosophy has traditionally been more contemplative and less active in its investigative methods than science.

How does science relate to its newer imitators, the social sciences? Again the key differences are subject-matter and method, although the contrasts are a little more subtle. The social sciences typically address issues on which it is difficult to operate within a consensual framework. Much more is done by argument between rival approaches, and less by development of rigorous chains of reasoning from generally accepted starting-points.

The natural sciences differ from the humanities and social sciences in that they deal merely with the external aspects of phenomena and do not need to consider in their subject-matter the complexities of meaning that pervade human culture. The disciplines of humanities and social science are said to be *hermeneutical*. To the extent that the present approach puts the observer back into the subject-matter, there is a place for hermeneutics in natural science too. The thought and actions of an active observer help to make meaningful the messages nature is constrained to give us. However, as scientists are not very self-conscious about these points, the complexities of meaning are nothing like as great as in most aspects of social life.

The humanities and social sciences are often said to deal with more complex subject matter than the natural sciences. For example, in the hierarchy of the sciences constructed by the positivist, Comte, sociology was seen as the last and most complex science, in that its subject-matter is also subject to the laws of all the other sciences.

On the view taken here, the simplicity of the natural sciences is at least in part a result of the methods by which scientists actively make their immediate realities simple. Confusing factors are actively excluded, at least until the conceptual system is able to cope with them. To do the same thing in the social sciences would often raise moral issues, for we cannot experimentally simplify aspects of the social world without changing normal social interactions in morally objectionable ways. (Just think what could be done to establish the interaction between genetics and environmental factors in human development if only we could do controlled experiments using such genetically identical sets of individuals as identical twins and clones.)

The earlier chapters of this book provide yet another reason for seeing the natural sciences as having a lower order of complexity than the human sciences. The immediate realities through which we come to our understanding of other people in social science is changed by their actions and by the beliefs that inform them. The reflexive features of immediate reality mean that there can be no final answers in social science, as long as it is in someone else's interest to act differently from the manner provided for in our answers. It is therefore foolish to try to model the human sciences too closely upon the natural sciences. For example, the problem of reflexivity makes the prediction of future actions feasible only in the short term, until knowledge of the prediction leaks out. Therefore it is not appropriate to make successful prediction a general criterion of good human science. Explaining past human actions is far easier, though ambiguities in interpretation of the evidence make it more likely that one's explanation is wrong. The doubts I have expressed about the traditional regard for scientific laws in natural science hold more trivially in social science. Descriptive laws of human behaviour hardly ever hold for uncoerced people who have an interest in acting contrary to the law, and if they know of the law, they might easily invent such an interest. I argued in section 5.5 that unrestricted natural laws are reports of the limits of human ingenuity in generating exceptions. Such laws can be found in social science, but there are not enough of them to explain much of human action. One such unrestricted law at present is that all human beings are mortal, but perhaps, one day, we will even find a way around that.

My own view of social science is that it is best regarded as a form of critical thinking. To the extent that it is concerned with precise *facts*, these are historical facts (that is, about the past, both recent and remote). Factual claims about the present and future are inextricably tied to questions of value. Therefore, social science is as much philosophical and historical as it is scientific.

Science is also closely related to technology as forms of intellectual activity. In 7.4, I emphasised the extent to which technology dilutes its use of scientific knowledge with knowledge from other sources. The overlap is also a feature of the two as ways of creating new knowledge. In the modern scientific age, the tradition of practical invention (as institutionalised by the system of patents for the right of initial monopoly over the exploitation of new inventions), has been greatly affected by science. Before the scientific revolution, invention was a hit-or-miss affair in craft practice. By the industrial revolution, inventors had ways to seek invention. They could systematically try variations on existing practice, looking for improvements. They could look for possible new practical phenomena of commercial significance by systematic variation on a range of possibilities. The increasing importance of science in technology was emphasised by the rise of science-based technological training. Eventually inventors could also trawl through the sea of scientific literature with nets set to catch exploitable possibilities. The creation of new technological knowledge now ranges from accidental improvement of craft practice to the self-conscious construction of new technological possibilities on scientific principles. The blurring of the two forms of knowledge creation is greatest in the areas which have attracted most research funds for science in the hope of practical benefits. However, the criteria of success remain different for technology and for science. In technology, success comes with a product that works and in particular, that works commercially under present market conditions. In science, in contrast, the criteria of success is not that it works, but that it is accepted as true.

7.8 Conclusion

In this section, I list a series of respects in which I believe it will be most productive to seek whatever distinctive features science has. Many of the important features of science are shared by similar intellectual activities and are not listed here. For example, many other intellectual practices besides science strive to draw upon experience, to be rationally co-ordinated and to be as explicit as possible. However, I have listed some more widely held features which are especially important in science. Other features are not listed because I have argued that they belong to images or ideals of science that are unrealistic or unrealisable.

1. One distinctive feature that arises from the anchoring of science in the combination of experience and the explicit use of reason is that a *science sacrifices completeness for rigour*. Even though perfect intellectual rigour is never obtained, it is always sought. That is, a scientist can never deflect a criticism accusing him or her of sloppy work by commenting that there is no need to be so rigorous in this science. In striving to be rigorous in its empirical grounding and in its rational representations, science in general tends to fragment into particular sciences, each of which is limited to some branch of phenomena investigated by practitioners using a characteristic repertoire of practical and conceptual skills. Science, as a whole, builds up as a collection of narrowly defined systems of more rigorous understanding which are only weakly linked to one another. A marked tendency of the search for rigour in experimental sciences is to construct artificially simplified (or isolated) domains of experimentally controlled phenomena in which explicit descriptions and explanations can be *made* to work, simply because an apparent failure is treated as a failure to control the phenomena adequately. The interface *between* the knowledge constructed in domains linked to different experimental protocols is not usually so rigorous as that *within* each domain.

The sacrifice of completeness for rigour is in contrast with the requirements of a technology, which must provide the best available solution to a problem even if available knowledge is known to be incomplete. It is also in contrast with complete world views.[2] Religions very often function as complete world views, in that, even when they do not provide direct answers to important questions, they offer the hope that the answer is contained within them. In contrast, science always has a place for human ignorance in its structure. If science cannot explain the meaning of life, the universe, and all that, it is not obliged to try, for Douglas Adams's question (to which the joking answer is 42) may not have a serious scientific answer. Extrapolations of science into a

---

[2] In the sense of the German term, *Weltanschauung*, which has been used quite freely in history of science.

complete world view inevitably go beyond science proper.[3] Among the areas in which the sacrifice of completeness for rigour in science is especially striking is in the refusal to make a choice of moral values (for how can one be rigorous about the choice of values?).

2. In section 7.4, I stated that *science is publicly shared, and scientific knowledge is in principle open to public criticism.* This is a feature of most didactic forms of knowledge, especially those occurring within the modern university tradition. It is often missing from less open systems of belief, such as those which flourish in secretive and totalitarian contexts. It is in decline in those commercially relevant parts of science which blur into technology. In some parts of molecular genetics there is a prospect of quick commercial exploitation of even the most fundamental discoveries, so that patenting precedes publication, and the art of protecting a patent monopoly often involves keeping some of the relevant knowledge secret.

3. *The content of science is subject to change.* As a consequence of the last point, the cognitive content of science is continually being augmented and revised. New science is constructed out of old science, and as a result some of the old science has to be discarded. *Current* science is that which survives *current* criticism. *Future* science (also) has to be able to survive *future* criticism. We expect future science to reject a significant part of what is now acceptable. This feature of science makes it quite unlike tradition-based belief systems. As was noted in the comparison of science with common sense, science does not *require* declarations of faith or commitment from its practitioners (even though they may be prepared to make such declarations). You can not be expected to have undying faith in doctrines you anticipate will undergo change in the future. The very institutions of science presuppose that its content will change.

4. I noted in 7.4 that *science is sustained in a distinctive institutional framework.* Within this framework, success in the systematic search for empirical novelty and the rational assimilation of that novelty is valued and rewarded. Rational criticism, social recognition, and appreciative exploitation of new work are complementary aspects of the same institutional processes. It is this institutional framework that sustains the rational, social and technological routines of science, transmitting them to successive generations. It is this framework (referred to as 'orthodox science') which allows the continuity of social support for science and the cumulation of perceived success. And

---

[3] Sometimes such extrapolative answers that try to provide a whole approach to life in the spirit of science are known as *scientism* (Cameron & Edge, 1979).

it is this framework which obliges the scientific practitioner to persist in scientific investigation, even when it is in conflict with established knowledge.

5. *The distinctive features of science are relative to its changing institutional form.* The importance of its institutionalised procedure means that what constitutes acceptable science is, to some extent, dependent on the social context. For example, the institutionalised practice of science has changed over time. The rituals of displaying the connection of one's work to that of one's predecessors (published work should cite relevant earlier publications) have become more elaborate over time, as have the processes of peer review of work seeking research grants or publication. In Dolby (1987) I defended the view that modern Creation science is acceptable science by the standards of the late seventeenth and early eighteenth century, even though it is not acceptable science by the more demanding standards of the late twentieth century. Similarly, the institutional pattern of science varies somewhat from place to place. In historical periods when the science of separate regions was not in full communication, different regional standards applied. To some extent also, variant standards apply in the institutional practice of different scientific disciplines.

6. There are, then, many local variations of the institutional framework of science. *Claims for scientific status may arise at any cognitive level, but are distinctively scientific only if they can survive at all cognitive levels.* Scientific knowledge requires the interaction of the individual and social levels. For example, although new scientific findings and insights may be constructed by individual geniuses, by discussion between collaborators in a research team, by graduate students following an educational routine, or even by machinery employing the latest principles of artificial intelligence, such creative activity does not constitute science until it is exposed to and survives criticism in the public domain. Idealised views of the content of science require it to be universal. If a claim were to stick at the level of its origin, it would not be science. There is nothing in the nature of the creative impulses or the organised reasoning that go on in a person's head that make them characteristically scientific. The creative and critical processes are not fundamentally different from, say, mathematical creativity or literary production. Nor are the creative and critical processes that go on in a specific local setting necessarily distinctive. Something may be accepted as science in a local (institutional) context, but unless it is able to become accepted elsewhere, in other institutional contexts it is only local science, not science proper. There is a great deal of local science which never becomes science proper. Other cases of local science become controversial science. (For example, I will later draw upon the argument of Sulloway (1991) that Freud created a special

institutional framework to sustain the psychoanalytic knowledge linked to psychoanalytic practice. Because this institutional setting was cut off from scientific and medical orthodoxy, and because of the way it was organised, Sulloway's account helps account for why Freud's doctrines thrived in some contexts while being regarded as discredited in others.) To a lesser extent, one can say that what is accepted as science at a particular time may not meet the standards of later times. However, here the custom is to say that the earlier science was indeed science, it is now old science, and no longer accepted as true, rather than non-science. A single person or a local practice may be able to anticipate the wider and more demanding requirements of wider regions and later times, but cannot have the authority of the later developments.

The precise details of the levels of cognitive activity as described in chapter 2 are not essential to science. A scientific instrument which provides a read-out in a human language can replace the perceptual system. Some development of artificial intelligence might replace individual human reasoning. Society might be structured in a quite different way, perhaps with communication processes sustained by computerised machinery, human beings, afraid of infection, never coming together to exchange ideas. But, unless the cognitive activities that replace those now occurring at these levels are allowed to affect one another, working towards cognitive conclusions sustainable at all levels, it is not science.

7. Crucial to the distinctive nature of *science* is that it *is focussed on the generation and rational assimilation of novel facts*. The last chapter explained the procedures involved. Facts are socially negotiated intellectual constructions, and as dependent upon the culture in which they are generated as they are on the aspect of external reality which they are claimed to represent. It is also part of the institutionalised procedure of science that the co-ordinating network of ideas is required to make sense of the novelty of what is newly accepted as fact. This feature of science can be dually represented in social terms and in rational terms.

8. *New scientific findings aim at coherence with existing science*. It is vital to the assimilation of facts that a way be found to connect them to existing knowledge. Ideally, facts should be represented within a coherent system of explicit rationality. Claimed facts which do not cohere with existing knowledge are not facts but puzzles, the assimilation of which may require their rejection or the modification of existing understanding. When a pattern emerges in a range of novel factual claims, and that pattern works against the wider informal patterns of coherence in science, there are similar problems in its assimilation. Evidence for human powers of telepathy and clairvoyance, striking though they appear in their own context, have yet to be

assimilated into the general patterns of science, and could only be assimilated by revolutionary changes in science. Just think of how much more misleadingly congenial immediate reality would be if merely thinking about an experiment made it more likely to succeed. Would we ever get the same result in an experiment done by believers and disbelievers in the theory under test?

9. Another distinctive feature of *science* is that it *is progressive*. However, I do not think that its progressiveness lies in the cumulation of facts, because the scientific content of an observation report changes with change of context and of scientific theory. Nor is science demonstrably progressive in the sense of cumulating general knowledge, for scientific knowledge is radically reworked in scientific revolutions. The way we assess the cumulation of scientific knowledge also changes as society changes.

The progressiveness of science comes from the technology it generates. Through technology, science provides a cumulative power of intervention in the part of the world we have learned to shape and control. It is an ever-growing resource to guide action.

This claim that science is progressive is taken as innocuous by most defenders of science, is regarded with suspicion by those attracted by anti-science, and is dismissed with scorn by those who regard scientific knowledge as a purely social construction, relative to the society producing it. As social relativists see it, science changes as society changes and for no other reason. I would offer the following as a litmus test of whether the reader's own views are compatible with the doctrine of the progressiveness of science. If you, the reader, are prepared to take seriously the idea of giving up science and starting again, on quite different principles, then you doubt that science is progressive. If however, whatever doubts you have about present-day science, you have the feeling that it is too valuable a resource to be abandoned, then, even if you think that it could be greatly improved, you intuitively accept the idea of scientific progress.

7.9 Summary

There are many philosophical accounts of the distinctive rational nature of science, each of which may be used to generate criticisms of the others. In this context, none has gained universal acceptance.

Scientific knowledge is ideally based on experience and rationally co-ordinated. It is explicit and publicly shared and although it may contain values, it does not *depend* on any. *The scientific process* ideally follows a rational methodology sustained in a special institutional framework. Scientific *discovery* involves psychological and social processes that have not been shown to be significantly different from creative processes elsewhere. The *justification* of discoveries involves knowledge proofs which aspire to absolute rational standards but also depend on special support varying with time and place.

Orthodox science differs from common sense in that it does not stop with the first satisfactory solution to the problems it seeks to solve. It differs from the quasi-scientific enthusiasms of popular culture in that it employs close technical reasoning from facts rather than building its support by appeal to widely held interests. It has much in common with the academic intellectual practices of the humanities and social sciences. What social science can do best is to provide further tools for critical thinking about social matters. While academic knowledge-constructing strategies simplify natural phenomena, they do the opposite to human phenomena, for people subsequently use the newly provided conceptual tools to complicate social interactions.

Scientific knowledge is not a complete web of belief but the outcome of a limited procedure for the generation and rational assimilation of novel facts. It cumulates novelty by the repetition of this process, so that its cognitive content often changes. Eventually, changes are forced in the wider belief systems within which it is embedded. Its continued practice cumulates power to intervene in the world. It has features at many cognitive levels and would break down into a less distinctive cognitive form if it were to lose its openness to perceptual surprise, to private thought, to collective local enthusiasm, to orderly cumulation, and to public criticism.

8

How is good science to be distinguished from bad science?

In this chapter, I will consider knowledge-producing activities which are claimed by their proponents to be scientific. The best practices are properly scientific, the worst are pseudo-scientific; in between lies a continuum from good science to bad science.

Our immediate judgements of pseudo-science are very often prejudgements, coming *before* careful open and rational deliberation. To turn the prejudgements into explicitly reasoned judgements involves identifying and scrutinising many hidden issues. Consider, for example, your own thinking in evaluating an area of popular belief which you are *not* inclined to take seriously. It might be the idea that putting a pyramid over a razor blade will keep it sharp indefinitely (pyramidology*), or that the Bermuda Triangle* poses a genuine danger to Caribbean shipping, or that psychic surgery really does involve removal of internal organic material, leaving the skin unbroken. If you dismiss any such case with a chuckle, rather than after careful investigation, you are presumably relying on the feeling that the very idea is ridiculous; you are, perhaps, fitting in with the conventional establishment attitude, which is to scorn such things. Your attitudes are not the result of your own considered thought, and it may be quite difficult to turn them into explicitly reasoned conclusions.

If our distinction between good science and inferior or pseudo-science is not just to be a label for value judgements with no explicit rational basis, we should identify criteria which can be applied to any case without prejudgement to help us come to clear conclusions. The criteria offered in this chapter will be developed further in the next chapter, which offers a theory of the pathologies of science.

8.1 Science in relation to pseudo-science

'Pseudo-science' is, literally, false science, or that which seems to be science or is professed to be science, but is not really. The term 'pseudo-science' has been used since the mid-nineteenth century as a pejorative term to debunk claims to knowledge on the grounds that they fail to meet the standards of science – they are foolishly or fraudulently conceived, in error, supported by irrational arguments, etc. Typically, such labelling has been done by defenders of the scientific establishment, and the targets of such labelling have been marginalised and discredited rivals to science.

The prophets, pioneers, and apologists of the scientific revolution of the seventeenth century and the popularisers of science-based rationality of the eighteenth-century Enlightenment, had comparable labels for what they regarded as false or sacrilegious knowledge. Such enterprises as Aristotelianism, astrology*, alchemy*, hermeticism, magic, cabalism, and so on were variously abused as dogmatism, quackery, mountebankery, trickery, and necromancy. Modern defenders of scientific enlightenment still attack as pseudo-scientific the traditional doctrines of astrology*, magic, and mysticism and also the latest intellectual fads of popular culture.

The attacks appear justified if claims of scientific status for the proponents of the marginal systems of belief can be shown to be undeserved. Therefore, if the label 'pseudo-science' is to be used satisfactorily, it should be possible to demonstrate the error or incompetence or fraudulence of the intellectual enterprise to which it is being applied. As this chapter will show, such a demonstration is not easy. In practice, most of those who have attacked intellectual practices as pseudo-science have regarded it as intuitively obvious that something was wrong and have then searched around for a way of legitimating their intuition. They have very often done this without careful study of the target of their attack, regarding their viewpoint as justified by educated common sense. Once the target is identified as pseudo-science, the critic looks for dishonest or foolish motives of the proponents or to the irrationality of their methods. Perhaps the same critical enthusiasm could have found similar targets in accepted science. There is a lack of symmetry between the way in which we use the language of 'establishing the truth' to account for our findings, but attribute the errors of our opponents to the intrusion of non-rational factors (Gilbert & Mulkay, 1984). Even such an eminent philosopher as Sir Karl Popper gave the appearance of being quite sure that astrology* is a pseudo-science (Popper, 1974) long before he had made clear how his analysis of pseudo-science legitimates his dismissive attitude.

The concept of 'pseudo-science' requires two separate attributions, very often by two different groups. The first is the claim of scientific status, and the second is the judgement that this claim is unmerited. Each of these attributions is likely to change over time in historical cases. For example, both Creation science* and evolutionary theory have been claimed as scientific by their respective proponents since the early nineteenth century. However, before Darwin's *Origin of Species* (1968, originally 1859), Creationism was the scientific orthodoxy and evolutionary theory was marginal. Today, in contrast, evolutionary theory is accepted as the scientific orthodoxy, and modern Creation science* is marginal to science and cognitively acceptable only in the Fundamentalist religious groups which sustain it.

8.2 Philosophy: science versus pseudo-science

The philosophy of the demarcation of science from pseudo-science is a topic with its own history. Here are some highlights. During the scientific revolution, prophets, pioneers, and apologists for natural philosophy were concerned to give legitimacy to their preferred form of knowledge and to distinguish it from other intellectual practices, with which it might be confused. Natural philosophy was separated from speculative philosophy, from theology, from the occult, and, in some contexts, from practice-oriented forms of knowledge such as medicine and arts and crafts. Special attention was given to demonstrating the contrast between proper knowledge and forms that were to be discredited, such as astrology, alchemy, and natural magic.

The idea of a distinction between acceptable and unacceptable claims to knowledge became part of the armament of those who campaigned on behalf of science in the centuries that followed. When, in the Enlightenment of the eighteenth century, appeal was made to rationality in undermining the traditional authority of church and state, reason based on systematic experience became a new kind of authority. Religious and scientific bases for knowledge were slowly disentangled.

After the French Revolution, early nineteenth-century positivism operated in a world in which science was regarded as having displaced religion. Positivism was particularly concerned to distinguish between the positive method of a proper scientific approach and the more primitive forms out of which it developed – theological and metaphysical thinking. Auguste Comte identified the three stages (theological, metaphysical, and positive) as stages of human history, of individual maturation, and of the development of each branch of science. The theological phase looked for explanation in terms of quasi-human or divine agency and the metaphysical phase sought explanations in terms of hidden causes such as essences. In contrast to both, the positive phase limited itself to collecting the facts and inferring the relationships among the facts. Comte rejected any aspect of science which retained residues of theological or metaphysical thinking in his *Cours de Philosophie Positive* (1969, originally 1830-42).

The positivist conception of science did not have it all its own way in the nineteenth century. Romanticism incorporated a rejection of the Enlightenment tendency to reduce everything to reason. Rival romantic conceptions of science briefly flourished in the early nineteenth century, and echoes of them still persist. Positivism contained within itself trends which moved away from scientific orthodoxy. In particular, the idea of giving science the functions of religion encouraged efforts to extrapolate science to cover all aspects of human life. Positivism affected the rise of psychical research into the phenomena of spiritualism, for in this way some religious ideas might be given a scientific basis.

By the early twentieth century, the flourishing of so many variants to scientific orthodoxy made the pseudo-science issue increasingly significant.[1] How could science proper be distinguished from the unjustified extrapolations and the false imitations? That twentieth-century question is the concern of this chapter.

Its discussion may be seen as a topic of applied philosophy. Philosophical problems of this kind do not always progress satisfactorily through intellectual discussion. Firstly, like any philosophical tradition, it is likely to lose its way as a result of its own intellectual processes, and, secondly, it is likely to be corrupted by the pressures of its biggest market, the ideology of the legitimation of preferred forms of science. I will briefly explain these two risks in turn.

It is a general problem in the history of philosophy, from which the philosophy of the demarcation of science is not exempt, that endless generations of discussion may actually be to the detriment of the subject rather than to its benefit. In the late nineteenth century, one commentator on the history of logic saw it as representative of how the later stages in the development of a subject can pervert its original aims.

The convenience of the teacher, the love of symmetry, the love of subtlety; easygoing indolence on the one hand and intellectual restlessness on the other – all these motives act from within on traditional matter without regard to any external pur-

---

[1] Though not everyone has been convinced. See for example, Laudan (1982).

pose whatever. Thus in Logic, difficulties have been glossed over and simplified for the dull understanding, while acute minds have revelled in variations and new and ingenious manipulations of the old formulae, and in multiplication and more exact and symmetrical definitions of the old distinctions.

(Minto, 1899, p. 2)

The second difficulty is that intellectual interest in the distinctive nature of scientific knowledge and scientific method is inevitably affected by the ideological needs of people to legitimate their preferred form of science. Even philosophers who write for one another cannot ignore the wider impact of their work. The popularity of Popper's falsificationist conception of science serves as an example. The simplest and most dogmatic version of Popper's philosophy (a version which it is easy to find criticised in Popper's own writings) declares that, while it is impossible to verify a scientific theory completely, a single publicly accepted observation can falsify it conclusively. Such a view provides a very sharp criterion of demarcation between the sciences, which are concerned to falsify their (testable) conjectures, and the pseudo-sciences, which avoid putting their speculations at risk of falsification. Such black-and-white judgements are very helpful to those defending specific doctrines as science. All that need be done is to show that an opposing view is unfalsifiable, or its defenders actively avoid trying to falsify it. (It will also be necessary to distract attention from the point that falsifying one's own theory may not be straight-forward either.) As we will see, it is very hard to sustain such a simplistic version of falsificationism against criticism. More readily defended versions of Popper's views make the demarcation between the scientific and the non-scientific less sharp, and do not allow 'instant rationality' in evaluating scientific arguments (Lakatos, 1970, p. 174). The greater ease of application of oversimplified forms of Popper's philosophy leads to their greater influence in ideologically committed strands of discussion and to a degenerative pressure upon philosophy.

I see the ideological uses of the philosophy of demarcation as important in practice because so many writers who invoke demarcations between science and pseudo-science do so in a way which shows that they have made their judgements of which is which before they start. They know what conclusions they will reach, and are looking for ways to induce others to accept them.

These critical comments are intended to suggest that one should not accept the philosophical approach to demarcation too naively. However, neither should the criticisms be accepted without question. Those disciplines (sociology in particular) which often wish to deflect philosophical criticisms of their own practice find it all too easy to dismiss philosophy as *merely* an ideological instrument. I propose that the question of the proper role for a philosophical approach should not be prejudged, but should be an open issue (initially, at least).

After these preliminary remarks, let us look in more detail at one case of how the philosophy of demarcation of science from pseudo-science has been discussed: the work of Karl Popper.

8.3 An example: Popper's demarcation criterion

Sir Karl Popper (1901-94) has been one of the most influential twentieth-century philosophers of science. He grew up in Vienna at the same time as the Vienna Circle was becoming the centre of logical positivism. Although he disagreed with them over many fundamental issues, he often sought to answer similar problems. He objected to the idea of rejecting all non-scientific statements as meaningless, and to the idea of using verifiability as a criterion of meaningfulness or of science. But he was concerned with the distinctive nature of scientific knowledge. He sought to avoid Hume's problem of induction by arguing that rather than arrive at general laws by induction, we may devise hypothetical law statements in any way we like (provided that they are compatible with past experience) and we should then seek to falsify them, so that we can learn from our mistakes.

Accounts of Popper's ideas by himself and others are widely available (Chalmers, 1982; Magee, 1973; O'Hear, 1982; Popper, 1959, 1974, 1979). A brief exposition and critique is included here, as Popper's arguments are especially suitable as an introduction to the philosophical demarcation of science from pseudo-science.

The essentials of Popper's criterion of demarcation

Assume that we have deliberately made it our task to live in this unknown world of ours; to adjust ourselves to it as well as we can; to take advantage of the opportunities we find in it; and to explain it, if possible (we need not assume that it is), and as far as possible, with the help of laws and explanatory theories. If we have made this our task, then there is no more rational procedure than the method of trial and error – of conjecture and refutation: of boldly proposing theories; of trying our best to show that these are erroneous; and of accepting them tentatively if our critical efforts are unsuccessful.

(Popper, 1974, p. 51.)

When seeking to build up an explanatory system of universal statements from an evidential basis of particular observations, Popper reminds us of the logical point that, while no finite sequence of observations can conclusively verify a (non-tautologous) universal statement, if an observational prediction is deduced from a universal statement, then an observation which negates that prediction falsifies the universal statement. Because of this asymmetry between verification and falsification, we can, he argues, learn more about the world by organising inquiry around the search for falsifications rather than verifications. For example, if our conjecture is 'All ravens are black', then we should proceed by attempting to find a raven that is *not* black rather than by looking for supporting instances.

In order to turn this asymmetry into a way of learning about the world, Popper requires that, in the face of a contradiction between theoretical prediction and observation, we should always reject the theory and not the observation. For this to be sensible advice, we must be able to specify in advance the kinds of observations which are unambiguous enough to have such powerful status in our reasoning. Clearly, a fleeting glimpse of a white bird that appeared to be a raven should not in itself lead to the abandonment of the conjecture that all ravens are black.

The universal statements we employ should be capable of falsification, indeed, they should be formulated in such a way as to make them as testable as possible. In particular, we should not define away the possibility of conflict between theory and new observation, or our theories will be empirically vacuous. Universal statements which are not empirically testable are metaphysical.

We should always proceed in science by seeking falsifications and rejecting universal conjectures which have been falsified. To act otherwise while claiming the status of science is pseudo-scientific.

When a conjecture is falsified, we must devise a new as yet unfalsified conjecture to replace it. Popper gives little rational guidance. We should not, however, produce the new conjecture merely by adjusting the original so as to exclude the falsifying instance (*ad hoc* modification). It would not do, for example, to revise 'All ravens are black' after discovering Sam, a white raven, by formulating the new conjecture, 'All ravens except Sam are black.' In general, our new hypothesis should be unfalsified by past experience and be at least as testable as the previous conjecture. To permit anything else would be pseudo-scientific.

Those conjectures which have survived attempts at refutation can be given some of the status of scientific knowledge. (Popper says that they are well corroborated.) Scientific knowledge will always remain fallible, for it is always subject to subsequent falsification. Although they remain liable to subsequent refutation, well-corroborated theories can used for practical purposes.

Typically in science, universal conjectures cannot be tested completely independently. Background knowledge must be assumed – for example, in the theory of the instruments used or about what is affecting the state of affairs under study. Normally, this background knowledge should consist of well-corroborated theories. Predictions are deduced from the conjecture in conjunction with the background knowledge. An observation which negates the prediction actually refutes the combination of background knowledge and conjecture. The refutation may be resolved by adjusting the conjecture or some aspect of the background knowledge. However, for the purposes of the test, Popper suggests that the background knowledge be regarded as better established than the conjecture, so that it is the latter which should be rejected.

Popper's criterion of demarcation applied to psychoanalysis

Freud's theory* and therapy of psychoanalysis have had an immense cultural impact, but have always been controversial. At least part of the controversy has been because Freud demanded the status of science for his ideas, but resisted the pressure to follow normal scientific procedures. He attempted an interdisciplinary synthesis, making assumptions which soon lost their disciplinary basis elsewhere (Patricia Kitcher, 1992). Rather than try to update the disciplinary connections, he constructed a new institutional milieu, the psychoanalytic movement, in which the authority for the empirical bases of his claims was located (Sulloway, 1991). His arguments had such rhetorical force that an immense number of clever people have found them irresistible. But the new way of thinking about human nature was inspirational rather than critical. Freud's philosophical critics accused him of providing the psychoanalytical movement with few mechanisms for the discovery and elimination of errors apart from the exercise of his own judgement. Even during his lifetime many rival developments from his original ideas emerged. After his death psychoanalytic doctrines multiplied, restrained only by institutional pressures for conformity. Eysenck (1985, p. 18) claimed that there were then more than 100 competing schools of psychoanalysis in New York. The discussion which follows does not attempt to deal with the general cultural significance of Freud's work, but is limited to how Freud's claim to be doing science can be evaluated by criteria related to Popper's demarcation proposal. The essence of Freud's theory and of the critical reaction to it is given in the appendix.

Popper developed his demarcation of science from pseudo-science with many supporting arguments and illustrations. His favourite illustration of good science was Einstein's Theory of General Relativity which, he believed, had shown its credentials by risking (and surviving) challenging tests. One of his favourite examples of pseudo-science was Freud's* theory of psychoanalysis. This was fashionable in early twentieth-century Vienna, and readily confirmed by countless observations, but very difficult to refute.[2]

The problem immediately arises that it is not possible for Popper to demonstrate empirically that such a hypothesis is completely irrefutable, for that would require showing that none of the infinite number of logical consequences is testable.[3] All Popper can argue is that no testable consequence has yet been found, which might be due to the lack of imagination by philosophical critics in thinking up suitable tests. In the case of psychoanalysis, there appears to be adequate evidence that many of Freud's conjectures in psychoanalysis were initially formulated so as to be refutable, and that Freud's critics often set out to refute them.[4] It is rather easier to argue *ad hominem* that Freud was pseudo-scientific in the sense proposed by Popper because he claimed the status of science for his ideas and yet did not look for demanding tests, instead adopting stratagems which made it harder rather than easier to falsify his key ideas. Cioffi (1970) has discussed Freud from this point of view.

We should note also that psychoanalysis is not a pure science, but a theory closely linked to, and in part justified by, the claimed effectiveness of psychoanalytic therapy. Popper's conception of science judges the epistemological value of a theory as coming from its survival of bold attempts at falsification, while the tests of a theory in an application such as therapy are not normally at all searching. Clearly it is not in the patient's interest to be treated by an analyst who is trying to maximise the likelihood of his theory failing. In practice, the culture of psychoanalysis is like that of engineers designing, say, (competitively costed but safe) bridges, who do not regard it as their role to question the theory they apply. They are more concerned to use theory to get successful practical results.

We should note, thirdly, that psychoanalysis is not a single universal conjecture, but a combination of a large number of ideas. Ideally, they should be testable in combination and separately. In practice, many scientific theories contain constituent ideas that are not independently falsifiable, such as claims

---

[2] Grünbaum (1984, chapter 11) argues that Popper is wrong in claiming that Freud's theory is easy to verify but hard to refute. On a more sophisticated reading of Freud's theory, it is quite hard to verify. He argues that whatever we conclude about Freud's theory, we are not entitled to infer an underlying weakness of the inductive method.

[3] cf., Grünbaum (1984, p. 113) and the review, by P. Kline (1987).

[4] Grünbaum (1984, Part I, chapter 1) shows that Freud's own work discussed at least two prima facie refutations. Eysenck, (1985, p. 14) argues that Freud's theories are refutable and sets about refuting them.

of the existence of unobserved or unobservable entities. Freud's claim of the existence of the unconscious is comparable in this respect to such claims in physical science as the existence of free quarks or of a preponderance of dark matter in galaxies. Claims that such entities actually exist cannot be directly verified (at least when proposed), and may be said not to have scientific status outside the scientific theory within which they were proposed. Eventually new testable implications of such untestable additions may be found. In the case of Freud's theory, however, many conjectures are offered which look as if they should have independent significance but do not appear to be falsifiable.

For example, a part of Freud's theory of dreams is the conjecture that all dreams are (disguised) wish fulfilments. Very often dream wishes come not from the conscious mind or from frustrated desires of the previous day, but from the unconscious. Unconscious wishes in dreams may represent repressed infantile desires. The disturbing content of unconscious wishes is disguised so as to allow the dreamer to continue sleeping. Freud's critics such as Cioffi argue that he deliberately made this conjecture impossible to falsify.

There are undoubtedly difficulties about any scientific study of dreams, for the dreamer is asleep at the time and can only report what happened on the basis of his or her memories on waking. Subsequent retelling of the dream in the presence of an analyst tends to modify the dream report still more. The dream report might be restructured in ways which are not true of the dream and there would be no way to discover the discrepancy.[5] If one tries to decide whether a recent dream fulfilled a wish, there might be a temptation to elaborate the dream report to fit.

It is not just the problematic nature of dream reports that makes Freud's hypothesis difficult to attack. We are told that the dream is normally the disguised symbolic expression of an unconscious desire. We cannot directly establish the existence and nature of an unconscious desire any more than we can directly establish the existence of the unconscious itself. Freud offers a route to its discovery: free association on the content of the dream report during psychoanalysis. The analysis may reveal the disguised wish to the analyst. We should note that such a 'verification' does not have a corresponding falsification. At what point in the analysis can it be concluded that no unconscious wish was present? One is tempted to think that Freud would have regarded an analysis which had not unearthed a disguised wish for the dream as incomplete. However, at least once, Freud did draw such a conclusion. The dream of a particularly recalcitrant patient resisted all analysis.

---

[5] Eysenck (1985, p. 130) reports how he confirmed the work of A. Luria who, in the early 1920s, had hypnotised subjects to have specific dreams and then analysed the realtionship between the original dream and subsequent accounts of it by the subject.

Finally, Freud concluded that the wish in question was to frustrate the pet theory of the analyst that all dreams are disguised wish fulfilments.

Another stratagem Freud used to keep his theory alive was to invent new classes of unconscious desires. In particular, he coped with the difficult category of shell-shocked soldiers in World War I who kept dreaming of dying by proposing that we all have a death instinct (thanatos) which becomes more important as we age, and that the abnormal circumstances of the soldiers was leading to this surfacing in their dreams, in spite of their youthfulness. Unlike the infantile wishes of the unconscious, the death wish manifested itself to these soldiers in undisguised form, and did wake them up.

Other examples of specific Freudian doctrines have similar difficulties. For example, Freud proposed the theory that all psychoneuroses have their origins in past sexual life. In its original form, the provocative idea that the cause of hysteria was sexual abuse in early childhood was extremely bold in Popper's technical sense of being readily falsified. But, in the face of criticism, Freud quickly modified this idea to the far less testable conjecture that the cause lay in the sexual fantasies of children (Masson, 1984). As with the theory of dreams, the only route to verification lay through what was said in psychoanalysis, and no empirical refutation seemed possible.

If we accept, then, that Freud protected many of his specific conjectures by making them testable only within psychoanalysis, the question remains of whether psychoanalytic theory as a whole is testable. Perhaps the most common argument in favour of Freudian psychoanalysis is that there must be something in it because it works. I argue elsewhere that this criterion is notoriously unreliable as an indicator of truth, as perceptions of what 'works' are usually situated in a less demanding cultural context than a full-fledged experimental test. Zande magic worked for the Azande, in Evans-Pritchard's classic study, but that did not make it true (see chapter 4). In the case of psychoanalysis, the fact that people who undergo psychoanalysis are pronounced cured means that both analyst and patient are satisfied that the analysis has come to its natural end and that the patient is now sufficiently able to face up to the problems that brought him or her to the analyst. But people very often recover naturally from the kinds of conditions which might have led them to analysis. And it is very difficult to be sure that simple suggestion was not a factor in the recovery under analysis. Placebo effects are very powerful in medicine, and, in spite of Freud's precautions not to invoke simple suggestion (in the manner of hypnotherapy), it is inevitable that the patient becomes aware at the appropriate point that now (s)he ought to think of him or herself as cured.

Attacks on the idea that psychoanalysis is true because it works have taken two main forms: The first is the counter-claim of Eysenck and others that, statistically, psychoanalysis produces only marginally more recoveries than putting people on a waiting list for several years, and that other therapies work at least as well (Rachman, 1963; Eysenck & Wilson, 1973; Eysenck 1985). The second attack is by A. Grünbaum (1984) in terms of what he calls 'the tally argument'. This is based on Freud's claim that there are cases which only psychoanalysis can cure. The evidence in such cases need not be put in statistical form with control data. Freud did not think that statistical evaluation of his theory was appropriate. Let us briefly consider each of these attacks in turn.

In 1952 and subsequently, Eysenck argued that the efficacy of psychoanalysis had never been properly tested, and set about a critical analysis of what evidence there was. He argued that the evidence was that the average duration of neurosis was one to two years and that two-thirds of patients have spontaneously recovered untreated after two years. Later, more detailed evidence supported the general conclusion that the spontaneous remission of patients continues to cumulate, so that about 90 per cent have recovered within five years. A few studies of psychotherapy which had included control groups tended to show that those who underwent psychoanalysis reported that they had benefited from it, but that their recovery rate was roughly comparable with the untreated cases, or worse, if those who broke off treatment are counted as uncured. Eysenck was concerned to discredit psychoanalysis in favour of the alternative of behaviour therapy, and claimed that the statistical evidence in the latter's favour was significantly better (Eysenck, 1985). Other studies have found that the recovery rate in a wide range of therapies is roughly similar. Perhaps, then, it is not the theory guiding such therapies which is working, but some factor all the therapies share, such as the 'therapeutic bond', or the placebo effect (Lambert et al., 1986). In the terms of the discussion of immediate reality offered in chapter 5, psychotherapeutic practice is a congenial immediate reality in which whatever theory is used is all too easily found to work.

However, evidence that psychoanalysis is not uniquely effective at bringing about cures has not been regarded as conclusive by defenders of psychoanalysis. Freud himself had made it clear that not every patient is suitable for psychoanalysis. The case must be correctly diagnosed before it is put into the statistics. Furthermore, the patient has to want to recover. Someone whose life is more tolerable while ill has an incentive to stay ill. In addition, the classic form of psychoanalysis requires the patient to work at overcoming resistance to facing up to repressed material, and relies on building up a bond between analyst and patient to do this (transference). If this bond is insufficient then the analysis may not progress. Furthermore, the defenders of psychoanalysis argued, the standard by which cure is judged in psychoanalysis is much higher than in untreated cases, in which it is the ability to cope adequately with the pressures of ordinary life. The self-understanding provided by psychoanalysis strives to enable the patient to be far more fulfilled than merely coping.

There does seem to be evidence that up to 30 per cent of patients with neurotic disorders fail to undergo spontaneous remission. Perhaps the claim that psychoanalysis works applies to some of these cases. Grünbaum's discussion of the tally argument becomes most relevant here.

Grünbaum claims that Freud's use of the tally argument makes him a more sophisticated scientific methodologist than is allowed for by such critics as Eysenck. The tally argument is, in essence, that when, with some guidance from the analyst, the patient comes to recognise that the presence of unconscious material is having an effect on his or her actions, this has a lasting curative effect on the neurosis. The benefit is because it is true. The inner conflicts will only be successfully resolved and the resistances overcome if the anticipatory ideas given in analysis tally with what is real in him (Grünbaum, 1984, p. 139). There may be some cases in which mere suggestion, as in hypnotic suggestion, produces the lasting suppression of symptoms. But, in genuine neuroses, mere suggestion does not produce this benefit. Grünbaum's careful and critical analysis of this argument leads him to the conclusion that in so far as the evidence for psychoanalysis is held to derive from patients under analysis, 'this warrant is remarkably weak' (ibid., p. 139). If Freud's theories are to be validated, the tests will have to come from studies that have not yet been done.

I have only covered a few of the main points of the criticisms of psychoanalysis so as to show the issues in trying to use a criterion of demarcation such as Popper's. There is a vast literature attacking and defending Freud, and, although Popper's criterion has been used, most writers have found it inadequate, and have gone on to develop their own techniques for criticism. But works such as those by Cioffi (1970), Eysenck (1985), and Grünbaum (1984, 1993) show that Popper's criterion comes fairly close to identifying the nature of a widely perceived weakness in Freudian doctrine.

In terms of the view of science constructed in Part I, the essence of Popper's critique of psychoanalytic theory should be accepted. In my view, psychoanalytical theory constructs a conceptual world which, although moderately open to cultural confrontation, is highly protected against direct evidential challenge. My discussion of immediate reality in chapter 5 suggests that part of the difficulty may be Freud's choice of empirical material to connect to his theories. He regularly chose what I described as excessively congenial immediate reality. Theories of such material often work for reasons other than those given. Psychoanalytical theory relies on such indirect forms of empirical support that the truth or error of any specific Freudian idea cannot be made to depend directly on a new factual finding. Once reality can no longer force changes in a coherent set of ideas, the empirical phase of knowledge construction is over, and the intellectual practice ceases to be empirical science. It may, like mathematics, continue to undergo internal development as a set of concepts. Its practical uses may be useful in ordinary social life, especially if we manipulate our immediate reality to match its conceptual system. It may also have cultural value in its confrontation with other sets of ideas. But it is no longer empirical science.

Criticisms of Popper's demarcation criterion

Although Popper's demarcation criterion of falsifiability has been very influential, it is not difficult to criticise it in its own terms, or from the viewpoint of other philosophers. Here are a number of points of criticism.

1. Every student of elementary science knows that his or her own scientific observations are untrustworthy. To propose to reject the textbook theory in the face of a contrary student observation could only be a joke. As we build up skill, our observations become more reliable. We should only agree to reject theory that conflicts with observation once it is established that the observation is reliable. The most straightforward way of doing this is to show that it is reproducible. It is therefore plausible to argue against Popper that observation statements require justification and that we normally require prior inductions to establish which factual reports can act as potential falsifiers.

2. Popper's rhetoric of justification of his methodology tends to refer to universal statements. 'All ravens are black' is falsified by the observation of just one white raven at any point in time. Therefore the universal statement should be rejected at its first falsification. If we recast scientific generalisations as probability statements, quite different implications follow. Consider the statement, 'It is very highly probable that any raven we find is black.' If we had found that 99.99 per cent of the 10,000 ravens seen in the past were black, we would not have reason to abandon the second statement. Nor would we need to abandon our prediction that the next raven we see will be black. Popper concedes that

probabilistic hypotheses are needed in science. Once we agree on statistical criteria for their refutation, they can be treated as testable hypotheses. However probabilistic hypotheses remain less desirable because they are not so readily falsifiable.

3. There is normally some theory presupposed in falsifying observations (such as the theory of the instruments used). So falsification is normally a conflict between higher-level theory and lower-level theory. We may not be able to establish in advance the relative reliability of the two theories, and always to prefer the lower-level theory may be bad advice.

4. In general, an observational prediction can be deduced from a theory only with the aid of much background knowledge. We can normally reconcile a conflict between observation and prediction from theory by making an adjustment to the background knowledge assumed (the Duhem-Quine thesis). In particular, it is often plausible to reject the *ceteris paribus* clause of the theory (the assumption that nothing else is interfering with the phenomenon explained by the theory) by proposing that, in making the prediction, we had assumed the world to be simpler than it actually is (Lakatos, 1970).

5. As Lakatos argued, (Lakatos, 1970) such stratagems allow a theory to be kept alive indefinitely. Perhaps all we can claim for a theory is its long-term record of predictive success.

6. As Kuhn argued, we can never afford to abandon our more powerful theories (which continue to work except for problems producing crisis) until we have something better to put in their place. We cannot do without a theory to guide research.

7. The Popperian doctrine applies best to theory-led science (such as much of physics), and not to observation-led science (as much of the rest of science is), in which the techniques of systematic observation generate novelty, which is assimilated by classification and explanation only after it is discovered.

8. In less mature sciences, we would learn more by falsifying cautious theories than by falsifying bold ones (Chalmers, 1973).

9. A new theory may be rather too easy to refute, simply because it has not been fully worked out. It may be best to allow new theories a period for development before insisting that refutation should lead to rejection. As Popper himself argued, intellectual enterprises which fail by his criterion should not be rejected out of hand. (Falsifiability is not intended as a criterion of meaning.) Unfalsifiable theories may develop into sciences by the natural elaboration of their own internal rationale. For example, ancient atomism, which was untestable metaphysics, was succeeded by

the powerful heuristic framework of seventeenth-century corpuscularism, which, in turn, was replaced by the nineteenth-century variants of atomism in chemistry and physics. The early twentieth-century successor to nineteenth-century atomism was a new science of atoms, which at last had a vast range of testable consequences.

10. Popper gives us minimal positive guidance on how to create new hypotheses to replace old falsified ones. (How to do this is a problem in some areas of science – hypothesis-starved science.)

11. Popper's criterion gives all testable but unfalsified theories the same status, even if they are inconsistent. In a fertile area we might easily generate an indefinitely large number of equally viable variant theories. We would need additional criteria if we have not the resources to investigate them all simultaneously. (Hypothesis-overwhelmed science.)

12. Popper's criterion works best for narrowly defined theories, viewed realistically, under conditions in which we seek to discover the limits of present ideas. It is of less value to the applied scientist who wants to make existing knowledge *work* reliably.

13. Popper's criterion does not apply directly to those who regard themselves as applying a body of doctrine rather than developing it further. Just as engineers cannot be expected to be forever testing the mechanical theories they employ, so practising astrologers cannot be expected to be forever testing the astrological doctrines they apply.

14. Popper's criterion does not apply to broadly defined programmes. Just as it would be silly to ask whether physics as a whole should be falsifiable, so it may be silly to ask, for example, whether Marxism as a whole should be falsifiable.

15. In practice, the social system of science is institutionalised around recognition of priority for positive discoveries (such as demonstrations of observational phenomena, experimental effects or strongly argued theoretical doctrines) rather than refutations. Acceptance of Popper's view would require a major and counter-intuitive reconstruction of the reward system of science.

16. Why should we prefer Popper's criterion to any other? Clearly it is not self-justifying, for it is not itself falsifiable. (And to argue for it inductively would be to give up falsificationism.) Its justification must lie outside the domain of science proper. If a system of science were offered with rival conceptions of scientific method and of the demarcation of science, then the debate between it and Popper's conception would have to be conducted at least partly according to non-scientific rules. For example, Marxism has been defended in terms of the view that science

is to be defined in terms of dialectical materialism. For Marxism, pseudo-science is that which is distorted by the values of an objectionable ideology (Rose & Rose, 1976).

Popper's methodology has a special significance for the view of science developed in Part I. Once systems of thought incorporate sufficient defensive systems to be able to evade criticism, they risk becoming closed, and, if they are in harmony with the form of life which sustains them, they may then avoid internal pressures to change even though the truth their ideas express may be purely local. In an open society, we seek wider truths by allowing the confrontation with opposing systems of ideas. In a rational society we strive to maximise the logical coherence of our thought. In a system of knowledge which claims an empirical scientific basis, we should always be seeking perceptual surprises and in other ways widening the number of ways in which reality can challenge our thoughts. Scientific knowledge must be subject to perpetual revision. It must remain uncertain knowledge. This view is essentially Popperian.

8.4 The history of science and pseudo-science

Let us look more closely at what the historical study of science can show about the relationship between science and pseudo-science.

The history of science is relevant to the present study because the origin of several modern sciences in the sixteenth and seventeenth centuries appears to lie in what are now regarded as pseudo-sciences. Astronomy was once closely linked to astrology. Chemistry grew out of the more practical aspects of alchemy. The incorporation of mathematical approaches into natural science grew out of rather mystical speculations on the relationship of mathematics and the world. In and after the scientific revolution, alternative approaches to the study of nature that competed with the new orthodoxy were marginalised and discredited.

The history of subsequent science records many individuals who proclaimed themselves as revolutionary thinkers but were subsequently rejected as cranks, as well as many failed attempts to generate new sciences out of material uncongenial to a scientific approach. Most scientific disciplines have undergone revolutionary changes, and occasionally the scientific orthodoxy and pseudo-science have exchanged places, most famously with the acceptance of the doctrine of the evolution of species with Darwin, and with the idea of major horizontal movements of the continents around the early 1960s. Accounts of all these changes have been given by historians of science, who record their findings about such matters as rich and persuasive narratives. Yet, as we will see, there is no generally accepted narrative account of the relationship between science and pseudo-science. In this section, I will treat the history of science as a useful source of information and insight, but a problematic one.

8.5 An example: the scientific revolution

The sixteenth and seventeenth century was a period of major transition in the study of the natural world. At the beginning of the period, the world picture which dominated the centres of learning was a fusion of Aristotelianism and Christianity. The unity of the scholastic world picture was beginning to break down with the appearance of mystical, occult, and speculative practices such as alchemy, hermeticism, cabalism, Neoplatonism, magic, and astrology. In the sixteenth century, a number of dramatic intellectual developments occurred, most especially in mathematics and astronomy among the physical sciences, and in anatomy and physiology among the medical sciences. The most spectacular transformation among the new sciences was in natural philosophy. A mechanical view of nature originating in the thought of ancient Greek atomism had developed into a doctrine that all the phenomena of the physical world can be explained in terms of the movement, interaction, and properties of the corpuscles out of which everything is composed. Among those who most successfully developed this view were Galileo, Descartes, and Newton. Galileo Galilei (1561-1642), used the newly discovered telescope to produce new and more powerful arguments for the argument of Copernicus that the Earth is not the centre of the universe but one of the planets orbiting the sun. Galileo went on to develop a non-Aristotelian terrestrial physics. René Descartes (1596-1650), in addition to making major contributions to philosophy and mathematics, developed a cosmology in terms of the mechanical world view which treated all space as composed of variously sized particles acting upon one another through direct contact. Among the less creative but still important contemporaries of Galileo and Descartes was Marin Mersenne (1588-1648), who built up a correspondence network that spread information between all the key individuals of the new natural philosophy, and disseminated very much the same conception of the philosophy, method, and social form of science as that which became the orthodoxy. Isaac Newton (1642-1727), later in the same century, produced the most highly regarded accomplishments of the new natural philosophy. Newton modified the mechanical philosophy by adding the idea of action at a distance. His systematic mathematical basis for mechanics, building around his concept of force, and incorporating the theory of an inverse square law of universal gravitational attraction, and his experimental studies of light, provided a framework for physical science. Although the scientific content of this framework gradually moved away from Newton's explicit concerns, its empirical limits were found only in the twentieth century.

Achievements such as these helped make the new natural philosophy into a major cultural force. Its social basis was secured by the creation of new scientific institutions, in particular, the Royal Society of London (founded 1660) and the French Académie des Sciences (founded 1666). By the early eighteenth century, it was widely accepted that modern knowledge had far outstripped what was known from the ancient world in many branches of natural philosophy. Even mathematics, in some branches of which the ancient Greeks had excelled, could boast major new achievements, and offered the prospect of many more. In such a context, natural philosophy was regarded as a valuable resource (Jacob, 1976), and much effort was devoted to assimilating it as an unthreatening part of the intellectual and religious orthodoxy. Nevertheless, by the mid-eighteenth century, Enlightenment writers popularised the rational and empirical methods by which a person is able to make his own cognitive judgements. In France, in particular, the new image was offered as a liberating force in attacks on the old centres of power of church and state.

When, in the eighteenth century, the Enlightenment *philosophes* wished to argue for the progress of reason, they presented a picture of the cumulative discovery and elimination of sources of error in human thought. The natural sciences were seen as a model of this process. This image guided most of subsequent historical writing on science until quite recently. Science, on this view is 'the edge of objectivity' (Gillispie, 1960). Scientific change is to be understood as the working out of an ever more efficient rationality. Scientific progress comes from the rational elimination of error and the cumulation of knowledge. From this viewpoint, it is quite easy to see how the sciences might grow out of the ancient errors of Aristotelianism and competing speculative intellectual practices. It can do so by the cumulative identification and elimination of sources of error in our knowledge and our methods of production of knowledge.

But the conception of scientific progress we have inherited from the Enlightenment is at least partly an artefact of our perspective. When we study the intellectual activity of an earlier age, we most readily see it through modern eyes, evaluating it as a contribution to present-day problems. If we are guided by past historical writing, it has very often grown out of the ideologies of winning parties. In such a perspective, the beginnings of science appear as faltering steps along the same path as was more clearly traced by later work. But it is clear that just as we can have no idea of how effectively our present science contributes to the science of three centuries in the future, our intellectual ancestors could have had no more idea of the science of our own time. The early sciences should be understood in terms of the ideals of their own age. Each intellectual generation exploited the accomplishments of its predecessors by improving them to meet the needs of an ever-changing context as judged by the standards of the time.[6]

Although a simple idea of the progress of science from false belief to true science is untenable, modern scholarship has regarded the 'scientific revolution' as a fundamental transition in the history of thought. If there is indeed a qualitative difference between modern science and its Renaissance precursors, what is their historical relationship? In particular, what was the role of practices like alchemy, hermeticism, cabalism, Neoplatonism, magic, and astrology? These rivals to Aristotelianism were simultaneously enthusiastically pursued and heavily attacked. They would now regarded as pseudo-sciences but were not so labelled at the time, as the concept of science had not yet taken its modern form. Four answers can be extracted from the discussions of historians of science on the historical relationship of astrology, alchemy and similar practices and the new sciences.

(a) *No connection*: the modern sciences had virtually nothing to do with astrology, alchemy, and the like. Modern natural philosophy, for example, emerged in the late Renaissance from the fusion of Greek philosophy and a thriving and innovative craft technology in a literate and mathematically qualified milieu.

(b) *Negative connection*: the competing forms of belief which flourished in the Renaissance were a negative stimulus to the new natural philosophy, which emerged as an alternative to them, its methodology being self-consciously worked out to avoid their difficulties.

(c) *Positive connection*: Astrology and other contested traditions were 'proto-sciences', which led directly to the new natural philosophy. There was continuity in subject-matter and in key historical chains of personal influence among the alternatives to the old Aristotelian orthodoxy.

(d) *Both science and pseudo-science were epiphenomenal to the process of change*: the change from the Renaissance intellectual practices to the orthodox knowledge of the seventeenth century is best understood as the effect of developments in the wider society. Changes in science in the seventeenth

---

[6] The mistake of seeing past science as trying and only imperfectly succeeding in becoming present science is an aspect of what is known (from Butterfield, 1931) as the Whig interpretation of history.

century were not primarily a reaction against earlier knowledge or a development of it. Rather, they were an example of the way the intellectual resources of the past were shaped into new forms by the pressures and opportunities of a new age.

I will say something about each of these views, arguing that each of them is more or less correct within its own frame of analysis. Then I will try to show how they can be integrated by fitting the frameworks together.

(a) No connection

If we limit the scope of our history of science by assuming that ever since the creation of the first of the sciences as a part of ancient Greek philosophy, they have been the forms of inquiry into the natural world practised by the intellectual elite of society, then the pseudo-sciences fall outside the scope of our historical investigation. This is the judgement of the pattern of early professional history of science influenced by Alexandre Koyré.[7] On this view, the fact that modern science arose only in Europe and not in India or China is to be explained by the special importance of ancient Greek thought. The origins of modern science lie in the Renaissance recovery of Greek philosophy and the rebirth of its spirit of inquiry. The philosophical impetus behind the revolution was the way the recovery of Platonic thought encouraged the mathematical representation of the world, in contrast to the strictly qualitative nature of Aristotelian physics.

The concern with mathematical representation was being taken up in a European culture in which a section of the learned were beginning to pay serious attention to the problems and accomplishments of new craft technologies. They were able to generate from them new lines of thought within the framework of mathematical representation. It was an age of geographical exploration and of conquest. Magnetism was being applied to the magnetic compass in the former, and gunpowder was being applied to cannons and other military applications in the latter. Much of the interest in new craft technologies was highly practical and opposed to bookish learning, but there were clear influences upon the learned. For example, William Gilbert's *De Magnete* (1958, originally 1600) was built upon earlier treatises of navigation using the magnetic compass, but is most famous for its use of a large spherical lodestone to investigate experimentally the idea that the Earth is a giant magnet. Similarly, discussions of the way in which projectiles move through

---

[7] Examples included Koyré (1957); Gillispie (1960); Hall (1983).

the air occurred in medieval Aristotelian physics, but it was Galileo (and his contemporaries) who actually produced a law of fall in the seventeenth century. Galileo's arguments included an appeal to experimental demonstrations. The proof of the law of fall was produced at a time when understanding the motion of cannon-balls was no longer just an interesting philosophical question but also a practical problem. The revolutionary synthesis of theory and practice showed a new readiness to turn the Greek form of naturalistic philosophical questioning and mathematical reasoning to matters for which answers could be checked by experiment and measurement (Crombie, 1952, p. 274; cf., Hall, 1952).

The texts in the history of science that develop the 'no connection' view of the role of pseudo-science in the scientific revolution tend either not to discuss astrology and alchemy at all, or to refer to them casually and dismissively. C. C. Gillispie (1960) noted the existence of the mystical jungle of alchemy. A. R. Hall (1983), with the opportunity to revise a text originally published two decades earlier, summarised his view of debates he had had in the previous decades with defenders of the 'positive connection' view. Hall conceded that several pioneers of the scientific revolution did dabble in pseudo-science, but dismissed this as 'a kind of atavism of which the progressive thinker himself may be unconscious'. For example, he did not believe that 'Newton's name is immortal because he read alchemical authors' (Hall, 1983, p. 2).

(b) Negative connection

While the 'no connection' view idealises the sciences by concentrating on the most elite sector of the activity of natural philosophy, the 'negative connection' view concentrates less on actual practice and more on the public presentation of the new natural philosophy – the manner by which writers sought to create an image of the new knowledge that showed its legitimacy. This view, like the last, has inherited its attitudes from the Enlightenment, particularly as they have been carried by positivist philosophy of science.

On the 'negative connection' view, by the end of the sixteenth century, there was something of a crisis in the intellectual community because of the number of competing systems of belief about the natural world. Although it was not a tolerant age, conflict between Protestant and Catholic religion being a major stimulus to war, there were enclaves in which a higher degree of independence of thought was tolerated; variants of religious, mystical, magical, medical, and speculative philosophical viewpoints flourished locally, to be disseminated effectively by the new technology of printing. Although some of these views appeared to the sceptical to be the product of fraud and charlatanry, many more were well worked out views that might have survived as the dominant tradition in a simpler 'closed society'. But in a society with many local factions, partially opened up to one another by the easier dissemination of information through printing, we can understand why those who looked outward from their local orthodoxy acquired a heightened sense of scepticism. A number of thinkers sought to develop sceptical arguments which could be used to dismiss the bulk of the competing systems, while leaving a favoured set of beliefs unscathed. In a Christian age, those favoured beliefs should include all of Christianity, or perhaps some essential subset of Christian beliefs which all the competing religious views shared.

In seeking to construct a new and authoritative image of knowledge, the prophets, pioneers, and apologists of the new approach set it in clear contrast with the competing systems which they rejected as unsatisfactory. I will consider especially Francis Bacon (1561-1626), Galileo Galilei, René Descartes, and Marin Mersenne.

In his *Novum Organon* (1620), Francis Bacon expressed his scepticism in terms of the errors and distortions produced by the four idols: the idol of the tribe, the idol of the cave, the idol of the market-place, and the idol of the theatre. These referred respectively to the errors resulting from general characteristics of human nature, the particular prejudices of individuals, the corruptions of language, and the dogmas of traditional theories.

Bacon was among those who attacked Aristotelianism in the name of the new form of knowledge. Although Aristotelianism was a thriving tradition which dominated the universities and had been absorbed into Christian orthodoxy by the synthesis of Thomas Aquinas, it was often attacked in the general intellectual culture. Bacon (1620, Book 1, aphorism 63), for example, criticised Aristotle's physics as corrupted by the arbitrary definitions and distinctions of his logic and without adequate recourse to experience. Contemporary Aristotelians were guilty of the idol of the theatre in their reverence for the authority of their tradition. Nevertheless, Bacon's own thought retains many features which were later to be criticised as Aristotelian. Bacon's attacks were not limited to Aristotelianism. He thought that experiments were important in the new learning, but regarded alchemy as a tradition which revealed the dangers involved. By generalising prematurely from a few obscure experiments and ignoring common notions, alchemists produced dogmas of a deformed and monstrous nature (Bacon, 1620, Book 1, aphorism 64). He was as scornful of the natural magicians. Their explanations in terms of sympathy and antipathy were no more than idle conjecture. They might occasionally produce novel effects, but only by chance, and, since their interest was in the wonderful, they were usually without any usefulness (Bacon, 1620, Book 1, aphorism 85). Bacon accused magicians of fraud, of craze for genius, and of megalomania; he criticised magic's non-progressive, non-cooperative methods and especially its attempts to replace human effort by a few drops of elixir. He was convinced that only infinite patience could unravel the 'riddles of nature' (Rossi, 1978, p. 31).

Bacon is not entirely authoritative as a representative of the new approach to knowledge. He did not contribute significantly to the content of natural philosophy and antecedents can be found for most of his methodological and epistemological suggestions. He sometimes dismissed the work of natural philosophers whose work has subsequently been accepted. For example, he dismissed Gilbert as one who based a system of nature on a single experiment. Perhaps his most significant contribution to the new natural philosophy was his vision of how to organise collective scientific investigation in *The New Atlantis* (Bacon, 1974, originally 1624), which influenced the early Royal Society later in the century.

Galileo, too, attacked Aristotelian cosmology and physics. For example, he ridiculed the fictional Aristotelian, Simplicio, in his *Dialogue Concerning the Two Chief World Systems* (1967, originally 1632). Galileo thought Aristotelian explanations were too often merely verbal, and sought to substitute quantitative forms of reasoning in natural philosophy. He was also scornful of the more mystically inclined of his contemporaries. For example, he criticised Kepler's belief that the moon influences the tides as the acceptance of 'occult properties and . . . such puerilities' (Galileo, 1967, p. 462).

It is Descartes among the pioneers of the scientific revolution whose relationship to the sceptical philosophical tradition is best known. Popkin (1979, originally 1964) argued for the special importance to Descartes and modern philosophy of the revival of Greek scepticism in sixteenth-century Europe. Descartes had been a student at the famous Jesuit school, La Flèche. He was exposed to the very best of existing knowledge, yet seemed to have been impressed by sceptical arguments, and to have eventually decided that a new approach to the production of knowledge was required. Descartes tells us in his *Discourse on Method* that, in spite of his diligence while a student, the main benefit had been the discovery of his own ignorance (Descartes, 1954, p. 9). For Descartes, mathematics was the branch of existing knowledge which most clearly reached the kind of rational certainty that he desired, and he set out to reconstruct knowledge with mathematical rigour. In his *Discourse of Method* and in his *Meditations*, he employed the method of extreme scepticism which led him to doubt even his own senses. He has been thought to have relied too strongly on the power of a priori reasoning in his system, which he attempted to construct from an excessively narrow base. In his earlier scientific researches, he had been obliged to supplement mathematical reasoning by introducing such weak forms of inference as analogy.

One substantial attack on the traditions in competition with natural philosophy in the early seventeenth century was by Mersenne in several large works published between 1623 and 1625. The basis of his lengthy critiques of such movements as sorcery, cabalism, alchemy, astrology, and animism, and of their recent exponents, was not simply sceptical. He sought to defend orthodox Catholic Christianity against them, in particular by keeping a place for miracles (Le Noble, 1971; Dear, 1988).

Mersenne put the methods of the new natural philosophy more nearly in the balance that later became dominant. Because he was at the centre of a scientific correspondence network that spanned Europe and beyond, he played a key role in the convergence of views of natural philosophers. In response to the earlier sceptical tradition, Mersenne thought that the new knowledge should limit itself to the world as experienced. He was appreciative of Bacon's programme for the new learning, but felt that his Idols were too sceptical in their implications and that Bacon had been too ambitious in hoping to discover the essences of things, especially within such an extreme empiricist philosophy. Mersenne, like Galileo and Descartes, advocated a mechanical philosophy. He thought that a mechanical philosophy of the world-as-experienced could avoid heretical dangers of earlier suspect traditions and yet provide an answer to scepticism and atheism. The mechanism of nature is constrained by quantitative laws, imposed by God and discoverable by man. Even man's body is a mechanism. Mersenne was an enthusiastic disseminator of Galileo's work especially the astronomical discoveries of his telescope. In the balance between quantitative experiment and mathematical reasoning he was, however, doubtful about the certainty Galileo claimed for his reasoning, and suspicious as to whether Galileo had made all the careful experiments he claimed. Mersenne insisted that experimental methods should be carefully described, repeated several times, and the actual results reported. This was the procedure he followed in his own work on acoustics.

The leading figures of the scientific revolution were anxious to show that, unlike some of its Renaissance competitors, the new knowledge was complementary to, and supportive of, established religion. God created the mechanism of the universe and it continues to operate only thanks to His divine providence. Bacon was of Protestant upbringing. In spite of his conflict with the church over Copernicanism, Galileo tried to be a faithful Catholic.

Descartes' care to avoid making his natural philosophy heretical was reinforced by his knowledge of Galileo's experiences. Mersenne belonged to the Order of Minims, and although there was some opportunity for independence in French Catholicism, he devoted a great deal of effort to constructing an image of knowledge fully supportive of orthodox faith.

(c) Positive connection

This view was at the peak of its influence in the historical study of science in the late 1960s and early 1970s.[8] According to this view, there was a considerable continuity between the new natural philosophy and the competing forms of knowledge of the late Renaissance. Ideas and techniques were taken over when they were thought to be true or helpful; furthermore there are continuous chains of personal influence. The new investigator of nature can be seen to have developed from the older role of the magus – for example, both sought control of natural forces.

In the Middle Ages, magic and alchemy had been conflated with demonic arts, necromancy, and the like. It was only with the Renaissance and a new evaluation of man's significance in the world that more respectable forms of magic could be reinstated as worthy of mankind to be practised without shame. Magic in the Renaissance became an intellectual achievement, praised by many important intellectuals. It retained this status until the time of Kepler, Bacon, Gassendi, and Descartes.

Many writers in the Renaissance attacked supernatural forms of magic, but admitted the importance of natural magic, which proceeds according to natural law, practitioners of which may exploit hidden powers of things to produce spectacular and unexpected effects. A similar contrast was made between true and false alchemy (sophistical alchemy was an intermediate form). False alchemy made all sorts of impossible claims, like curing disease instantaneously and bestowing eternal youth. Alchemy had become debased by the unscrupulous.

Let us consider alchemy in more detail. Traditional alchemy was a combination of a mystical attitude to the world, given expression in elaborate forms of symbolism (expressed in complex language or symbolic diagrams), and a set of manipulative practices involving the transformation of matter. The material changes produced by the alchemists' art reflected the higher realities of the universe. Not all alchemists merely sought some get-rich-quick device for turning base metals into gold (Kearney, 1971). In the mid-sixteenth cen-

[8] Examples include Debus (1965, 1972); Pagel (1958); Righini Bonelli & Shea (1975); and Yates (1964).

tury, Paracelsus, a controversial figure combining genius, showmanship, and fraud, had developed a medical side to alchemy – arguing for the importance of mineral remedies for illnesses. Paracelsus, both as mystical magician and as physician, had tremendous influence. There was a rapid rise in interest in iatrochemistry, the medical application of alchemy, around this period. At the end of the sixteenth century, there was also a great enthusiasm for the mystical side of alchemy, along with other occult sciences. Alchemy was something to be taken seriously. The mechanical philosophers who were to become the core of the new natural philosophy of the scientific revolution were not usually mystically inclined. But they were concerned with the ultimate nature of matter, with the kinds of transformations that could be achieved by the experimental manipulation of matter. They took the practical side of alchemy seriously, and tried to represent alchemical ideas in their own atomistic terminology in which all was reduced to matter and motion. Many of them accepted, for example, the possibility of transmutation (as with the transmutation of base metals into gold), and showed how it could occur within their viewpoint. This phase of the relation between alchemy and natural philosophy lasted until late in the seventeenth century. Such leading natural philosophers as Robert Boyle and Isaac Newton took alchemy seriously in this way. Those investigators of the seventeenth and eighteenth centuries who were most interested in reproducing the practical claims of the alchemical tradition were also pioneers of the new science of chemistry. They concluded that the more grandiose claims of the alchemists were in error. They eliminated approach after approach to the problem of transmutation, for example.

There was one striking difference between the new mechanical philosophers and the older alchemists. The alchemists had a secret tradition. Such knowledge as the transmutation of base metals into gold was dangerous in the wrong hands. So their public utterances were made obscure by concealing truths in allegorical form which only the initiated would understand. The mechanical philosophers on the other hand increasingly adopted a conception of knowledge as something for the public benefit. Everything should be published and publicly scrutinised. This is a striking difference between the later science and the earlier practices. Obscurities and mistakes were not exposed and eliminated from the secret tradition as easily as in a tradition of critical public discussion.

It was, of course, hard for the new chemists to be sure that they had shown an error in some much repeated alchemical claim. The process took more than a generation. It was made easier by a second point of connection between the earlier occult arts and modern science, the continuity of personal influence. The flowering of occult arts at the end of the sixteenth century was rich in mysticism. But within the movement was a widespread interest in social reform. J. V. Andreae, for example, combined mystical alchemy with a hope for universal reform. His *Christianopolis* was published a year before Francis Bacon's *New Atlantis*. One especially important figure in the rise of British natural philosophy was Samuel Hartlib. He was heavily immersed in the tradition of educational and religious reform and the group that formed around him in the mid-seventeenth century played some part in the rise of the Royal Society. Hartlib acted as an information centre, with a wide correspondence network. This was later taken over by Henry Oldenburg, the son-in-law of one of his assistants. Oldenburg later continued the same information-sharing activities as secretary of the Royal Society. The Hartlib group, among which was Robert Boyle, were very interested in alchemy. This is only natural, considering the combination of interests of their intellectual antecedents. The Hartlib group collected and circulated publications and manuscripts on alchemy.

One figure in this story is especially interesting – Isaac Newton. Newton is the most famous natural philosopher of the seventeenth century – and perhaps of any century. His achievements were taken by subsequent generations as paradigms of scientific research. Newton's published work was disciplined in method – nothing was claimed which had not been argued for by a combination of induction and deduction, the latter modelled on the deductive method of geometry. In his private life, Newton was deeply interested in alchemy. Although he did not publish on it, he took copious notes on books and manuscripts which he apparently got hold of by his friendships with Cambridge contacts of the Hartlib correspondence network. He did prolonged and detailed experimental studies (Dobbs, 1975, 1991; Westfall, 1980).

Religion was an important motivation in all of Newton's work. Understanding of God's creation, the world, can be informed by interpretation of the knowledge revealed in ancient sacred texts. He thought that, in earliest times, God had given the secrets of natural philosophy and of true religion to a select few. Much of the knowledge was subsequently lost, but partially recovered later to be incorporated in fables and mythic formulations where it would remain hidden from the vulgar. Newton thought that his own experiments would be most productive if they proceeded in conjunction with the study of records of the ancient past. As a part of his natural philosophy, he was interested in the interactions of very small particles, knowledge of which he believed to be hidden in allegorical form in alchemical writings. It was the *most* mystical alchemical writings which he thought were the most important to study – as these were thought to represent the oldest part of alchemy, especially the work attributed to Hermes Trismegismus. In his studies and experiments, Newton had important insights into what was and was not possible by alchemical manipulations. But he never published anything directly concerned with alchemy. He never seemed to have decided that this work was misdirected. It was inappropriate to publish when he had not got to the bottom of the problems. If Newton had been successful, he would still have been cautious about publishing, for in this context there may have been some residual effect of the often repeated warning of the alchemists that their knowledge was dangerous and should not be revealed to non-initiates. Although he had not established any obviously dangerous conclusions, it was possible that there were unseen dangers in what he had done. His unorthodox religious views were another reason not to publicise this work.

Newton was, then, doing alchemy for reasons that were natural for someone in *his* time and place, even though they do not fit with later images of Newton the 'man of science'. He was not directly following the alchemical tradition, though his beliefs led him to take it seriously and therefore to seek to extract information of value from it.

(d) Both science and the pseudo-sciences were epiphenomenal to underlying change processes

In the 1970s, some historians of science argued for the importance of 'external' factors in scientific change, according to which the changing historical form and content of science is the outcome of processes in the rest of society.[9] This form of argument had been developed by Marxists. For example, the Soviet writer, Boris Hessen (1971, originally 1931) offered Western historians of science an account of the origins of Newton's *Principia* in terms of responses to the basic economic problems of the day. His argument, while interesting, had seemed too extreme, for there was little sign in Newton's published and unpublished writings of a significant connection between such matters and his natural philosophy. Later Marxist writers of the mid-century sought to develop more historically sophisticated ways of assimilating the development of science into materialist views of wider historical processes. As non-Marxist historians of science also became more interested in the social dimensions of science, externalism versus internalism briefly flared up into an intellectual issue. (A later account of the issue is given in Shapin, 1992.)

The possibility arises, then, that perhaps the scientific revolution was the result of the emergence of new attitudes, social roles, and interests emerging

[9] The issue of internalism versus externalism had emerged in the late 1960s. See, for example, Kuhn (1968).

more generally in the rapidly changing society of the seventeenth century. Recent history of the science of this period has tended to construct explanations in terms of the categories used by the historical actors. These actors normally constructed intellectual boundaries between the task of constructing new knowledge of nature and other concerns. Although they did not agree precisely where the boundary should be (Shapin and Schaffer, 1985), one effect of the boundaries being created was the assumption of a certain degree of autonomy for the new sciences. The historical actors did not see the new sciences as a product of wider processes.

Marxist writers had assumed that unconscious forces could be at work on historical actors whose thought was always shaped to some degree by the false consciousness of the ruling classes. This is an attractive assumption, but it is difficult to establish a satisfactory methodology for historical processes which work outside consciousness in an intellectual domain like science which is so dominated by the flow of conscious ideas.

Perhaps, however, we *can* identify some wider historical processes which caught up those concerned with the sciences. For example, it has been argued by O. Hannaway (1975) that the transition from alchemy to chemistry can be linked to the appearance of a new form of didactic alchemy, particularly in the *Alchemia* of Andreas Libavius (1597). Hannaway contrasted this new didactic chemistry with the chemical philosophy of the followers of Paracelsus. Libavius gave expression to a new conception of chemistry as a discipline, and presented a quite different conception of the significance of language in the explication of nature. In contrast to the Paracelsian, Croll, whom Libavius criticised, and whom Hannaway chooses as an example of the opposing tradition, Libavius' didactic chemistry was committed to the values of open and clear communication. Croll's epistemology was individualistic, seeing knowledge as created within the individual by a process of individual attraction between the objects of the world and the microcosmic representations of them in man. On Croll's view, knowledge was not dictated by reason, and could not be read straightforwardly from nature or even from books. The didactic role was a deliberate attempt to apply the principles of humanist pedagogy to alchemy. It was a viable possibility in the context. But why did didactic chemistry displace alchemy? Paracelsianism continued to thrive well into the seventeenth century. Christie and Golinski (1982) suggested that our problem has become, not simply how and why chemistry displaced alchemy, but to understand the more complex extended dialectical process by which didactic chemistry established and maintained a distinct disciplinary identity and also responded to extrinsic forces from other spheres of discourse. The change in the rhetoric of chemistry is part of a wider cultural change.

Attempts to explain the key changes of the scientific revolution (including the transition from pseudo-sciences to science) in terms of wider social processes, remain programmatic.

It may be interesting to consider a sketch of another way in which this programme might be applied. The argument which follows employs an even more general explanatory scheme than Marxist economic determinism, one which can also bypass the conceptual categories of the historical actors. The account looks in particular at the processes by which modern society slowly came to welcome open-ended forms of social change.

Today, we live in a society in which history is accelerating (Hann (ed.), 1994). Some kinds of change are out of control. Although there are many institutions which regulate and hold back undesired initiatives for social change, there are also many kinds of change which are valued and encouraged. We welcome new artistic ideas with few limits. We welcome new kinds of knowledge with only moral scruples and financial caution over research costs. The capitalist system encourages competitive confrontations which might lead to economic growth. Even though we do not all welcome new military tactics or technology, we have great difficulty in holding them back because of our fear of external military threats.

The reason I say that change is out of control is because many of the changes we encourage are 'multipliers' (cf., Boulding, 1978) – they lead to chains of further changes, the effects of which we are unable to evaluate. We contrast our society with earlier societies which seem to have resisted change or to have had very specific visions of the kind of change to be sought. Our culture has been enthusiastic about the kinds of changes labelled as economic growth and/or as social progress since about the mid eighteenth century.

The creation of new knowledge through the pursuit of science is such a multiplier of historical change: it has played a significant and increasing role in the generation of historical change over the last two centuries. To some extent, our attitude to science arises because the knowledge it offers is not empty of practical consequences but repeatedly generates technologies which deliver desired economic results. We see an essential continuity in the form of science since about the seventeenth century. So, surely, any discussion of the question which heads this section must reveal how we moved from forms of science which did not work in generating economic change to ones which do?

Let us consider, schematically, how it was that our society moved from having little concern with change into one in which so much of social change is the result of socially approved multipliers of change.

For many reasons, late Medieval and Renaissance European society found its circumstances changing rapidly and became receptive to internal changes in order to adjust. A non-human pressure was the dramatic drop in population caused by plagues. An external human factor was the contact with other societies to the South and East, including the recovery of additional ancient learning conserved in Byzantine and Arab culture. An internal factor was that as power became concentrated in small nation states, there was a continuing development in the technology of war, which forced further social changes. By the sixteenth century, European society was changing very fast, and under these circumstances, resources which could help cope with change were likely to be favourably received. Some of the ways Renaissance society adjusted, however, were a systematic source of further changes. The systems which built up trade facilitated further exploration which led to new external contacts. The assimilation of gunpowder led to a sequences of changes in military technology. Printing led to a revolutionary increase in the communication of the written word. It became more difficult to control those who sought to profit by sharing knowledge widely. Not every adjustment to changing circumstances made in this society produced further effects, but some had dramatic repercussions. Exploration, invention, and easier communication could all be seen as 'change multipliers' in that society.

Sixteenth-century European society was changing so fast that it was beginning to restructure itself so as to adapt to circumstances characterised by continuing change. Increasingly, the changes occurring were due to the multipliers of change already in place. One effect of this was the rapid growth in competing traditions of natural knowledge which we have already noted. They could be seen as a form of social adaptation to the pressures of sustained change, for mechanisms which sustain diversity are an adaptive response to new challenges.

In the seventeenth century, times remained unsettled. England, for example, was greatly affected in mid century by civil wars. However, as more of the changes were produced from within society, the idea of welcoming or resisting internally generated changes was coming to be appreciated as a political issue. Resources for resisting some forms of change became more influential. For example, the attempt grew to construct an acceptable form of knowledge to replace the old Aristotelian Christian orthodoxy with a synthesis of the new knowledge with the less divisive features of Christianity. The new sciences of the seventeenth century are, on this view, a product of the way the intellectuals of the period sought to resolve the tensions between the pressures for and against further change in the domain of knowledge.

The new sciences of the seventeenth century sought characteristics appropriate for a society seeking stability, where their predecessors had features that could cope with changing circumstances. The key seventeenth-century developments were especially marked at the end of the English civil war, when the urge for recovery of stability was especially strong. The following principles stand out: elitism (Greek standards of intellectual rigour were appealed to and a new elite institution, the Royal Society, was founded); orthodoxy (in particular, the new knowledge should be supportive of religion in non-divisive ways); non-secretiveness (its practitioners competed publicly in displays of their virtuosity). The new knowledge avoided the populist, the heretical, and the occult. On these principles, the material selected for inclusion in the new sciences naturally included all those strands which were perceived as having produced significant accomplishments. For example, Copernicanism had been developed in different ways by Bruno, Galileo, Kepler, and Descartes. The Newtonian synthesis of astronomy and terrestrial mechanics selected and built upon what appeared to be the best of diverse variants of earlier astronomy. It did not simply follow steps laid down in some tightly defined tradition of earlier forms of mechanical philosophy.

By the early eighteenth century, it may have seemed that the sciences, at least, were now under control. They could safely build up knowledge without endangering society by stimulating further change. But the very form of authoritative knowledge they represented was to become the most powerful multiplier of change of all. The mode of rigorous technical reasoning, abstracted from its wider social context, facilitates change if widely applied. If tacit craft knowledge is represented verbally in powerful systems of thought, it becomes easier to think about its improvement and also to teach it to others. The new knowledge could more easily transfer to new social niches within societies which shared a culture of formalised knowledge. Soon, many of the Enlightenment philosophers began to use this form of knowledge as an instrument of political liberation. They argued that we should be sceptical of what the state and religious authorities tell us, putting our trust in what we can establish by our own senses and reasoning.

Simultaneously, the emerging capitalist system was bringing about economic change by giving increasing encouragement to invention. New centres of political power, allied to commerce and manufacture, were generating further change, reaching a climax with the industrial revolution. Emerging political groups saw science as a resource that could lead to the kinds of changes they were seeking. Their views may at first have been more ideological than practical. But, by the nineteenth century, science was becoming a major resource for technological change, a trend that has continued without any sign of an end in the twentieth century.

In this sketch I have sought to show that the growth of variant forms of belief about the natural world in the sixteenth century was a response to widespread social change. A number of multipliers of change were produced as change internal to society was tolerated in the context of the need to adjust to changing times. In the seventeenth century, attempts were made to slow the multiplication of forms of knowledge by consolidating the new learning into a stabilising instrument. This was to have new standards of intellectual rigour, tied into existing religion which could reduce the fragmentation of society by producing universal facts which are true for all. But it was too late. The authoritative form of natural philosophy which resulted was taken up by new groups who sought to make the sciences into a tool for the kinds of change that they could exploit. Other sources of change produced new problems for the new learning to address. By the nineteenth century, science had become one of the most powerful generators of social change.

The overall process as described here would not have been seen in these precise terms by the historical actors. Their own purposes were conceptualised in other, more specific ways. The process was more pervasive than can be appreciated by study of their own rhetoric. Society was restructuring itself to be able to cope with changes which increasingly came from its own internal structures.

Integration of the explanations

The scientific revolution should not be considered as a completely self-driven cultural process. It is part of a bigger cultural change that also affected commerce, technology, medicine, art, music, literature, and religion. Perhaps it can be tied to changes in the underlying political and economic system. Marx thought that the political changes were a result of changes in the economic system. Weber argued that the economic changes (the rise of capitalism) were the result of religious changes (Protestantism). In the modern era, however, we are finding that new religious forms in traditional cultures (such as cargo cults) are often generated in societies undergoing rapid change such as that produced in by contact with the West. My intuition is that all these changes are interlocked in some broader process. My suggestions about the role of multipliers of change in European society are an attempt to give substance to this feeling.

From this viewpoint, the other approaches discussed in the section are at best partial explanations. Each is looking selectively at parts of a larger process. Mid twentieth century history of science tended to build upon Enlightenment views of the rise of modern science, which put at the centre of the stage the late seventeenth-century Newtonian synthesis, its anchoring in the intellectual activities of the Royal Society and the philosophical support given by Locke (Locke, 1979, originally 1689). Earlier science was considered in terms of how it led to these developments. Later twentieth-century culture became increasingly preoccupied by perceptions of the pervasiveness of social change and by doubts that such changes can be fitted into an Enlightenment scheme of social progress. Postmodernist sensibilities encourage historians of early modern science to consider the diversity of beliefs of the sixteenth century in relation to the local circumstances and to doubt whether the imperfect sifting out of a more limited orthodoxy in the seventeenth century deserves to be termed a 'scientific revolution' at all.

The 'no connection' analysis concentrates on highlights of the raw material of the sixteenth century out of which the dominant seventeenth-century versions of natural philosophy were constructed. It does not really explain why this variant construction displaced all others, for it concentrates on an elitist view of knowledge. In its idealised view, the most trustworthy forms of experience were to be processed in terms provided by the most rigorous forms of mathematical expression. The actual historical process was more diverse than such an ideal allows for. The 'positive connection' analysis points to continuities between sixteenth- and seventeenth-century intellectual life ignored in the 'no connection' approach. To make a sharp distinction between the sixteenth-century intellectual practices which passed into seventeenth-century scientific orthodoxy and those which were marginalised or rejected requires imposing the standards of the later century upon the earlier one. The 'negative connection' analysis reveals something of the process by which practitioners of seventeenth-century natural philosophy set out to define a distinctive practice. They constructed boundaries between their own activities and the old orthodoxy of Christianised Aristotelianism and also competing occult, heretical and mystical practices. These boundaries were also made to define the limits of legitimacy and orthodoxy for forms of knowledge construction.

My conclusion is that the science and pseudo-science issue in the scientific revolution is to be seen as growing out of attempts to bring the process of knowledge creation under a greater degree of social control. Efforts were made to manage it for the general good of society. However, that is not precisely how it was seen at the time. There were a great number of rhetorics each suited to the expression of particular perceptions. An important one was religion. True knowledge must be in harmony with true religion. False forms of belief can lead to heresy, from which society must be protected. Although antagonism to heresy appears to have a theological basis, it involves an attempt to restrict a certain kind of change – religious change. As society became more secular, as in eighteenth-century France, this particular rhetoric declined. But the science and pseudo-science issue did not go away, for false knowledge can still lead people astray in a society seeking to hold itself together.

The science and pseudo-science issue can, then, appear in many different rhetorical guises. My discussion suggests that societies in the crisis of needing new answers to the problems of change are sometimes prepared to tolerate more diversity of belief than other societies. Perhaps our own society has, for that reason, moved towards cognitive pluralism. Those people who see crises looming out of present political, economic, military, ecological, religious, and artistic trends are more prepared to lower their critical standards of belief. Those who see no such crises vigorously object to the decline of intellectual standards. The way we understand the scientific revolution is, then, a function of our own society as much as of the past societies we study historically.

8.6 Summary

There are many variant forms of cognitive practice, claimed as science by their proponents and rejected as pseudo-science by their critics. Is there a rational demarcation criterion between science and pseudo-science?

The philosophical criterion offered by Popper is discussed and applied to psychoanalysis. Popper's criterion requires the scientist to formulate observationally falsifiable hypotheses. (S)he should seek refuting observations, rejecting a hypothesis when it is falsified. Not to do this is pseudo-scientific. Some of the critics of psychoanalysis have pointed out that in spite of its great cultural impact, it was formulated and defended in a way that systematically evaded the possibility of falsification. Popper's criterion turns out to be too crude to demonstrate that psychoanalysis is a pseudo-science, but Popper's criterion does nevertheless identify a widely perceived weakness in Freudian doctrine.

In the rise of modern science, modern disciplinary forms often emerged in close association with flourishing practices such as astrology, alchemy, and the occult arts, now regarded as pseudo-sciences. What was their relationship in the scientific revolution of the sixteenth and seventeenth centuries? Four accounts are reported and it is argued that they are all partially correct, each applying to limited aspects of the overall process. It is true at the level of content and method that (a) the new natural philosophy combined the best of the recently recovered Greek philosophy with a new appreciation of craft practice. It is also true at the level of justificatory rhetoric that (b) the defenders of the new knowledge sought to sharpen the contrast between the rigour of natural philosophy and the untrustworthiness of the pseudo-sciences. At the level of the people involved (c) there was a continuity between the practitioners of the pseudo-sciences and the virtuosi of natural philosophy, strikingly illustrated by the case of Isaac Newton. However, to explain why natural philosophy displaced the pseudo-sciences, we must consider how the sciences fit into the wider social changes occurring in society. The general explanation offered here concentrates on how society moved from merely responding to the challenges of change to incorporating 'change multipliers' and to surfing on the wave of internally generated change, under the illusion that change was under control. Modern science was a product of this process but did not become a major change generator until the nineteenth and twentieth centuries.

9

A theory of the pathologies of science

9.1 Pathological science

In this chapter, I present an account of some of the ways in which we can distinguish good scientific procedures from bad. It is tied to the theory of science presented in chapter 6. The term 'pathology' is being used here as a generalisation from the classic paper by I. Langmuir, 'Pathological Science' (1989, originally 1953). Irving Langmuir, an eminent physical chemist (Nobel prize in chemistry, 1932), often gave talks on the subject of pathological science. A transcript of one such talk, given in 1953, was finally published in 1989. Langmuir's discussion (Langmuir, 1989, p. 44) is summarised in terms of six symptoms of pathological science:

Symptoms of pathological science

1 The maximum effect that is observed is produced by a causative agent of barely detectable intensity, and the magnitude of the effect is substantially independent of the intensity of the cause.

2 The effect is of a magnitude that remains close to the limit of detectability, or, many measurements are necessary because of the very low statistical significance of the results.

3 There are claims of great accuracy.

4 Fantastic theories contrary to experience are suggested.

5 Criticisms are met by *ad hoc* excuses thought up on the spur of the moment.

6 The ratio of supporters to critics rises up to somewhere near 50 per cent and then falls gradually to oblivion.

Langmuir was concerned to show how, in some circumstances, well-intentioned scientists could convince themselves and their immediate colleagues that they were seeing and measuring something which, with hindsight, we can be sure was not there. Langmuir described as pathological science, 'cases where there is no dishonesty involved but where people are tricked into false results by a lack of understanding about what people can do to themselves in the way of being led astray by subjective effects, wishful thinking or threshold interactions'.

In my development of Langmuir's concept, I see pathological science as science which begins with some illusion, fallacy, or unnoticed interference with the immediate reality under study, that produces a bold or surprising result. However, instead of the deviation being picked up as an error and quickly corrected, it is sustained or even increased by some failure of the subsequent cognitive activity of science at one or more levels. Very often other practitioners criticise it as departing unacceptably from the usual intellectual standards of science (though they may only do so with hindsight). A common pathological pattern is for people to find it difficult to acknowledge that they have made an error. Believing they have the truth, they try to manipulate the orthodox procedures to give greater credibility to their claims. It is a pathology of science for a scientist to be able to control how other people judge his or her claims to knowledge. The typical development of pathological science, then, involves an uncorrected exaggeration of the accuracy and significance of the phenomenon found, to produce the pathological effects that Langmuir lists.

Langmuir's examples included N-rays* which I discuss next. It is not difficult to think of other cases in the history of science to which it might be applied (e.g. Goethe's theory of colour*). The published version of his talk was readily applied to subsequent discussion of the cold fusion* debacle of 1989 (e.g. Morrison, 1990).

After an account of the N-ray case, I will discuss how pathological science can arise at each stage in the general process of scientific activity discussed in chapter 6. My aim is to show the diversity of ways in which orthodox science can go wrong. My argument is that such failures are not so much failures according to absolute standards of rationality, but according to the institutionalised imperatives of the immediate context of the cognitive activity. Behaviour that would be quite reasonable in other contexts, becomes often pathological only when it occurs as a part of science.

9.2 The N-ray story

In 1903, the French academician René Blondlot, a physicist at Nancy University, 'discovered' N-rays. He had been studying X-rays by their effect of making brighter a faint spark jumping across a small air gap. He found that the same detection method was also picking up another kind of radiation with rather different properties. The new radiation could, for example, be produced by a very hot wire in an iron tube with a thin aluminium window in it. The radiation penetrated aluminium, black paper, and wood, but was absorbed by lead or water. Blondlot reported his discovery to the Paris Academy of Sciences on 3 March 1903, and over the next few months made further reports, describing the properties of the radiation, which he called n-rays (later N-rays, after the town of Nancy in which his university was based). He presented fifteen papers to the Academy between 1902 and 1904 (reprinted in Blondlot, 1905). His research programme was based on conjectures inspired by the earlier study of X-rays. The dramatic scientific discovery of X-rays in the previous decade had revealed clearly and unambiguously the existence of a form of radiation which penetrated all but the densest matter. Although, like any new discovery, the first insights into the nature of the N-ray phenomenon had to be exploratory, the impression that Blondlot's new discovery was under proper experimental control built up very quickly. He regularly confirmed his conjectures about new properties for the radiation. All his early methods of detection of N-rays and a second form, N_1 rays, involved subjective judgements that a dim light had been made brighter – whether an electric spark, a small flame, a phosphorescent screen, or a dimly lit piece of paper. He soon found that when N-rays are directed onto the eyes, vision becomes more sensitive. He was therefore able to use his eyes as a rough-and-ready N-ray detector simply by judging whether or not he could see the time on the wall clock in an almost dark laboratory.

It might seem that the detection of N-rays involves judging a sensory threshold, inevitably with a high level of error. Blondlot found that individuals vary in their skill. (*Revue Scientifique*, 1904, p. 546). Some people could see the brightness changes immediately, while others needed considerable practice. Once seen, however, consistent results were obtained. Although the pioneering work had to be done by the simple methods described above, Blondlot was working towards a more objective measure using photographs to compare the brightness of a spark with and without N-ray illumination. However, the spark tended to vary in brightness spontaneously, and the methods Blondlot used to average out the variation allowed major errors to creep into the timing of the photographs. These errors tended to make the immediate reality excessively congenial.

In spite of (or as we now think, because of) the difficulties of the sensory discriminations, Blondlot made many exciting 'discoveries'. N-radiation turned out to have a wide spectrum of frequencies, which he was able to separate using a prism and to measure their wavelengths by two different methods. The disagreement in the data was less than 4 per cent. Many of his students and associates were soon helping him on the new project. Six other authors at Nancy presented papers to the Academy of Sciences. One of the most prolific was A. Charpentier, professor of medicine at Nancy. He discovered that contracting muscles, active nerves, and the central nervous system all emit N-rays. His observations were subject to priority disputes[1] (an ironic twist if you accept that the phenomena were spurious). Two Paris-based scientists Jean Becquerel and A. Broca, went to Nancy to acquire the appropriate skills, and with their students were soon publishing observations on N-rays. Other scientists were less successful, reporting marginal results, or failure. No one outside France published positive results. In 1904, for example, eight letters were published in *Nature*, reporting failures to observe N-rays.[2] It was pointed out in such publications that the observations were not nearly as simple as Blondlot's publications implied. Judgements of the brightness of dim light can be affected by such factors as the increasing adaptation of the eyes to the dark, the lower sensitivity of the eye to dim light when viewed directly than when viewed peripherally, and the hypnotic effect of prolonged staring at a small relatively bright object. Uncontrolled physical influences could also complicate matters. The brightness of an electric spark could be affected by placing a metal plate close to it, as Blondlot did when introducing his lead screen. And small changes in the temperature of a dimly glowing phosphorescent screen produced large changes in its brightness.

In the standard story,[3] the last of the letters to *Nature*, in September 1904, is seen as settling the matter. The American, R. W. Wood (1904) had visited Blondlot's laboratory and the phenomenon had been demonstrated to him. He had not been able to see the N-ray brightening effect, a common experience of Blondlot's guests, but in the darkened room he had been able to make a few unnoticed changes in the apparatus. In particular, his hosts were still able to detect maxima and minima in their N-ray spectrum even after Wood

---

[1] See, for example, D'Arsonval (1904), which rejects calims for priority over Charpentier's 14 December paper to the Académie, naming five people as being among those who had made such claims.
[2] Swinton (1904a, 1904b); Burke (1904a, 1904b); Rudge (1904); Schenk (1904); McKendrick & Colquhoun (1904); Wood (1904).
[3] For example. Firth (1969); Langmuir (1989); Price (1961), pp. 81-91; Rostand (1960), pp. 11-51. For more historical depth, see Nye (1980).

had secretly removed the prism supposedly producing it. Wood also criticised the methods by which photographs of N-rays were being produced.

Although Wood's letter is given the key role in discrediting N-rays outside France, the story went on a little longer. The French were not immediately persuaded that Blondlot had systematically been deceiving himself rather than making a one-off error during Wood's visit in the use of a very difficult technique. On 22 October, *La Revue Scientifique*, which had been reporting the main developments in N-rays, translated Wood's letter, and the following week, gave a conscientious review of the developments since Blondlot's first paper (Langevin, 1904). The journal also instituted a survey of the opinions of established French physicists (and a few non-French individuals who had been directly involved). Fifty-seven replies were published, including Blondlot's reply to Wood's letter. Eleven physicists refused to offer an opinion, either on principle, or claiming ignorance. Twenty-eight had tried to confirm some aspect of Blondlot's work but had failed. Most (seventeen) of these avoided inferring that N-rays do not exist, pointing out that they may not have been careful or persistent enough in what was clearly a difficult and skilled task. A minority (eleven) stated, more or less forcefully, their doubts about the existence of N-rays – on the basis of their own experience and through general arguments. Of the seventeen respondents who believed in the existence of N-rays, nine were from outside Nancy. Seven of the latter had not detected the radiation personally, basing their judgement on the integrity and competence of Blondlot and the high quality of his previous research. Henri Poincaré, for example, admitted that his was a blind faith in the existence of N-rays, since Blondlot had explained to him how defects in his vision were preventing him from detecting the subtle effects of the new radiation (Poincaré, 1904, p. 682).

The opinion survey showed clearly that the only people committed to N-rays were those who had detected them for themselves, initially with Blondlot's guidance, together with those who were prepared to trust Blondlot's expertise. Neither of these considerations had much appeal to scientists outside France.

The survey could not, of course, resolve the problem as if it were a political referendum. However, attempts at more objective tests for N-rays were not able to overcome the increasing level of suspicion, and by 1906 scientific publications of the subject had ceased. At the end of 1904, the Leconte prize of the Academy of Sciences, for which Blondlot had been chosen some months earlier, was given to him, not for his discovery of N-rays, but for the work of his whole career (Académie des Sciences, 1904).

I will return to the N-ray story at several stages of this discussion. What happened fits Langmuir's model of pathological science extremely well. In the language of the discussion of this chapter, the judgement of hindsight is that there were failures in the manner in which nature was made to generate meaningful sensory messages. Indeed, the messages that nature was supposed to be producing, through subjective discriminations of brightness levels by Blondlot and his colleagues, are now thought to be illusory.

9.3 Pathologies in the production and negotiation of scientific facts

The difference between sciences building upon illusion and those that do not is obvious in traditional epistemology, but the contrast is worth probing more deeply in the context of the epistemological relativism that haunts the work of microsociologists investigating the social construction of scientific facts.

I see the process of producing inscriptions as going wrong when it becomes clear to critics that the messages alleged to be from nature are too ambiguous and no way is available to reduce the ambiguity. This ambiguity can result from:

1. The choice of phenomena uncongenial to the science approach. Ideally, it should be possible to produce measurable effects as and when required by the direct action of the scientist. In the extreme case of rare transient phenomena that only untrained people have seen, severe problems for the science can result.
2. Inappropriate or inadequate procedures to control the phenomena. This most readily happens through lack of relevant practical experience or when scientific understanding is misapplied – leading to temporary confusion.
3. The use of unreliable methods in the generation of messages from nature. The trend in science has been to replace methods involving difficult sensory judgements by objective instrumental recording of data. But newly discovered phenomena cannot always immediately be investigated to the sophisticated standards of established science. This was an aspect of Blondlot's problem.
4. The imposition of unreliable interpretations upon laboratory results. If we receive ambiguous messages from the phenomena, we are more dependent upon skilled scientific interpretation. But even experts can fool themselves. If we can, we use our power over nature to generate

messages which are easier to interpret. If we cannot do this, as with telepathy, there are grounds for suspicion about what is really going on.

The ideal practice of science in this context is a counsel of perfection which cannot normally be achieved. Scientific communities institutionalise techniques to minimise ambiguities in a manner appropriate to the research situation. Failure to follow such procedures results in pathology.

It seems obvious that the N-ray case involved collective self-delusion, which was eventually exposed by those who were not subject to it. Rostand (1960) argues that the recent discovery of so many forms of radiation made it likely that there were new forms of radiation to be discovered and gave guidance in how new discoveries were to be made. Blondlot and his colleagues were insufficiently cautious in such an intellectual climate.

If the N-ray phenomenon had existed and better instrumentation to display it had been developed, it would have been amenable to scientific investigation. The paradigm of X-ray research proved to be easy to follow. In retrospect, however, we judge that the appearance of control was illusory. Right from the beginning, it was clear that the methods by which N-rays were being detected were near the margins of acceptability. Blondlot unwittingly overstepped those margins. His supporters came to accept that his methods were inconclusive and his critics that they were open to self-deception. He tried to introduce instrumental elaborations of the method and especially, without full success, to represent the brightening effect photographically. In hindsight, it is easy to see that illusion had led to self-delusion.

I think that the account of the last paragraph is a tale of a mistake which any ambitious researcher could have made. The story becomes much more clearly pathological when we consider how Blondlot acted so as to minimise self-doubt. As it became clear that not everybody could detect N-rays, Blondlot made out that it was a highly skilled task, but one that could be learned in his laboratory. His defence could be seen as a holding action, until photographic methods could be made more objective and reliable. But, at least in the short term, observation of N-rays relied on interpreting an obscure message from nature, rather than on reading a clear message correctly. Blondlot had set up a situation analogous to non-scientific interpretations of natural signs. When witch-doctors read the entrails of a sacrificial animal, they interpret nature as giving them signs. But we, of another culture, are likely to judge that, whatever the state of the entrails, the cultural meaning assigned is constrained rather more by the witch doctor and his society than it is by natural phenomena. So it seems to have been with Blondlot. However, in Blondlot's case, it was not the cultural norm that led him to make a skilled mystique out of seeing N-rays, rather it was a defensive strategy to stave off criticism.

Phenomena uncongenial to the scientific approach

Science is not suited to the study of all kinds of phenomena as they are initially encountered. Science has built up effective methods of treating specific forms of material within a number of disciplinary approaches. Some potential objects of study, like, say, the nature of extraterrestrial life, lie outside this range. There are many historical examples of initially uncongenial branches of phenomena being found which were turned into suitable objects of scientific study, being so transformed by the process that they no longer had the same significance, to non-scientists at least. One such story (Westrum, 1978) is of how stones that fell from the sky were regarded with great suspicion by the eighteenth century scientific orthodoxy. Stories of such stones were generally dismissed as the delusions of the ignorant peasants who saw them and found them, or sometimes as the result of lightning hitting the ground and melting the immediate spot (thus explaining any associated light and noise). When, after a particularly heavy fall of meteorites at l'Aigle (70 miles from Paris) in 1803, the French academician Biot investigated, and drew the closely reasoned conclusion that they had indeed fallen from the sky, the stones themselves became the focus of interest. It was found that they had distinct mineral forms, which led to a properly scientific study of meteorites. Clearly it was more acceptable to natural scientists to undertake a mineralogical investigation of a particular class of stones than it was to investigate the reports of low-status witnesses.

New popular cultural enthusiasms quite often encourage the study of uncongenial phenomena which prove resistant to such a transformation into a scientifically researchable form. This is particularly likely when the transformation into closely examined reproducible procedures under experimental or statistical control threatens to strip the phenomenon of its initial cultural significance. (Psychical research* is a relevant example)

There are many examples of popular science devoted to uncongenial phenomena. Dealing with uncongenial phenomena is normally only a transitional problem in orthodox science – the pathology would be to persist when attempts to transform the phenomena fail.

Pathologies in the generation and interpretation of messages from nature

It is only in experimental sciences that direct control of the phenomena is possible. The skill of the experimenter lies partly in the ability to make things happen as claimed and partly in observational skill. There is a component of tacit knowledge involved. However, for the purposes of science, things must happen for the reasons given. A conjuror can make surprising things happen, but he misleads us as to how it was done. When the phenomena observed in science occur because a scientist has cheated, discovery of this results in the breakdown of institutionalised trust, and a pathology of science results.

Langmuir did not choose to include fraudulent experimentation in his examples of pathological sciences, as he limited his discussion to honest scientists. Fraudulent experiments certainly occur (Broad and Wade, 1982). Some cases involve people who have convinced themselves of the existence of some phenomenon and yield to the pressure to produce interesting results by inventing them. It is unlikely that science is led in the wrong direction by fraudulent experimental results. Nondescript fraud is easy to get away with, but results which attract attention are not. If others wish to build on surprising results, they are under pressure to generate results of comparable interest. In the absence of collective conspiracy, fraudulent novelties can quickly be isolated and, exposed. A celebrated case was that of William Summerlin's patchwork mouse* in 1974. An older case in which similar suspicions were involved was that of Kammerer and the mid-wife toad*.

Those sciences which must *find* their phenomena rather than produce them are slightly more vulnerable to fraud, for it is not always possible to make direct checks. For example, in astronomy, unpredicted transient phenomena, are somewhat uncongenial for science, since it is very difficult to check up on errors, much less fraud. Most astronomical phenomena continue over such a long timescale however, that they are as readily checked as in experimental sciences. Nevertheless, pathological examples have occurred. Perhaps the most famous was the episode of the Martian canals* in the late nineteenth century.

In some sciences, such as palaeontology, the concern is with found objects such as fossils. Especially interesting fossils are likely to be unique (to begin with, anyway), so that dishonesty is hard to prove. Cases such as Piltdown Man*, in which a unique set of fossil fragments was only established as fraudulent forty years later, and other cases in which the accusation of fraud has been made, but not established, show that there is a special problem here, apparently because any who succumb to the temptation to cheat, know that they might well get away with it. In the case of Piltdown Man, the fossils were kept with such care by the enthusiastic professional custodians that it was difficult, for a generation, to establish the fraud by examination of the sacred relics themselves. Equally spectacular was the recent case of the Indian geologist, V. J. Gupta, who was accused of confusing research in Himalayan geology for decades by publishing reports of 'salted finds' and other dishonesty, but had become so senior that it was very difficult to discredit him He was finally found guilty by his university in 1994.

Failures in the negotiation of facts

The simplest such pathology might be the individual who fails to involve other people at all in his or her findings. I discuss this possibility in the next chapter.

Another possible pathology of the negotiation of factual claims would be attempts to convince others by deceitful methods. To his critics, the sixteenth century iatrochemist Paracelsus was such a flamboyant charlatan in his methods of persuading his audience of the truth of his claims and the efficacy of his mineral medical remedies. The establishment critics in the 1780s of Mesmer* regarded him similarly. (In both cases, however, they were not *simply* charlatans.) We are prepared to believe that some modern advocates of unorthodox ideas and products in the popular domain are also charlatans. It is difficult to be sure which of the showmen of orthodox science are best accused of charlatanry, at least until their deceit has been exposed. On the whole, modern science only seems to attract (or expose) really flamboyant charlatans in areas that attract very generous research funding.

The N-ray story gives the impression that Blondlot was scrupulously correct in the manner in which he presented his claims to the scientific world. He communicated fully, and although he may at first have made his observations seem much simpler and unambiguous than he later claimed they were when he was under attack, this is not a rare feature of scientific presentations. When the interested parties came to his door explaining how hard they had found it to produce N-rays, he was happy to show them the practical technique. If he was very defensive after Wood's visit and exposé, that is not, perhaps, surprising. It is not in itself a pathology.

The case of cold fusion Cold fusion,* which hit the headlines in 1989, seems to have been made much more pathological by the negotiation process. The discoverers, M. Fleischmann and S. Pons, of the University of Utah, regarded themselves as being in a race with S. Jones at nearby Brigham

Young University, and used a press conference to announce their finding, rather than a properly refereed scientific paper. They became increasingly secretive about the discovery, which would have been revolutionary if true, if only because they were under pressure to maximise the commercial potential of their result. The natural caution of scientists in evaluating such a radical claim as cold fusion was swamped by the use of uncritical systems of rapid communication, in particular, the popular press and computer networks. People around the world were soon reporting that they had managed to produce confirmatory evidence. However, these quick tests soon turned out to be unreliable, because those making them were not accustomed to using the (widely available) instrumentation in such a non-standard way. The positive claims kept having to be withdrawn. Critical work took a little longer, and within about a year influential critical reports were available. The life-cycle of the process was very much as Langmuir described as characteristic of pathological science. (Morrison, 1990) The institutions of science were not obviously damaged in the long term, although the careers of some of those caught up in the episode may have been.

The normal way of describing what went wrong in the negotiation of scientific inscriptions in the cold-fusion example is to say that the critical stages of peer review were bypassed when Fleischmann and Pons held their press conference. And yet, the process was little different from the discovery of 'high temperature superconductivity' a little earlier, in 1986, a non-pathological case of an exciting new development in physical science. The original work on a superconducting ceramic material was more carefully prepared than that of Fleischmann and Pons, but, after the initial publication, the story was too urgent and too important to be held back by leisurely scientific publication. As in the cold-fusion case, even more dramatic findings were reported from around the world. Criticisms were also made. However, in the case of superconductivity, quick and unreliable tests of the phenomenon were not difficult to turn into persuasive demonstrations (such as the party trick of making a magnet float above a superconducting material). As with cold fusion, a more restrained communication system might have reduced the number of positive claims that later had to be withdrawn, but it would also have slowed the rate at which successful positive work was done, which resulted in rapid improvement in the ease of manufacture of new ceramic superconductors and higher levels of performance.

9.4 Pathologies in the reconciliation of fact and expectation

I expect that somewhere there is a Popperian hell in which perfectly adequate theories are constantly being discarded because people pay too much attention to spurious facts. (Perhaps it lies in some of the animal models on which potential human medicines are first tested.) But Popper thought he lived in a world in which there was more of a problem about people who produced pathological defences of cherished theories to shield them from the good facts which challenged them.

The discussion which follows explores the extent to which Popperian principles actually apply at this stage of the science process, allowing us to identify a specific source of pathological science. My development of Langmuir's concept holds that a faulty intellectual standard which is immediately identified as such and corrected should not count as pathological. But if the practitioners attempt to sustain their beliefs by deflecting normally effective criticism, then that may be pathological. Popper's work suggests that suitable candidates for such pathological science are the defenders of a doctrine who systematically refuse to expose it to the risk of refutation or to accept prima facie refutations when they are produced. Freudian psychoanalysis and popular astrology approximately fit the specification. In both cases, the theory could be tested, but its defenders surrounded it with defensive buttressing. Lakatos argued, however, that similar strategies are employed in healthy scientific research programmes, the hard core of the theory being protected against a contrary run of facts by adjustments to a protective belt of auxiliary hypotheses. We see, then, that scientific practice which is pathological according to Popper is perfectly acceptable according to Lakatos, the distinction between good and bad science only being possible, according to the latter, in long-term retrospective judgements.

Furthermore, as Kuhn suggested (1970a), in the absence of a satisfactory alternative, the heuristic value of having a theory (paradigm) at all is so great that scientists naturally keep the theory they have. They may move into a state of crisis, in which they look for a better theory, rather than immediately rejecting it because it cannot cope with the current run of factual evidence. Kuhn regarded crisis as a standard though transitional phase of science, rather than a pathological state.

Although there can be no hard-and-fast rule, it would appear that extreme examples of explaining away contrary evidence *are* regarded as unsatisfactory in orthodox science, in the sense that people feel uneasy when it is done. A famous contemporary example is the non-observation of the free quark. The quark is one of the universal constituents of the subatomic particles of ordinary matter, but, whenever a hypothesis was produced which predicted free quarks in a particular context, none was found. The protective belt of matter theory has been elaborated so that we no longer expect to be able to see quarks. The general theory in which this difficulty is embedded goes from triumph to triumph, so that a little unease can be tolerated. But, if ever a *whole* body of theory were to find itself in such a situation, people would not hesitate to label it pathological. This happens in some examples of unorthodox popular science.

9.5 Pathologies of the assimilation of revised hypotheses

Theories of conspiracy as pathology of science?

There are corners of our culture in which people are inclined to look for, and tend to find, conspiracies, secret groupings of people who use fair means and foul to attack and undermine contrary interests. The extremes of politics are especially prone to such thinking – the general level of anxiety is high enough to produce enough conspiratorial counter-conspiracies, which makes conspiracy theory self-fulfilling. It is therefore, perfectly reasonable to look for conspiracies in such a context. A culture of conspiracy theories legitimates itself. The temptation to look for conspiracies is, therefore, sometimes but not always appropriate. It is very common for the defenders of unorthodox views in the cultic fringes of society to buttress their beliefs with a conspiracy theory. The same tendency can even occur in orthodox science.

Looking for conspiracies is a pathology of science because it signals a breakdown of the mutual trust on which the institutional system of science builds, because the freedom to view all counter-evidence and counter-argument offered by other people as part of a conspiracy is an extremely effective way to defend one's own ideas against contrary tendencies in the network of understanding, and because, once proposed, it is very difficult to be persuaded to abandon a conspiracy theory. If Blondlot thought that Wood was conspiring against him in secretly removing his N-ray prism during the experiment, then doubtless the thought was justified. But, once one thinks in terms of conspiracies, the restoration of normal science is extremely difficult.

Secret science as a pathology of science?

There is nothing irrational about deciding to keep scientific findings secret within the group that produced them. The knowledge produced could be too dangerous to tell the world – as some of the precursors to science regarded their forms of knowledge (for example, it would debase the currency if every-one knew how to turn base metals into gold). It could be a competitive secret – as when only one's own state or company should know. Military and commercial secret science abounds. One of the early stages of the growth of the modern form of intellectual property in science goes back to when sixteenth-century university mathematicians challenged each other to solve problems and prove conjectures. The trick was to have a secret method for solving a class of problems which no one else could solve, and which could not be worked out from the answers the challenger had given to earlier problems in the class. This kind of secrecy continues in science – scientists are tempted to keep quiet about the techniques by which they reached credit-worthy results, which are themselves freely publicised (Hagstrom, 1965, espe-cially chapter 2).

Scientific activity, then, is an interplay between the public sharing of credit-worthy results and keeping secret that which is to one's personal or factional advantage. And yet, the effective functioning of science would seem to require the free exchange of information and ideas in an open discussion. Under what circumstances is secrecy pathological?

Secrecy may well have been pathological in some of the sixteenth-century precursors of modern science. Traditions which, like alchemy, disguised their findings in ambiguous forms which only initiates could understand, may have sought to avoid the corrupting effects of uncontrolled communication, but in their secret traditions, corruptions and distortions of knowledge were easily introduced and hard to correct. In particular, traditions which subscribed to the view that knowledge cannot be taught but only gained as a personal insight of the suitably initiated were far less effective at correcting their knowledge than didactic traditions in which knowledge is passed to the next generation by open, explicit (and critical) teaching.

When the normal process of scientific scrutiny fails for secret forms of modern military and commercial research, I do not think that the result should be labelled as pathological science but as technology. Then the appro-priate standard is the weaker technological one of whether or not a process or product *works* as required. There remain questions about the special interests of those who judge whether secret technology actually works, but that is a problem for the critical evaluation of technology.

Some secret features of science can persist without serious consequences, provided that they are linked to an open system of scientific communication. Only if what is secret is critical to the process of science would the effect be pathological. The present system of orthodox science is so open and so redundant in its communication system that, for any matter of widespread social interest, keeping some small part secret can be of only passing consequence, as very soon the same result will have been reached in another more public way. Even closely guarded military secrets are only secret in the short term. I suspect that there is no longer any great difficulty about using publicly accessible documentation to make a working, if merely marginally effective, atomic fission bomb, including the especially complex task of producing suitable fissionable material.

External pressure on scientific conclusions as a pathology of science?

In chapter 4, I looked at science in other social systems, and in particular totalitarian systems. The argument by the apologists for Western science was presented that, when science is subject to strong external pressures, it is liable to produce spurious results. If that argument is accepted, then Lysenkoist science is an example of pathological science. The opposing analysis was offered that the knowledge produced in science is *always* adapted to the external social setting. No science is an island. The imposition of the external views of one social system will appear pathological to defenders of the science of another. To Stalinists, Western bourgeois capitalist science was as pathological as Lysenko's Stalinist science was to Westerners.

The resolution offered there and in this chapter is that although science is always affected by the social context in which it is constructed, we may also take into account the social contexts through and to which it is successfully transferred. Very often, a preliminary form of science flourishes only in the localised context of its creation. Some forms later come to flourish, suitably modified, in a wider range of social settings. The scientific ideal of the universality of knowledge requires that this should happen, and science that can only survive with the support of special local conditions increasingly appears as pathological. Since all new variants of science start by being local knowledge, it is not appropriate to say that all purely local candidates for scientific knowledge are pathological. But if, in the long term, a knowledge claim has proved unacceptable to all outside a given locality, then it would become pathological to defend it as science. In the co-evolution of knowledge and of society, such pathological forms will tend to die with the societies which support them.

Controversial science as pathological science?

In chapter 6, it was suggested that controversy is not merely a passing phase of Kuhnian science in crisis, but an endemic feature of the intellectual activity. However, it was also argued that controversies are normally resolved or suppressed. It may be that if a part of orthodox science is subject to sustained and damaging controversies then that condition is pathological. Such a claim would itself be controversial, because some views of the intellectual process assume it to work through the dialectical tension of contradiction. Furthermore, never-ending controversies are a feature of many social sciences. However, in orthodox natural science, at least, there are powerful mechanisms for controversy resolution. A science in which those mechanisms are ineffective may be in a pathological state. A science in which disagreements can neither be legislated out of science proper, nor tied down to the outcome of specific empirical inquiry, is likely to generate lasting disagreements. However, the best examples of sustained inability to agree lie outside orthodox natural science.

9.6 Failures in inferences from received knowledge

It was claimed in chapter 6 that the pathologies of scholarship in science are not greatly different from those of the use of intellectual learning in other contexts. The requirements are more institutional than rational, an especially useful organising concept being the social importance of identification of other people's intellectual property and the careful delineation of one's own intellectual claims in relation to the property of others.

Plagiarism

One of the most widely noticed intellectual failings is plagiarism. In intellectual systems (such as theology) in which novel additions are less important, and it is thought that most things worth saying have been said many times already, intellectual property is less important. In literature, originality is of great importance, the issue usually turning on the precise wording used, so that the legal institution of copyright can be applied. In science, the most blatant forms of plagiarism sometimes involve directly copying someone else's publication; however, it is also a serious offence to steal an idea, even one that has only been partly developed in prior research.

Plagiarism is a form of dishonesty which sometimes causes problems in institutions employing such a scientist, for they may not have the competent researcher they think they do. Within science, researchers will be less trusting in their communication with one another if plagiarism is perceived to be widespread. However, scientific knowledge is not led astray by plagiarism, for the only truth it subverts is that of correct authorship. Those individuals who fear that an associate may be a plagiarist learn to exercise a higher level of caution around him or her. Those institutions which need to monitor the quality of their manpower can do so by tightening up the management of science on principles common to all knowledge-based activities.

The pathology of urgent applications

Sometimes the urgent needs of everyday life can ask more of science than it can provide satisfactorily to its own standards. The result can look like a pathological science. As an oversimplified example from science fiction, consider what is to be done on discovering that the world is going to end tomorrow. Let us suppose that it has been found by some properly scientific process, such as observing that a newly identified asteroid 11 km across is on a collision course with the Earth. If true, crisis action may salvage something from the global catastrophe. But there might be a mistake in the observation and reasoning; we cannot be certain until it is too late. Many natural and social sciences find themselves in less extreme versions of this millenarian dilemma of apocalyptic revelation. For example, this was the perception of such environmental alarmists of the 1960s as Commoner, (1970, originally 1963) and Ehrlich (1971, originally 1968). Sometimes we judge that the need for taking some action is more desperate than the need for great rigour in the science, at other times we doubt that any special action need be taken and confidently wait for the error in the prediction to be exposed. It would seem that, from a properly scientific point of view, we should always be sceptical about those who promise that the world will end tomorrow, only the faithful being saved. But the rationality of actual life is not always identical to the rationality of science. Whether the matter is the best way to educate our children, or the likelihood of our souls surviving our bodily death, we gamble in hope and in fear before all the evidence is in. Most of ordinary life could be seen as pathological science.

9.7 Concluding discussion

In this account of the pathologies of science, I have tried to show that the ideal cognitive processes and products are not always achieved, but that the normal feedback loops connecting the many areas and levels of cognitive activity should eventually pick up and correct whatever is (socially) identified as an error. Where there is disagreement about proper procedures, the higher levels of negotiation are drawn on. And, because of the peculiarly effective methods of controversy resolution and controversy elimination in science, the whole process works quite well, at least in the long term. The more systematic the evasion of normal procedures for detection and elimination of error, the more pathological the resulting science. In the absence of such pathologies, the good science that results is a distinctive form of knowledge.

9.8 Summary

This chapter identifies a number of ways in which science can fail to optimise the production of knowledge. Since it has been argued that all science is open to error and subject to revision, the discovery of error is not in itself a sign of pathological science. That comes with attempts to immunise the practice against the discovery and elimination of error.

Some phenomena are unsuited to the scientific approach and should be avoided or transformed into a more suitable form. Every effort should be made to avoid ambiguity in the messages we make nature write upon our instruments and in the methods we employ for their interpretation. Our negotiation of factual claims should avoid secrecy or deceit. The manner in which we reconcile fact and expectation should allow perceptual surprises to force change in existing ideas. The manner in which successful new theories are assimilated should be as open to criticism and counter-argument as possible, for example by avoiding secrecy and conspiracy against one's rivals. The use of existing knowledge should follow the rules institutionalised more generally for learned practice.

Part III

Changing science in a changing world?

10

What are acceptable variations of present science?

10.1 Introduction

The nature of the practice of science is changing along with its content. In the past, the changes have been on a curve of continuing growth, as perceived successes are repeated and society finds new uses for science. In the present day, the main driving force is economic investment in science as an overhead on science-based technologies. As a result, knowledge creation is now mostly in the context of knowledge use. With the recent decline in its rate of expansion, science now grows too slowly for new component activities simply to be added to established ones. Change involves loss as well as addition. In addition to the practical limits to growth, the problems produced by science-based technologies are providing negative feedback. Science struggles to adjust itself to deflect criticism and loss of support.

Our understanding of science, then, should take into account its changing nature. The present-day preoccupation with short-term economic value is producing effects disliked by both scientists and their public. Should we, perhaps, try to influence the way science changes?

All the options involve change. We cannot move back. A more slowly changing pre-scientific society could not sustain the science-based technology on which our survival at present population levels depends. Since past science was always in a state of exponential growth, to keep science in its traditional form would involve it growing further in a society which cannot grow and change fast enough to accommodate it. To keep science as an activity without growth would change it into a more conservative practice, no longer affected by the influx of so many young people who have much to gain by challenging old dogmas. We cannot afford to abandon the effective creation of new knowledge, and yet we cannot sustain the established ways. Science must keep on changing.

We must, it appears, live with change in the nature of science, just as we have long lived with change in the content of scientific knowledge. If we leave science to change in its own way, we might expect that each level of cognitive activity would optimise the distribution of benefit at that level. The adjustments do not simply optimise the pursuit of science, for they are unavoidably dependent on the external social relations of science. For example, in the past, scientists have modified science to suit the social circumstances of the short-term crises of war. At present, in areas with strong anti-vivisection movements, scientists actively seek ways to reduce the need to experiment on live animals. Perhaps there are optimal ways to influence science through such external pressures? Perhaps we can make science better as it adjusts to changing circumstances?

We saw that efforts to adjust science in accordance with the dominant ideology in totalitarian regimes can rather easily make science worse. To influence science too much risks producing ineffective science or even pseudo-science. The purpose of this chapter, therefore, is to explore some of the changes to science which have been advocated and to see what the consequences might be. I will look at proposed changes to scientific orthodoxy, some gradual and some radical. The more radical alternatives that do not build on present science are unlikely to be as effective as present science in its current social setting. Because the alternatives could not compete directly, they are only likely to survive in isolated places or by finding other social uses for the 'knowledge' they create. Perhaps, for example, much of orthodox science is less suited as mass entertainment than some of the popular alternatives in present society.

I will presuppose in this chapter that the primary function of science is the optimised creation of new knowledge of the external world, and that any alternative form of science which does not seek to do this can be relabelled as non-science or as pseudo-science.

A number of possible alternative forms of science will be discussed, in the ascending order of the hierarchy of levels of cognitive activity. The initial discussion of how far computers will be able to do science, emerges out of the idea that increasingly we substitute instruments for human action in, and perception of, the world. The exploration of the possibility of a feminist science emerges out of a discussion of permissible variations in the role of individuals. The treatment of psychical research* as a possible science of the supernatural emerges out of a discussion of how science might be related in a different way to popular culture.

10.2 Science done by computers?

Computers are instruments used as tools. When used in science, they are scientific instruments. They are a form of electronic instrumentation. In this section I will discuss how instrumentation has already changed science and promises to change it further in the immediate future. I will then look at the more contentious issue of what these trends imply for the possibility of a variant of science carried on entirely by computers.

There have been three especially dramatic changes in the history of the practice of science. The first was the introduction of publicly defensible and teachable rational methods of investigation and argument as ways of establishing new knowledge of the world. This was especially associated with the scientific revolution in the seventeenth century. The second was the institutionalisation of training in research routines in the nineteenth century. By getting students to follow standard observational and experimental practices, the professor could simultaneously get his research done and train his students. His students had learned how to do research by becoming teachers with research students in their turn. This made possible a new scale of professional pedagogically based science. The third change was the systematic introduction into scientific research of instruments which observe, measure, and control the objects of study, and aid calculation and modelling within the related systems of conceptual representation. Instrumentation was always a feature of modern science, but the great change was with the twentieth-century revolution in electronic instrumentation.

Electronic instruments first emerged in the late nineteenth century by the exploitation of new effects produced by the passage of electricity through glass vessels from which much or virtually all the gas had been extracted. The new effects included fluorescence of gases, electron beams which made a suitably treated wall glow, X-rays, and so on. They were applied in incandescent filament light bulbs, fluorescent lights, X-rays as a diagnostic technique, radio valves, and television tubes. They were also applied in a number of electronic instruments, among the first of which were X-ray machines and oscilloscopes. As the industrial significance of telegraphy, telephones, radio, and television grew, so did the general availability of the technology. It could be used to construct electronic circuitry to detect, amplify, measure, and record any phenomenon once a physical effect was known by which the phenomenon could be converted into an electrical signal. The versatility of electronic instrumentation was increasingly perceived in the mid-twentieth century. This versatility was dramatically increased after the discovery of transistors (in 1947) led to new generations of solid-state electronic devices.

What I describe as the revolution in electronic instrumentation came about with the rapid development and use of this kind of resource within science in the decades after World War II. At that time, much more money was being spent on science. Many authorities had been persuaded that science was a good investment by their perception of the military effectiveness of scientific accomplishments in World War II. The extra money was spent in many ways, but a great deal was spent on making or buying the new kind of electronic instruments, in the expectation that they would greatly increase research productivity.

One of the most dramatic events in this revolution in electronic instrumentation was the development of electronic computers. All three of the revolutions in scientific procedure that I have just described helped make it possible for computers to do aspects of science. By turning science from a haphazard business, requiring unrepeatable luck or genius, into a more reliable and manageable activity, they made it more amenable to computing. The seventeenth-century revolution in scientific rationality has enabled us to construct routines for the manipulation of symbols which, in the twentieth century, can be delegated to computers. Now a large proportion of routine calculation is handed over to machines. The nineteenth-century revolution in the social organisation of science has given us the capacity to turn a research idea into standard investigative routine that even a research student can follow. In the last quarter of the twentieth century, computers have become cheaper than graduate students so that we have every incentive to pass the less demanding of such routines over to computers. The transition has been managed by assigning graduate students the task of constructing the computer procedures which do what earlier generations of research students had to do themselves. The revolution in electronic instrumentation has provided the economic resources, the technical means, and institutional opportunity to make these changes.

The trend in the use of computers in science is for more powerful uses to be found for them. Could computers eventually take over science? Could it be that the end of the exponential growth of science in the late twentieth century, described by Price (1963), is just the levelling-off of human involvement, while the contribution of computers goes on expanding without immediate limit?

This question is linked to long-lived debates between strongly entrenched interests over artificial intelligence. Are there capacities and competencies in human minds which computers could never imitate (Dreyfus, 1992)? Some sociologists of knowledge (e.g. Collins, 1990) see the social negotiation process of science as equally beyond the capacity of computers. In contrast, those who believe in artificial intelligence (AI) have seen every new computing development as opening up new vistas of opportunity for computers. Soon, we are told by the enthusiasts, we will live in a science-fiction world in which computers are not merely our equals, but our superiors.

Whether or not computers will ever be able to do all of science, we are clearly witnessing a process in which more and more of science is handed over to computers. The twenty-first-century revolution in science-based computing will surprise us by the extent to which science is done by computers and by the manner in which computer take-over has transformed science.

I propose to discuss the first stages of that twenty-first-century scientific revolution, its historical roots and its long-term prospects. I will look at the role of computers in observation, experimental control, calculation, conceptual modelling, cognition more generally, and the various stages of social negotiation.

Computers and instrumental aids to observation

Computers are one form of the electronic instrumentation that has already transformed the perceptual stage of science. Less and less of scientific observation and measurement is done by the raw senses of the skilled observer. From the beginning of the scientific revolution, observational judgements had to be standardised and reproducible to carry weight in interpersonal negotiation. Since the nineteenth century, they have also had to be capable of being made by students, and since the mid-twentieth century, also by assistants and technicians. Instrumental observation facilitates these requirements. When cross-checking confirms the agreement between diverse types of instrumental observations, scientists have all the more reason to trust their results. Very often, science has gone so far beyond the unaided senses that immediate sensory checks are of minimal value. We can, of course, check electron microscopes at low powers against optical microscopes and the latter at low powers can be checked against careful direct visual scrutiny. The trend towards interpersonal reliability of observation has been at the expense of richer forms of sensory experience. Instrumental observation usually tells you what you make it tell you rather than surprising you with a new perceptual experience. The surprise has to be cognitive rather than perceptual.

The revolution in the instrumentation of observation has led to revolutions in the content of a number of twentieth-century sciences. Its impact has been greatest in information-starved sciences, where the limits to what we could find out about the relevant aspects of the world were previously a major constraint on scientific understanding.

As an example, astronomy has always been limited by what we can observe. The ancient astronomy of direct visual observation, perhaps with sighting devices to aid measurement, had little to work on but the pattern of positional change in the sun, moon, stars, and planets. Perhaps early astronomy did so well because its central phenomena were so self-contained and yet revealed significant patterns. Copernican sun-centred astronomy and its development by Kepler (elliptical orbits to replace the compounding of circular motions) claimed to offer a clearer conceptual view of the heavens than its Ptolemaic Earth-centred precursor. When Galileo turned his telescope to the heavens, a new revolution in astronomy began. The Copernican view became demonstrably superior to the ancient more intuitive alternative. The greatest achievement of the new age of astronomy was the Newtonian account of universal gravitation. For the next few centuries, it was the modest capacities of the telescope as much as of gravitational theory which provided the limits to astronomy. The discovery of new planets and other celestial objects came with improvements in telescopes. The demonstration of chemical spectra in the sun and stars and the later discovery of the redshift of galaxies was due to the development of another instrument, the spectroscope, attached to the telescope.

When, in the mid-twentieth century, the revolution in electronic instrumentation occurred, astronomy benefited in two initial ways. Firstly, with the development of radio-astronomy, which allowed the observation of a range of non-visible radiation from the stars. Secondly, from new miniaturised instrumentation which allowed the more effective exploitation of ballooning and rocketry, which could lift the instruments above the Earth's atmosphere. Both these developments were extended by further similar instrumental innovations. We can now study the entire electromagnetic spectrum of the heavens and can also collect any heavenly matter that comes our way. Astronomy is crucially dependent on the new kinds of instrumentation in the present age of space exploration on a limited budget.

Astronomy has, then, always grown by the assimilation of novel discoveries made possible with new scientific instruments. The sudden rush of discovery of novel phenomena produced by the revolution in electronic instrumentation has provided has produced a twentieth-century revolution in astronomy. Pulsars, quasars, and other significant but puzzling entities in modern astronomy were only found through modern developments in instrumentation. The old ways of proceeding have not been discredited, but they are now tied into a radically different intellectual practice, dealing with more complex subject-matter.

Astronomy is an example of a science once starved of information, now transformed by modern instrumentation. Another example is the medical investigation of the interior workings of individual living human bodies. Where early modern medicine had to approach this question indirectly, through anatomy based on post-mortem dissection, and comparative physiology based on torturing animals, present-day medicine has a vast array of investigative techniques which enable us to find out what is going on inside an individual's body without doing a great deal of harm. X-rays were one of the very first electronic instruments. Today, we have ultrasonic scans, nuclear magnetic resonance scans, PET scans, thermography, various forms of video-aided internal spying and other surgical aids, and a vast range of biochemical assay techniques. If there is something you really want to know about the internal state of an ailing individual, instruments are now available to help you find out.

Scientific observation has been permanently changed by easy access to the benefits of ever more powerful instrumentation. Even the most qualitative of soft sciences has been changed. We can, for example, monitor the movement of a wild animal by attaching a transmitter to it, perhaps to be detected by satellite. We cannot go back to simple human observation, because there is so much we now rely on that would no longer be observable or measurable. People study digital readouts, or look at computer-generated images, or assimilate instrumental readings in other trustworthy ways. Only at transient frontiers of scientific exploration for which instruments have not yet been developed do we still rely primarily on raw sensation.

Computers as control devices

In the age of the computer, electronic instrumentation is not merely an aid to passive observation. It also assists us to make phenomena take the form we require for their effective study. The change has been most dramatic in areas of routine science where the same investigative process is to be used repeatedly. It is still easiest to use a human to do a single experiment to discover a novel effect. But if a team of scientists, thinking on a grander scale, wish to apply an investigative technique repeatedly, they may reduce the task to a bureaucratic routine, and then program a computer-controlled system to carry it out. The initial complexity of setting up the procedure is compensated for by the vastly greater efficiency with which it can repeatedly be performed. Computerised experimental procedures have gone furthest in the medical technology of diagnostic testing. It is by building on this large area of experience that computer control is being applied in one of the most dramatic routine research projects, the Human Genome Project. To read the human genetic code is a boringly repetitive piece of routine research, and well suited to being assigned to computers. It may even be that the task would be prohibitively expensive to do any other way. Perhaps computerised control systems will bring about cognitive revolutions through the findings of such large scale routine science.

Sceptics doubt whether the use of computers in this way will have a dramatic effect on science. Initially, at least, control procedures must be precisely specified if they are to be turned into computer operations. This involves a loss of the vagueness that allows the discovery of new effects by playing around with their variation. It takes longer to develop ways of using computers to improve the control procedures. The revolutions promised from computerisation are not yet from new insights into the control procedures themselves, but from their application on a larger and more systematic scale than is humanly feasible.

Computers as calculators

In chapter 4, I distinguished explicit rationality and informal rationality. Electronic computers have been of great help in dealing with the routines of explicit rationality. Since ancient times, we have used physical aids in calculations. Indeed, the rules of explicit arithmetic reasoning are tied to what works well in the manipulation of pebbles (as on a counting board), or beads on an abacus, or physical operations in related notation. Such rules that work for both physical and conceptual operations were readily transferred to computers. The extended powers of symbol manipulation that computers provide has become a significant stimulus to the development of logic and mathematics.

Computers have transformed the rapid processing of information in science. Computers were first applied in science to speed up time-consuming calculations, either because they were so complex, or because of the sheer number of calculations required. Because earlier transformations of science had built on the formal processing of symbolic representations, calculation was an important feature of prior practice. Mathematical processing of precise and copious measurements can easily get out of hand. Some sciences were, as a result, limited in their development by the overwhelming amount of information they had to deal with. Computers have had a revolutionary effect in such information-overwhelmed disciplines.

Consider meteorology, for example. There is an immediate application for attempts to model weather systems on the basis of the changing pattern in local measurements like pressure, temperature, wind speed, and precipitation, provided the calculations the model requires can be made faster than the actual changes in the weather being modelled. This only happened, (except for the very simplest models) with the advent of large computers in the second half of the twentieth century. Now the large amount of weather data collected can be processed sufficiently rapidly to make weather modelling (locally and globally) an effective procedure. There has been a revolution in meteorology, one which had to wait for computers before it could happen. (Perhaps weather forecasting could have had a different revolution if modern satellite pictures of changing cloud patterns had happened first. But that revolution has not taken place in a distinctive form, because powerful computers came before the satellite images.)

Other data-overwhelmed sciences, too, have undergone revolutionary changes as information-processing requirements suddenly became manageable. Virtually unusable conceptual techniques were turned into routine tools. Seismic surveying of underground terrain became more straightforward for geological prospecting. Working out molecular structures from X-ray analysis of crystalline organic materials became routine.

Computer modelling

One of the great changes in recent science has been the dramatic rise in the use of modelling. Classically, physical science sought to represent an aspect of the world in symbolic relationships. However, the rules for the manipulation of symbols can be arbitrary, and in difficult cases it can be difficult to judge in the abstract whether the consequences of those rules continue to correspond to the world. Nineteenth-century British physicists, such as Lord Kelvin, argued that we could trust mathematically formulated theories most fully if a simple mechanical model of the theory and thus of the original phenomenon could be formulated. The ability to check our thinking through correspondences between the original phenomenon, the mathematical theory, and the mechanical model made the process of scientific representation more reliable. In particular, a means was provided to transcend the limits of the particular mathematical formulation, for unused features of the mechanical model could be used to think up changes in the mathematics. Although to the distaste of continental physicists (most famously Pierre Duhem), this view of models was developed further in the twentieth century, and became a standard idea in philosophy of science (Campbell, 1920; Hesse 1963, 1980). A widely used elementary example was the ideal gas theory. According to this theory, such observed physical properties of gases as pressure and temperature are the result of the rapid free movement of the tiny molecules composing them. The mechanical model of an ideal gas is a system of tiny but perfectly elastic particles which behave like billiard-balls moving in three dimensions, bouncing off one another and the walls of the container. The properties of gases can be related to the behaviour of the model. The ideal gas laws do not apply exactly to real gases, and the model suggests how to revise them, as, for example, when account is taken of the proportion of the space through which movement occurs is occupied by the billiard-balls/molecules. At very high pressures, this becomes a significant correction.

Computer modelling has turned out to be far more powerful and flexible than the older notion of mechanical models. Once a theory has been formulated precisely, imaginary examples of the theory can be constructed which can be turned into computer simulations. Although it is not plausible to argue that the ability to construct a computer model makes a theory more believable, such a model certainly allows the scientists to work through how the theory would apply in any particular case. Before doing an experimental study, it is possible to work out how things ought to happen according to the theory by running a simulation. Since this is very often cheaper than doing the experiment, the researcher can then be clearer about what the experiment should show, and so reduce wasted experiments.

Because it is relatively cheap, great ingenuity is now devoted to finding new uses for computer modelling. Where an experiment is not feasible, we can use computer models to work out what we think would happen. In applied science, when we need to make predictions in terms of our theories, computer simulations help us. In cognitive psychology, we can develop theories of mental processes using computers to see which ideas are workable and whether they can model the limits of specific human capacities.

I suggested in chapter 5 that the process of modelling blurs our sharp conventional distinction between mental processes and physical processes, and therefore the distinction between words and things. Computers form an immediate reality controlled by the programmer. What the computer does can be made to correspond to working through an example in terms of the rules of a formal system. But computers are actually physical systems. Their electronic states follow one another according to clearly understood physical principles. A computer model is simultaneously a conceptual instantiation of a mental process and a physical system set up to parallel some other physical system.

Perhaps we have been mistaken in assuming that scientific thinking is conducted entirely inside our heads. Much of thought can be externalised, in a suitable context. All living systems, human minds included, function well only in a suitable context, stabilised by the interaction of its manifold components. Human thought requires all the prerequisites of healthy human biological functioning, together with appropriate social surroundings. Part of the functioning of thought may actually be going on in that sustaining context. Similarly, tools like computers only function effectively in a suitable context. The functioning is better regarded as an achievement of tool-plus-context, rather than of the tool alone. Much of the detail of human thought can be done outside our heads. Other humans can help us in defining our task, written records can aid memory, calculating aids can assist calculation. Perhaps there is very little in the role of the individual in the scientific process that cannot be performed in a suitably supportive context with the human brain being substituted by a computer.

The scientific routines of observation, measurement, calculation, and control are not yet done entirely by computers. There is still a place for observational description in disciplinary areas unsuited to quantitative methodology. Much of science is exploratory research. This often cannot be reduced to routine, as the constant redefinition of the problem domain never demands the same routine procedure often enough. However, the very way an individual scientist conceives of how to deal with an observational or experimental problem is through experience of it while using the tools of the particular scientific practice, and, if these tools are electronic instrumentation, they are used. If a research problem is formulated in a computerised environment, it may be done with computer-based design techniques, leading on to computer-controlled procedures. Some parts of the scientific world (especially those involving large machines) are already like this. Radio-telescope scientists have long seen the heavens through the computers which control their equipment and process and display their results.

When, in the seventeenth century, some parts of science proved amenable to mathematically formulated rational procedures, other parts of science did not stop, but they lost status and interest because they were slower to give interesting results. When, in the nineteenth century, some parts of science proved amenable to the social routine of one professor with many assistants and students, other parts of science did not stop, and indeed, in the hands of the cleverest people, they were more important in scientific achievement. But not everyone is a creative genius, and the more routine methods allowed less inspired people to be creative and productive. They increasingly displaced individual scientific creativity. Now, at the end of the twentieth century, computer routines are becoming cheap and easy to use. Other science will not stop, but the proportion of resources devoted to more old-fashioned kinds of science will surely decline.

Are computers able to take over informal rationality as well as explicit rationality?

Will computers remain the limited scientific tools they are at present or will they take over the cognitive activity of the individual mind? At present the question turns on the limits of explicit objective rationality. The success of science has come from our skill at finding ways of making the implicit explicit, at converting the intuitive into topics following clear and unambiguous rules. To the extent that we can do that, we can find ways to use computers in science. If we are to make computers do more, we must spell out the rules by which such rules are to be created.

For example, is *scientific creativity* a mysterious process, or is it something that people can be taught? If the latter, then perhaps the procedures for teaching creativity can be turned into principles of computer procedures. Present computers have been made our slaves, following our every bidding, and lacking in the initiative to answer back (would you tolerate a computer package that was rude to you and pointed out your errors to anyone looking over your shoulder?) Slaves perhaps, but why not creative slaves, provided their creativity can be kept under control? In at least some circumstances, we are prepared to coax creative initiative out of computers, even though we would not want it more generally.

It turns out that the great intellectual tool of constructing a cognitive map of concepts and their relationships within technical scientific language is something that makes certain kinds of computer creativity possible. Even though a computer lacks understanding, if it can be provided with a set of symbols and rules for their manipulation in a technical domain, then it can be set to discover more possibilities inherent in those symbols and rules. It can be provided with powerful algorithms which invariably solve suitable problems, and with weaker heuristic rules which give some guidance when the more powerful methods are inappropriate. Two phases of research on artificial intelligence have hinted at what is possible. The research of Herbert Simon and his collaborators on computational methods of problem solving, as with 'General Problem Solver' in 1957, looked for and claimed to have found general algorithms for manipulating symbols which could solve problems in any domain in which the answer to the problem was entirely supplied by the information available. Self-contained puzzles like the missionaries and cannibals problem or the tower of Hanoi problem,[1] can be solved by computer, even though humans can find them difficult to master. The key problem for humans appears to be 'back-tracking' – moving away from the apparent solution in order to make progress.

Later, the advent of expert systems (their period of greatest fashionableness was the early 1980s) allowed artificial intelligence to be applied to domains in which a large number of special facts and local rules were to be used. One of the earliest systems (developed in the late 1960s), DENDRAL, for example, could use rules culled from the practice of the relevant specialists to identify chemical structure from chemical data such as mass spectrographs. Another system, MYCIN, could diagnose blood and meningitis infections from answers to questions about symptoms. These procedures were not locked into a particular state of knowledge. New facts and rules could be added to the databases from which they worked. From MYCIN, an empty shell, EMYCIN was produced into which any database of specialist knowledge could be inserted. DENDRAL'S list of rules could even be extended by an addition to the program, METADENDRAL, the purpose of which was to discover new rules by looking at the spectra of known compounds. METADENDRAL appeared to be getting very close to scientific creativity of a limited kind.

In this context, there has been heated argument about the potential for producing computer procedures which are genuinely creative. A set of programs designed for scientific discovery were discussed by Langley et al. (1987). One set of programs, BACON, could use the original data of classic scientific discoveries and use procedures analogous to Mill's canons of induction to arrive at laws inherent in the data. For example, BACON could rediscover Boyle's law.

Critics of Langley et al., (e.g. Brannigan, 1989, p. 609) pointed out that the data fed into the computer process was generated in terms of concepts from which Boyle's law could indeed be obtained, and carefully selected to be true, in the sense that it was consistent with the final result. All this is easy in hindsight. Programmes like BACON would have a harder time being put to work on the present frontiers of knowledge, where we do not know what to

---

[1] These problems can give difficulty to those who have not learned strategies for solving them. The missionaries and cannibals problem involves working out how three missionaries and three cannibals may cross a river by using a boat which can only carry up to two people. It is required that at no stage may the cannibals outnumber the missionaries on either bank, lest they eat them. The tower of Hanoi problem involves three rings, one large, one medium, and one small, and three pegs. The rings are initially on one of the pegs with the largest at the bottom and the smallest at the top. The task is to find the smallest number of moves of one ring as a time which will transfer the rings to a different peg, leaving them in the same order. At no stage may a larger ring be placed above a smaller ring on the same peg. The use of these problems in the cognitive study of problem-solving is discussed in Kahney (1986).

scrutinise seriously, what to regard as suggestive and partly true, and what is merely spurious. All the criticisms of what is presupposed using simple forms of induction apply to BACON. However, if ever a situation were to occur in which simple induction was an appropriate rational method, then the latest successor to BACON could be run on a computer and do the final stages of the creative work. Computer creativity could become yet another computer tool.

The kind of creativity that I have been discussing so far is rather reminiscent of Kuhn's account of normal science (Kuhn, 1970a). Creativity has been reduced to the elaboration or extension of a formal framework of analysis by using techniques that are a part of that framework. And, since students can be trained to be creative in this way, it would not be too surprising to discover that the procedures can be spelled out even more carefully for computers to follow, slave-like.

A more extreme form of creativity is that in which we discover the limits of the present framework of analysis and modify it more or less drastically so that it can cope with what it failed to do before. In the most radical cases, those described by Kuhn (1970a) as revolutionary science, a significant number of the old conceptual rules are rejected and the whole domain has to be reworked. In less radical cases, it may seem that the creative extension of the system is a natural but surprising growth out of what was already there. If there are meta-rules about how and under what circumstances the present rules are to be extended, then perhaps future computers will be able to accomplish this kind of creativity too. The rules for the extension of present formalised systems of concepts are not so clear as those within the system itself. Humans, at least, make them up as they go along. Such rules for changing rules might, for example, be justified by the independent assessment of the results they produce, once those results are available and can be scrutinised. But any process which is itself rule-governed can, in principle, be assigned to a computer.

Those computer programs which use explicitly formulated rules to determine permissible sequences of causal steps are trapped within the prison of formal rationality to a much greater extent than humans. However, it is difficult, even for a human, to provide a communicable account of how we can escape explicit reasoning systems, since natural language communication is very like running a socially programmed system. However free our personal imagination may be, when it comes to putting our thought into words, we are guided by the standard relations of the terms available.

It will suffice in the present discussion to appeal to the two ways introduced in chapter 4 in which individual informal rationality goes beyond explicit rationality. Humans are adept at working in more than one formal system simultaneously. And, when dealing with a specific problem, humans can produce optimal partial solutions even when there is no known sequence of steps which will produce a full solution. In both cases, it is possible that *future* computers will be able to do this too.

There is a rich culture of how humans deal simultaneously with more than one coherent framework of thought. A metaphor can import meaning from one framework into another. One account of human creativity (Koestler, 1964) talks of the 'bisociative act', in which one frame of thought is used to overcome an obstacle in another. When Isaac Newton in the famous apple anecdote thought of the moon as falling like the apple, but also as moving tangentially to the Earth so fast that in falling it never hit the Earth but went around it, he was bringing together the frameworks of celestial and terrestrial physics. Perhaps Koestler is right, and the major resource in the creative extension of frameworks of thought is finding new ways to link them to other frameworks.

Surely, such a process could be carried out by a computer? We could find ways for the computer to operate two or more systems in parallel, each with its own database and rule system, and to program it with heuristic rules on how to use the legitimate moves it discovers in one system as a resource for looking for new legitimate moves in the other. It might for example, use a set of meta-rules by which it can identify identities, similarities, and differences between the two systems (Hesse 1961, 1963). Within such a system, new kinds of computer creativity would then be possible.

Such a computer system would remain limited to the manipulation of formal symbols. It would still miss out on what is so important to humans, the way we link symbols to reality. Our manipulation of the symbols of our various kinds of language are not purely syntactic but semantic. Searle (1980) argued this point very vigorously in terms of his Chinese-room example. A human who does not know Chinese might, like a computer, follow a set of formal rules which imitates a native Chinese speaker persuasively. But, just as the human cannot be said to understand Chinese, nor can the computer. Unlike the native speaker of Chinese, they are just following syntactic rules.

It seems to me that the sequence of electronic states of a computer could be made to model semantic processing as easily as they can already be made to model syntactic processing. Consider my suggestion of a computer which creatively relates two formal symbolic systems. One very basic application of such techniques would be to supply the computer with a perceptual language to summarise hypotheses about the immediate causes of data in its rich range of sensors, together with a higher-level conceptual system. If it could be induced to find correspondences between its percepts and its concepts, it would have the beginnings of a semantic base for its symbol system. When discussing the physical objects, tables, it could try to find the perceptual input to which the concept 'table' might relate. Perhaps we might see the beginning of this in T. Winograd's program, SHRDLU (Winograd, 1973), which could process commands and answer elementary questions about the children's blocks which it was able to manipulate.

Such a computer (we are now in the realm of futuristic speculation) might then be capable of operating moderately effectively at the two lowest levels of cognitive activity.

Just as computers could transcend operating within a single system of explicit rationality, I see no reason why they should not be able to operate my other example of informal individual rationality, the 'optimiser with zoom'. A human can holistically consider a wide or narrow domain of potentially relevant considerations and come up with an optimal but uncertain conclusion. There are many systems of formal reasoning in which optimisation procedures may be modelled. Probability theory is one example, and the new connectionist (neural network) learning systems are another. The main limit of such systems tends to be that they have to ignore everything which is still outside the system, and therefore the optimisation process cannot zoom all the way out from a close focus. The neural-network approach models the way neurons may be connected in the brain. For example, when set the task of visual recognition, the system is programmed to adjust the connections between its component nodes so as to optimise the recognition of general patterns in the input data. Its optimisation is limited to the immediate data input. If the higher levels of human thought are at all like this, the optimisation should also take some account of feedback from the rest of what is going on in the brain. To connect computing versions of neural network subsystems in some such way is within the scope of current research.

The connectionist idea of how human optimisation may occur, suggests how to implement a computer version of the task of relating two distinct frameworks of thought. If the two coherent cognitive structures go on in separate neural networks, it may be possible to construct supplementary connections between the networks which allows them to affect one another in the same ways that humans manage so well (analogy, critical comparison, creative extension of one by the other, etc).

From slaves to companions – computers and social negotiation of knowledge

There is a trend in present-day computing towards the development of computing devices which enhance the capacities and competence of individuals who make use of them. The interface between person and computer might, in the future be made very close, as in the science-fiction idea of the cyborg, a man–machine composite, or the cyberpunk vision of a computer chip that is socketed into a person's skull, enabling direct connection between the computer network and the nervous system. These visions are often thought to be horrific, even though hearing-aid devices which directly stimulate the nerves of the ear are now available. A less alarming projection is computer systems which remain external to the individual and yet interface in a way that ordinary people can readily cope with. Virtual-reality systems may be a portent of the future. By feeding the human senses with an electronically produced simulation of an artificial world in which there is sensory feedback from movement, actions can simultaneously make physical sense to the person and have appropriate effects within the computer system. Information could, for example, be spatially organised like a library which one could move through in virtual reality.

Another aspect of the trend to computers as companions is to help people to make computers do their bidding by designing the person–computer interface to simulate an ordinary interpersonal interaction. One could then ask the computer to do something and the computer would answer you appropriately. Further dialogue could clarify any misapprehension. This trend, an extrapolation of 'user-friendliness' will, it is hoped, make it easier to put computers to work in human terms. With this kind of interface, doing science with the aid of computers will be as easy as doing science within a cohesive group.

The present trend is to combine computers and humans in what might be called 'social intelligence'. However, more developments will be required before computers can display social intelligence in the absence of humans.

The knowledge that individuals construct should not merely be optimised with respect to perceptual experience (in the context of action), but should also be reconciled with the cognitive activities of the wider social domain. In part, this wider social process might be seen as analogous to what goes on in our own heads. By exposing our ideas to a range of people, with variant patterns of belief, in different immediate situations and with diverse interests, the social negotiations that go on provide a way to turn the ideas into an approximately optimised interpersonal belief. In a suitably institutionalised culture of negotiation, such interpersonal beliefs deserve to be called scientific knowledge.

If future computers are to be able to function in ways that enable them to substitute for individuals in the scientific process, perhaps they could also substitute for the negotiation process by modelling human interaction.

Certainly, any computerised version of the optimisation process I described might be able to incorporate in distinct subsystems the beliefs and viewpoint of more that one individual. The final knowledge output could then represent that which maximised acceptability to all the viewpoints incorporated. Have I, then, provided a complete sketch theory of a computer-based science? The problem with the vision just provided is that it projects the present too far. Perhaps it is too soon to abstract a hypothesised essential function of the social negotiation of knowledge. In the rest of this section, I will explore a role for computers appropriate to the *present* kind of social negotiation, leaving more mysterious precisely what such negotiation achieves.

An important part of science as it is done by humans is that the conclusions one individual reaches are made acceptable to others. To the extent that the evidence and argument can be made conclusive, even a computer could do what a human scientist does. Present computers can come very near what is required in argument, but problems remain over the authority given to evidence. If someone systematically doubts a computerised procedure for observation and measurement, the two most obvious checks are on how the procedure was carried out and on whether a comparable result can be reached in a different way. I suspect that the situation is not greatly different from that which now goes on in the use of automated tests of alcohol levels in car drivers. The right to use automated tests to convict for drunk-driving offences has been given in a number of legal systems, for example, the British. If the automated procedure is sufficiently exposed to scrutiny by competent people with a contrary interest, and they cannot make an effective case against the prevailing practice, then they must acknowledge the authority of the result. It is always possible to have an additional test done in some other way, in the hope that it will reveal some problem in the original procedure. If it does not, then the automated result is likely to keep its authority.

The validity of decisive computer-generated argument is, then, straightforward to establish. Since the computer is following clear rules, it need only spell out the sequence of steps by which a validly inferred conclusion was reached, so that a human may check them, perhaps using another automated reasoning process. The famous computer proof of the four-colour problem, previously discussed, provides an example. Although the computer proof was too tediously complex to be fully checked by an individual, its parts can be checked by other computer programs and by other individuals. In principle, if all who formulate a way to check their doubts find that their challenge fails, the proof becomes accepted, for its objectors will be discredited if they continue to make a fuss.

I see no problem, therefore, in computers being delegated authority for conclusions reached by rigorous methods that everyone is already happy with. However, in science, judgements sometimes involve conclusions which are not beyond conceivable doubt but are deemed more reasonable than alternatives. How could computers participate in such negotiations?

The problem is already with us. Expert systems take some self-contained body of expert knowledge that has been encapsulated in a database of facts and specific rules. They then apply an inbuilt logical procedure to provide answers to questions. However, it is a poor expert system which merely supplies answers. Suppose a doctor who is having a bad day uses a medical diagnostic expert system and acts on its advice without further check. If the computer turned out to be wrong, the fault would be the doctor's. We do not at present give authority to the judgements of expert systems, but make our own judgements in the light of the information they provide. For this reason, it is best practice to have expert systems justify their conclusions in ways that the computer user finds helpful or persuasive in deciding to take responsibility for the judgement. There may even be several levels of justification, for users of different competence and with different levels of uncertainty. The generation of such explanations by expert systems are a first step towards computers taking on the task of negotiation. If this facility were to be developed in the computers used in science, it might even come to be the case that computer-generated reasons were a standard resource for subsequent human negotiation.

If we return to the metaphor of computers as slaves, the negotiations that computers can undertake successfully are those that slaves can do. A slave has no authority as such, but might be delegated a specific authority in a particular context in which he or she had specific knowledge. Similarly, the computer can be delegated authority which can carry weight in wider cognitive negotiations.

For computers to function as more than slaves in the higher levels of cognitive activity in science, they would have to be granted some of the rights we give persons. The process would be easiest if computers were housed in mechanisms which function as persons do – the science-fiction robot. However, that may not happen, as it is more likely in the middle-term future that computers will be given capacities which resemble human ones (for we use human self-knowledge to guide us in creating such computer capacities) but which are developed in ways which complement human powers rather than merely copy them. Computers will enrich human society rather than competing with humans and eventually taking over directly. As computers are developed which can do things humans cannot do, the question arises of whether their negotiating role will continue to be analogous to that of slaves.

The main limiting factor in such a trend is not the capacities of computer technology, but what rights and responsibilities people will attribute to computers. For computers to take a significant part in the higher, social levels of science, they must be allowed sufficient power to be able to negotiate machine interests. People who at present worry about whether scientists should no longer be welcome in society and whether we should be happy to accept scientific knowledge merely on the authority of scientists, might have future counterparts who worry about whether powerful computers are welcome, and whether the science they produce might be contrary to human interests.

In this last fantasy, it might seem that I have gone much too far. Perhaps I have. But one trend points this way. Where previous computers have been locked into running on the narrow rails of their programming, so that everything they do is fully determined (even if the programmer failed to make the determinate process do precisely what was intended), neural networks are taught by a learning procedure which is not under such precise control. It may turn out that future neural computers are about as much under control as human children are under parental control. Perhaps that trend towards loss of control will not be allowed to continue, and perhaps it will never lead to computers taking initiative – for that would depend on the totality of input on which their learning is based.

There is a general economic argument about the effect of partial mechanisation of human labour. Whatever machines make cheap and easy is no longer so highly valued. In part this is because things which are in ready supply do not command as high a price. This applies within science. Mere calculation has become a low-status activity, so that we now may wonder why the great early nineteenth-century mathematician C. F. Gauss was prepared to spend so much time on routine but complex calculations (he was a lightning calculator) rather than on mathematical discovery. As observation is handed over to machines, it is the thought before and after the observation that becomes the more precious part of science. If, in the future, creativity or social negotiation is best done by computers, we will have to change our values once again. And, if science is ever handed over entirely to machines, the knowledge creation industry will be a declining asset, ever easier to do and risky as a long-term investment.

In conclusion, there is nothing in principle to stop computers doing science. The present changes are only the beginning. As more of science is done by computers, we can expect its nature to change radically, just as it has already changed.

Computers do not yet function as socialised individuals. A discussion of acceptable variations of science that deals with higher levels of cognitive activity in the more immediate future must still focus upon human beings. Since science has traditionally been dominated by men, it is worth asking what a science would be like which was constructed on different, feminist, principles.

10.3 Towards a feminist science?

Feminist discussions of science have pointed out the pervasive effects of male interests in past and present science. In this section, feminist critiques are briefly reviewed, but the main concern is with whether a future form of science can be constructed on feminist principles that is both viable and distinctly different from that produced within past patriarchal culture .

This section is concerned with variant forms of science which come from changes at the level of individual cognition. I will begin with a brief argument that by 'individual' I mean a properly socialised person. Modern culture has gone through a period which gave exaggerated importance to the possibility of an autonomous individual, capable of acting independently of all other people. Such an idea is quite unworkable. We are entirely dependent on growing up in society. This claim is supported by the few cases of children surviving in the wild. One such was the Wild Boy of Aveyron (see, for example, Shattuck, 1980) who was found in 1800 aged eleven or twelve. He appeared to have survived for several years alone in the woods, before he began to take food from the fields of outlying farms. Close study of the boy after his capture showed that, in spite of his alert interest in food, he was incapable of learning language and in some ways was little better than an idiot, although he cannot have been an idiot in all respects since he had survived by himself so long.

The notion that individuals can be intellectually self-sufficient in doing science has been disseminated further by the depiction of scientists in fiction. One popular literary stereotype of scientific motivation since Mary Shelley's *Frankenstein*, the scientific romances of Jules Verne and of H. G. Wells, is of the isolated scientist driven by non-social or anti-social needs. This is perpetuated in the pulp stereotype of the mad scientist of comic and film, who has produced great technical accomplishments and is ready to launch them on an unprepared and defenceless populace, perhaps as part of a wish to take over the world.

There have indeed been individual scientists who were driven mainly by personal satisfactions. Isaac Newton was undoubtedly secretive, being reluctant to tell the Royal Society what he had done without strong persuasive pressure. An even more extreme example of an individual who was so uninterested in scientific approval that he did not bother to present his ideas to anyone else was the aristocratic eighteenth-century natural philosopher, Henry Cavendish, who made lengthy studies of static electricity without bothering to publish them. (The work appeared only in the mid-nineteenth century in Maxwell, 1867) In some of his most individualistic work, Cavendish compared the effect of the discharge of Leyden jars through different kinds and amounts of electrical conductor, coming to an analogue of the later concept of resistance. He judged equality of resistance in two materials by deciding that the discharge of the same number of Leyden jars through each of them and also through himself gave him the same electric shock. This subjective and somewhat masochistic form of scientific measurement had been displaced by instruments before comparable results were obtained by other people in the next century. Cavendish seems to have obtained good results (Maxwell, 1867), but without communication and negotiation it was locked out of the normal scientific process.

Even in such an extreme case as that of Cavendish, we are considering an individual who had learned about science from others and had already learned how to do research and communicate it to the scientific community. His work was preparatory to publication, although he did not in the end publish it. Without the stimulus and support of the scientific community, it is increasingly difficult for an isolated individual to sustain the effectiveness of the research done. We will find it most productive, then, to consider the level of individual cognitive activity in terms of properly socialised individuals.

The explicit rationality of science has been indifferent to the precise nature of the individuals who practice it. For example, we might expect scientific rationality to be gender blind, and therefore no target for feminism. But gender-blind rationality has traditionally been practised by males in a male-dominated society. Our attention is drawn to the informal rationality of such individuals, and this may be open to feminist criticism.

The fact that past science has mainly been done by men may have affected the form and content of scientific knowledge. The general form of such an argument may be illustrated by a parallel argument over the effect on science of the age distribution of scientists. I will briefly describe these age effects in order that we may better see how to view the feminist case.

The age effect

At several points in this book I have briefly alluded to the idea that because science has grown exponentially for so many centuries, it has settled into an institutional form which makes it a culture dominated by youth. In a steadily growing social practice, those who are new to the practice numerically dominate. According to Price (1963), for two centuries, science had been doubling in size every twelve years or so. Therefore, if all scientists had careers of about forty years, then, over that period, the age cohort of new entries would have been about eight times the size of that of people nearing the end of their career. This numerical dominance by the young was magnified further by the close links of science with education. Many people work at science for a few years immediately after graduating and before going on to do other things.

Although science is numerically dominated by the young, there are also pressures which concentrate power in the hands of those longer established in their careers. According to Price (1963), the most creative and productive individuals tend to move to the institutions with the best research facilities where they benefit from the productivity of the best research students. There they become nodes in a privileged communication network which gives an advantage in exploiting newly emerged opportunities for research. Scientific elites are vastly more productive and more highly recognised than the lower orders of the scientific establishment. However, even in such centres of power, it is easy to slip off the crest of the wave of creation of influential scientific knowledge. The more numerous individuals of the next intellectual generation are constantly producing new knowledge which threatens to make old contributions out of date. Institutions in which older scientists can hold back the ideas of the young are likely to be bypassed for at least a generation.

The dominance of youth in scientific culture facilitates its receptivity to change, its orientation to the future, and reduced interest in conserving the past. The young, seeking to make their way, without a rich knowledge of the immediate past, are more receptive to ideas which promise to open new horizons. They compete with one another by displaying even newer observations and ideas. When radical approaches begin to gather momentum, the enthusiasm of the young can swamp the cautious responses of those who are set in their ways. At their most extreme, the enthusiasms of young scientists look analogous to the following of fashion. If such a fashion produces results, its intellectual leaders can gain rapid promotion under conditions of institutional growth, and by middle age can displace the power of the smaller number of the elderly. As a result, it takes only a decade or so to reorient science in a revolutionary new direction by change due to the arrival of new generations. All this is quite unlike more stable and traditional disciplinary areas such as, say, theology, in which the power of the oldest most senior individuals remains greatest. Scientific culture has traditionally been a culture of young adults.

In the last few decades of the twentieth century, the exponential growth of science has begun to tail off. The steady-state science which we are moving towards will only continue to be dominated by youth if scientific research is increasingly a youthful phase of careers of individuals who move on to other things. It will be interesting to see just how much the points just made continue to apply.

If past science has been shaped by being dominated by the psychology of the young, it has presumably also been shaped by being dominated by men. Feminist critiques of science claim this. I will discuss first the rise of the cultural acceptability of feminist arguments over science, the feminist critique of past science, and finally the possibility of a properly feminist science.

The acceptability of feminism

Modern feminism has something to say to everybody in its appeal to our sense of fairness. A sense of fair play is a powerful tool for change in present society. Children learn early in family life that it pays to campaign for a fair share of desirable items, and use the same principle throughout their lives. As the social roles of men and women have recently begun to overlap more than traditionally, the perception has emerged that it is only fair that men and women should have access to the same social opportunities and that traditional male-dominated society is unfair to women. It is only recently that this perception has been widely seen as persuasive.

In my discussion of women's brains in 4.4, I suggested that, in the nineteenth century, it was easy for men of science to discover that women were inferior, because they were indeed inferior in the social roles dominated by men – men made them so and women were not usually in a position to argue back effectively or to realise their potential. Such intellectual discussions were under the control of men, for whom it made sense to say that women should remain in their proper place. The question of fairness could not easily arise in a society which so sharply distinguished the sex roles. But, even then, the separation of the roles of men and women was changing in industrial society. The division of labour under capitalism was producing a society in which more people could afford to devote effort to matters other than sheer survival. Life expectancy was recovering from the initial problems of industrialisation. The changes had a major effect on women's roles. The main response to the changing circumstances was a re-emphasis on a woman's role within the family. However, eventually these duties did not keep all women permanently occupied. There were phases in the lives of many adult women which did not need to be devoted to childbearing, child-rearing, housekeeping, and, for the poorer, subsistence wage-labour. Many young women could afford to think about education, and many older women were able to turn to other activities after their children had grown up. There had always been a few women with ample leisure, though according to Wollstonecraft (1988, originally 1792) aristocratic ladies had minds enfeebled by false refinement (p. 7). She looked for new opportunities for women in the more natural conditions of the middle class (p. 9). As more women could afford to turn their energies in new directions, the new opportunities were mainly in activities dominated by men. In the women's movements of the early and late twentieth century, a collective consciousness was built up that the existing patriarchal society was unfair to women, and that something should be done about it.

Science, for example, had traditionally been done almost entirely by men, the contrast between the sexes being especially striking among the greatest achievers of science. This had seemed quite natural to the men, for few women were thought to have the intellectual qualities required.

In the changed social awareness of the late twentieth century, it is no longer appropriate to say that men have made women inferior to themselves. We now educate men and women to be intellectual equals. There was a transitional period in which it seemed that men might predominate at the extremes of genius and imbecility (Shields, 1982). But, apart from a number of X-linked genetic defects that reduce the intelligence of males more often than of females, it seems that that conclusion too was only a carry-over from a male dominated society in which social motivations to display intellectual excellence of a very high level were more important to men than women.

Undoubtedly, differences between men and women remain. In addition to differences dependent on reproductive function, men are somewhat taller and stronger (on average, not invariably). Could there also be some subtle biological basis for differences in intellectual functioning? When this possibility was explored in modern human sociobiology, feminists were well represented among sociobiology's critics. They pointed out that, in the absence of conclusive evidence, such a supposition appeared as ideological, legitimating the present inequalities of patriarchal society. And, since then, feminists have been suspicious of the search for innate biological differences between men and women in intellectual functioning. Since our biology is presumably fixed in a way our social nature is not, appeal to a biological basis for gender differences is regarded by most feminists as part of a conservative attempt to defend the established gender inequalities.

Modern feminists reject men's ideological use of biology and biological medicine to legitimate traditional male claims of innate superiority over women (e.g. Sayers, 1982). Feminists are therefore very careful in appealing to biology to support their own claims. The individual, as most feminists see it, is formed by society, and little girls are brought up in a different way from little boys. Very many social roles are linked to the sharply separated sex-roles of our society. Of course, individuals vary genetically, but it is the genetic differences which are given social significance which come to matter, and they matter in the way that society makes them. Our gender is important to us because we make it matter, while we usually remain blissfully ignorant of our personal inherited susceptibility to specific diseases (though perhaps this will also matter in the twenty-first century). The fundamental significance of the different biology of women is that they are naturally, but not inevitably, more concerned with the practical implications of the biology of reproduction, of childbirth and child-rearing, while men naturally, but not inevitably, find these things more incidental, and certainly do not deliberately build on such differences in the sciences they construct.

Modern feminism seeks to analyse the unfairness of the treatment of women in past male-dominated societies from a modern point of view. It points to the numerous social inequalities that still exist and seeks to accelerate their elimination from society. The changes in gender relationships are occurring more rapidly than the turnover of the generations, and many of the present inequalities at the top of society are a reflection of social patterns of the male-dominated society of older generations. The changes have not yet worked their way through to senior citizens.

The critical task of feminism may be self-limiting. If the social roles of each gender continue to converge, and society settles down in the new form, presumably the feminist critique would only apply historically. However, such a vision presupposes that there is no problem in adjusting to a new sense of what is fair for all. Presumably only women will continue to give birth and suckle babies. This demanding activity can be a distraction in many male-dominated roles to which women aspire. Perhaps women should encourage the construction of new social roles which are more in harmony with the biological necessities of women's lives, and somehow persuade men that the newly constructed roles should displace traditional male-dominated roles. Perhaps the new roles will be less competitive and more cooperative. Men may turn out to be more willing to co-operate with women than to compete with them. A science-fiction alternative would be to use an extended form of modern medical technology to manage the production of babies, so that men and women could have identical social roles unaffected by the now redundant differences of biological functioning (e.g. Piercy, 1979). Perhaps, however, the ideal of a gender-blind society is the wrong one. Some feminists would prefer an ideal society in which women dominate rather than men. Or should we see male–female relationships as just one symptom of a more general problem in our society, and recognise, with Marxists, the exploitative nature of the past organisation of both men's and women's work?

Given such difficulties, it is not surprising to find that, although feminists agree on the unfairness of the past, they do not agree on their ideals for the future. There are no past examples of society which are exactly right for feminists, and therefore the process of establishing the credentials of the new ideals is bound to be controversial. In their agreement on the unfairness of present society, feminists appear as a radical movement, seeking to criticise the status quo. Inevitably, their critical rhetoric echoes that of other contemporary radical movements. Other radical movements disagree among themselves, which accentuates polarisations within feminism.

These comments also apply to science. Feminists agree on the unfairness of past science, but do not agree on how a feminist science should be constructed.

The feminist critique of past science

Feminists like to quote Virginia Woolf, 'Science, it would seem is not sexless; he is a man, a father and infected too.'

Of course, science has traditionally been done by men. The word 'scientist' was invented in the early nineteenth century to replace 'man of science'. Women had sometimes been encouraged to take an interest in science since the eighteenth century, but were still a tiny minority in the nineteenth century. Only in the twentieth century has there been sufficient involvement of women in science for it to be appropriate to complain about them being kept in junior positions. (There is also widespread exploitation of the more numerous *young* male students by *established* scientists, but the young may eventually cumulate credibility, while women more often stay excluded.)

Accounts by feminists of the past images of women that were legitimated by science provide striking contrasts with our present-day images. The example of the involvement of women in sport today provides us with the view that the best women do almost as well as the best men and better than the non-elite males of their sport. The difference is greatest in displays of strength and least in displays of endurance. Yet, in Victorian times, medical science found women to be weak and vulnerable creatures who should be protected from all extreme challenges for their own good.

Much of the unfairness to women of past science has more to do with how knowledge was *used* (and abused) rather than how it was *created*. The past use of scientific knowledge has mainly been under the control of men, some of whom have been quite prepared to apply it in the sex wars. Feminist critiques of past science describe the way men have applied science in their own interest. In the present day, the appropriate corrective to such use and abuse of science might well be to apply science in women's interests, rather than to change science.

Throughout this book I have tried to distinguish between the creation and the use of scientific knowledge, even though I have conceded that the present economic pressures on science are producing institutions which blur them together. I deny that a knowledge claim should be accepted as science merely because it is the theory of a socially successful practice. At best that is technology. At worst it is the pseudo-scientific ideology of legitimation. Therefore, I see the problem of how feminism could change science as concerning the *creation* of acceptable scientific knowledge, and leave to the sex wars the problem of how scientific knowledge is *used*.

It is undoubtedly the case that modern forms of science emerged within social practices dominated by men. The social motivations out of which the knowledge, institutions, and tradition of science grew were based on competition for intellectual glory among restricted elites. Such elites were only open to outsiders once they had proved that they could excel at the prevailing social and intellectual games. Women were not normally eligible for membership of such elites. As science came to be practised within its own institutions, it became a more open system in which socially ambitious classes of outsiders frequently found social advancement was possible, while they were still blocked from other cultural practices. For they could make their way in science by their wits, with little need for wealth or prior social connections. Minority religions, racial groups, and immigrants have all benefited from the relative accessibility of a career in science and science has benefited from their involvement. Only recently has it been possible for women to benefit in the same way, and the opening of science to women has yet to be completed.

According to feminists, the content of past scientific knowledge has been shaped by the exclusion of women. The knowing agent always produces knowledge which is a function of his or her situation, and in particular of the power relations of that situation. Woman can often spot episodes in which men have failed to appreciate their interests. As there is no way to produce a science which is neutral to all interests, a satisfactory science should always include a representation of women's interests.

This general view of knowledge-producing activity has been applied to the analysis of images of science constructed in male-dominated ages. There is a tradition, going back into prehistory, well represented by the Pythagoreans, of structuring the world out of opposites: light–dark, good–bad, male–female, right–left, static–moving, and so on. The Pythagoreans took one side in each of these dichotomies, favouring the light over the dark, right over left, male over female. This way of thinking encourages the linking of corresponding members of different pairs, giving a rich set of associations to any specific pair, and in particular of male–female. Jacques Derrida is famed for criticisms of the way Western metaphysics is based on such binary oppositions. Some modern feminists, following Derrida (Grosz, 1990, p. 93), have attacked the role that such dichotomies play in the history of philosophy. For example, the Enlightenment view of scientific rationality operated in terms of such dichotomies as nature–culture, subject–object, knower–known. By linking the male–female dichotomy to that of science–nature, some writers, such as Francis Bacon, had found it appropriate to use sexual metaphors to describe the controlled manipulation of nature. This was still going on in the nineteenth century. We learn from Jordanova (1989, chapter 5) of a statue in the Paris Medical Faculty. It is of a bare-breasted female who is taking off a veil, and has the inscription, 'Nature unveils herself before Science.'

Feminists argue that the pervasive metaphorical structuring of the relation of science and the world in terms of the male–female distinction has had lasting effects, so that even where such metaphors are not used, they still work under the surface. The Enlightenment dichotomies concerning scientific rationality led to the conception of science as understanding and dominating the world. Feminists see such a scientific world view as androcentric. They seek to construct an alternative viewpoint.

Present science

The natural sciences in general, and the physical sciences in particular, still do not recruit women proportionately, and their higher social orders still reflect a time in which it was difficult for women to combine motherhood with careers demanding high levels of commitment.

Present-day science is adjusting to the changes in gender roles and gender relationships. Although some of the men who dominate science may not welcome the change, science does not provide a rationale for resistance, in the way that some Christian churches are able to use theology to legitimate resistance to female priests. The scientific ideals of the creation of knowledge by gender-blind forms of rationality offer no basis for opposing the idea that women should participate fully in the creation of knowledge.

Towards a feminist science?

I will now explore the possibility and viability of a future feminist science. Feminists do not agree on the constructive aspects of their campaign. Feminist critiques are often of past *images* of science. Actual science is more diverse than these images, so that it is possible to argue that feminism offers a new way of talking about science, rather than a new kind of science. Feminist critiques share many features with other critical studies of science of the present day, so that it is not easy to select a distinctively feminist alternative from among the manifold visions of how to change present science. These difficulties help to explain why there is still no clear image of a future more feminist science.

In spite of these difficulties it will be possible to come to an answer to our question. It is helpful once more to organise the discussion in terms of the hierarchy of levels of cognitive activity.

Perception As I pointed out, most feminists are not attracted to the idea of innate intellectual differences between the sexes, seeing this as no more than the conservative ideology of biological legitimation of the inequalities of the status quo. The key differences they find between the sexes lie in developmental and social psychology and in the structure of society. Therefore, a feminist science would not require a change in the perceptual equipment of humans. However, it might be used differently. Perhaps it is *men* who tend to focus upon the perception of the consequences of their own actions, who are so concerned to dominate their subject-matter, to control it according to their will, rather than relating to it empathetically, trying to get inside its situational perspective. Perhaps it is men who are inclined to represent the world in abstract and objective terms rather than through the richness of their sensory involvement. These effects are best dealt with at higher, more social cognitive levels. We will discuss them in the sections that follow.

Individual cognition Some feminist writers have developed the view that the way women think might be a valuable corrective to dominant trends in modern intellectual thought. Women are closer to nature, because a woman's biological functioning is more omnipresent in her life. There is a psychoanalytical argument in feminism that because girls remain closer to their mothers than boys, they do not learn to separate themselves so sharply from their nurturing environment. Boys develop the separation and ego-detachment defined as the central requirement of scientific method. The Enlightenment dichotomisation of the world into nature/culture, subject/object, knower/known led to a view of understanding as external domination. Feminists seek to construct a more participatory view in contrast to this androcentrism (Hekman, 1990).

I will not attempt to go through a full list of all the ways in which women might think and come to knowledge differently from men, for my discussion does not require it. (There is a good review discussion in Hekman, 1990.)

There must be many factors that work in various ways on individual cognition. For example, we all need food. It is an even more immediate feature of our biology than having and raising children. Not all of us need to think much about food. The scientific thought of those individuals who never bother about food, not worrying where it is coming from and not paying much attention to how it tastes, will be affected differently from those for whom food is of day-to-day importance. This contrast is likely to have some effect in the sciences which relate to food. We might well expect agricultural science to be done differently by those more preoccupied by food. Perhaps, one might even speculate that some of the great controversies in agricultural science, such as that between the chemical management of crop production and the 'organic' school (of muck and magic), have turned in part on the distinctions between the different ways in which people are concerned with food.

Food is only one example of the indefinitely long list of known and unknown factors at work on our thought. At the level of individual cognition, the thought of each individual is constrained by, and optimises, a great number of situational factors, some unique to that individual, some shared by many individuals. In addition to the factors that separate individuals as females or males, there are factors which divide them in other ways. These can play an important role in individual creativity, including scientific creativity. All of them matter, but it is not immediately obvious that any are uniquely privileged in the creation of new science.

Feminist alternatives in creativity One highly regarded work on feminist science is Evelyn Fox Keller's biography of Barbara McClintock (Keller, 1983). McClintock, a Nobel prize winner for her work on a previously unknown genetic mechanism (jumping genes) is reported to have made progress by rejecting the normal division of subject and object. Instead, she cultivated a 'feeling for the organism', in which she tried to work through the process, and her problem, by getting 'inside' the situation of the organism under study. Such a strategy, which is the orthodoxy in much of social science, especially that influenced by Max Weber (in his advocacy of the operation of 'Verstehen'), is less common in natural science, but far from unknown. In biology it is appreciated as being of some value, but also as risking projecting inappropriate anthropomorphic qualities on to the organism under study.

It is to be expected that, whatever the orthodox methods of scientific discovery happen to be, they are likely to build up a residue of recalcitrant cases, in which alternative modes of investigative thought may prove effective. If feminist styles of creativity can be more productive in some situations than establishment male-dominated science, they can lead to significant and appreciated discoveries.

For example, it has been suggested (e.g. Harding, 1986) that the traditional images of science fail to pay attention to the choice of scientific problem. I think this is correct. We do not have a rational theory of problem choice. The factors which are played off against one another include: the perceived worth of the project (including the benefits which might follow), the ease of conducting the research (which involves the requirement of gathering of economic and manpower resources), the timeliness of the work (for example, some things must be done now or someone else will do them), and so on. The optimal choice of research problem that results will vary with situational values. We find historically that the distribution of research effort varies greatly from individual to individual, from local group to local group, and from nation to nation. Science as a whole has not needed rationally to regulate problem choice. Many research enterprises will fail, but some will succeed dramatically. Science as a whole succeeds best by encouraging a diversity of approaches wide enough to provide a high chance of producing some successes. In this context, it is to be expected that women in general and feminists in particular will frequently choose different research problems. This is to the benefit of science. The more successful feminist strategies for choosing and creatively investigating research problems may even become the orthodoxy in specific areas.

Perhaps some feminists would like to suggest that some of the problems men choose would be better left uninvestigated. There are no good epistemological reasons for doing this, but arguments for not studying some problems have sometimes been made on moral grounds. If dangerous consequences are anticipated from research, or if the integrity of human life appears to be challenged, then research may be discouraged. Such bans are always problematic because they are made from a position of ignorance. We are never so clear about the consequences of future cognitive possibilities. Furthermore, in an open system of information exchange such as science, any local prohibition on research tends to be bypassed by the growth of neighbouring knowledge that has not been banned.

It is difficult to see, therefore, that any systematic difference would be made to the cognitive nature of science by the hints of feminists on problem choice. The uses to which science is most readily put may well change, however.

The justification of knowledge Individual creativity, situationally embedded as it is, must be capable of being built on by proving acceptable to other individuals in other situations. It is not clear at this stage that feminism has any alternative to offer other than the present kind of negotiation of objectified knowledge claims. The conventional way to do this is to objectify, to depersonalise each scientific claim, making it easier to be picked up by others with different interests in different contexts.

Suppose science were done by groups not dominated by men and that it turned out that certain styles of thought, widespread among women of present society, were especially productive. Would such a feminist science be different from, and/or preferable to present science?

For example, it has sometimes been proposed by feminists that women are by nature and culture, more co-operative, less competitive than men. This seems most plausible if we think about aggressive confrontations between individuals. Men are more likely to come to blows. However, it is not clear that such aggression makes a particular difference to the issue of competition and co-operation in science. Male scientists do not characteristically come to blows over their science. Science is somewhat like a mutual admiration society, rather than an open struggle for domination. There can, of course, be competition in the desire to be admired more than others. This kind of competition is often institutionalised by establishing rules that reveal a dominance hierarchy, a pecking order. Our present culture quite often agonises over the fairest rules for the pecking order of mixed groups. Should there be one pecking order for men and one for women, as in many sports, or can the two orders be integrated, as in educational systems? Should science have such a pecking order (presumably in terms of expertise and accomplishment on whatever topic is at issue), or should it seek to avoid it?

Perhaps feminists are asking us to imagine a new social order in which competition to establish dominance hierarchies no longer occurs. We are being asked to envisage a society in which there is no use for this aspect of human nature, so that it dies away. I am not sure if this would work as intended. Within groups, competition is suppressed in favour of co-operation as the group faces up to some external threat. The internal competition is restored when the external threat goes away. Doubtless, it would be possible to galvanise groups into co-operating when the only external threats were imaginary. This is a dangerous ideal, however, for, if people are to be manipulated so that they can work with a common and effective sense of purpose, we have a key ingredient of totalitarianism. For example, the competitive impulse is turned in science into the conflict between rival interests linked to rival views of the truth, and the co-operative impulse into compromising interests to produce and promote a shared vision of the truth. If competition is systematically suppressed in favour of co-operation at every interpersonal cognitive level, then the current vision is exposed to less checks. It moves towards being the only view – towards the kind of closed world view from which we escaped long ago and now fear in terms of the resurgence of totalitarianism. As long as science is uncertain knowledge, we must encourage competition between the alternative possibilities, and, if that involves competition for status between the advocates of those views, then so be it.

The most dangerous side of competition is that between groups in which conflicts of interest escalate into aggressive forms. If feminists or anyone else can find ways of preventing disagreement leading to war, present society would be very pleased. Science on the whole does not encourage this kind of escalation into conflict, because its practitioners must co-operate to be seen to be playing the game at all, and in the game of science there are procedures for the closure of argument by appeals to 'the facts'. However, in its external use, science amplifies the destructive capacity of technology, the most obvious danger to modern society.

In summary, then, the appeal of feminists to co-operation rather than competition may be misdirected when applied to science. It could be based on dislike of a kind of aggression which is only incidental to science. It could be dislike of the present tendency in society to manage processes through setting individuals and groups challenges which require them to compete with others. The alternative to this would involve a revolutionary change to the social order. Perhaps this alternative would also need a change in human nature resulting from the elimination of all those activities in which one individual finds it an advantage to set herself above another. The change in science would come as part of the wider change.

Institutional knowledge As a social system, science emerged with the feature that an individual could enhance his status (often in a particular social context) by displaying, not erudition, but new knowledge. This kind of competitive display was tied into the wider economic system. A doctor might, for example, acquire a favourable reputation through his fame in research. Furthermore, the capitalist system allowed people to sell their new-found knowledge and skills as much as their labour. Science proved to be an intellectually demanding activity, which, during the scientific revolution glorified its pioneers. Others sought the same kind of self-satisfying honour. Because so few could do it and even fewer do it well, it was relatively straightforward (though hard to achieve success) for those with sufficient talent to acquire a name in science. Science institutionalised rules for establishing priority in discovery, and a reward system for the easier identification of those most deserving of recognition, who could serve as role models for the young. The domination of this system by the young offered a promise of making one's way in the world rather quickly through science (perhaps to move on to other things).

The success of forms of science which offered social mobility through education was institutionally consolidated in ways which presupposed that men would be the primary recruits into science. For example, the results which were best rewarded were those requiring obsessional commitment over many years, which was difficult to combine with the career distractions of motherhood. The prospect of women using education as a stepping-stone to a better career could not at first compete with marriage as a route to social improvement. Those women who were first attracted into careers most often chose subject areas more traditionally attractive to them than science. The few who were drawn into science were often already in the professional classes, sustaining the cultural values of their origins.

The institutional form of a future feminist science would presumably have a different social basis, one making it attractive to the normal life plans of women. On the old model, science was a channel by which creative talent could fairly easily display its quality. Unless the demands of childbearing and early child-rearing are redistributed (which is quite possible in a technologically inventive age) women would have to do their science in early adulthood, part-time, or in middle age. The present kind of science is at the upper limits of the talents of the general population, and requires extreme dedication to doing something impressively well. It may simply be too hard for humans in their declining years, and part-timers may never rise out of the lowest ranks. Perhaps we can eventually make scientific creativity less intellectually challenging, but by then the knowledge production process might be done by machines. One way to make science attractive to women would be to construct a social system in which early recognition in science is seen as an advantage to ambitious women who can then move out of science, perhaps to build on their youthful reputations in later life. Unfortunately, the recent trend of our society has been to make science less attractive to the socially ambitious, male or female. There are easier ways to make one's way in the world than in science. If this could be changed and science could be to the taste of the young once again, then it could be attractive to ambitious young women.

The qualities that have made science so successful and influential emerged in a male-dominated culture and will change as the wider society changes. At present science is turning from being a vehicle for the ambition of individuals seeking social mobility into an investment device to keep technology productive in the long term. To remain successful when people of quality no longer clamour to do science at the level of investment available, science will have to modify its present institutional form. Perhaps enough social changes are being introduced to encourage ambitious young women to take up science in their own interests.

The most valuable suggestions of feminism at the institutional level of science arise from its critique of the way the institutionalisation processes in science can so easily suppress women's interests. This process very often occurs through pressures from the wider society.

Wider society Although it defines itself in terms of gender, a characteristic of human individuals, feminism has become primarily a criticism of society as a whole. Therefore it is in its relation to the wider society that we might expect feminism to be most interested in science. Certainly, male-dominated science can be seen as being quite as unfair to women as other institutions in society. The scientific and medical study of women in past societies is especially vulnerable to feminist critiques. The account given in chapter 4 of the argument for the intellectual inferiority of women is just a small part of that story.

Perhaps the most important feminist requirement for science is that knowledge be created and used in ways which sustain women's and feminists' interests. In this respect, there does not seem to be a profound difference between feminism and any other group that might use science. Every such group would like to maximise the extent to which new knowledge is created in its interests and existing knowledge is turned to its advantage. If past science does not seem to be suited to such purposes because it was created by men, then a feminist science should correct the deficit. The same kind of argument *could* be developed by the elderly, should they see science as having been primarily created by the young, or by organised criminals, if they see science as an instrument of the establishment. The same kind of argument *is* offered in the developing world, where science is seen as geared to the capital-intensive technology of the developed world. In these terms, a feminist science would be a science that remained as effective as possible at the generation of new knowledge, more effort being made to redirect it to serve a different set of interests.

In conclusion, this discussion recognises that science needs to change and is changing in response to feminist pressures. Science is adapting to these changes, as it has adapted to changes in its social basis in the past. I do not think, however, that these changes are especially revolutionary for the processes by which knowledge is created. The metaphysical dichotomies attacked by Derrida and his feminist followers, such as that between male science and female nature, do *not* seem to me to be constituent of science at the technical level. If such rhetorical flourishes disappear in an age more in tune with feminist sensibilities, science will go on very much as before. The impact of feminism may well be greater on the uses of science, where some male interests are being overthrown.

Women's interests can be expected to function as a corrective pressure within existing science, in particular by overcoming the exclusivity of past male-dominated forms. Feminist science might dominate specific research areas and social niches, while also joining the wider scientific community in engaging the intellectual issues of knowledge creation. Science as a whole can retain its ideal of being an open system of knowledge, but should represent within itself a wider range of social interests.

Some feminists may hold the view that feminism should lead to a much more radically transformed cognitive activity. This could compete from within a new social niche or take over in an intellectual revolution as a new orthodoxy. There are risks in both such radical strategies. A competing form of science must establish its own history of successes to be taken seriously, and the principles leading to these successes might be assimilated by present science, the advantage being lost. If radical feminist science becomes a new orthodoxy, which reworks existing knowledge in its own terms, and it rejects the received explicit rationality of science in favour of feminist informal rationality, it risks becoming less of a science and more like a pre-scientific culture, no longer so effective at generating new knowledge.

10.4 Psychical research

In this section, I will explore the possibility of an alternative science arising from a different social basis in society. There are many places in popular culture, in learned non-science and in the cults and sects of the fringes of society, in which alternative systems of systematic belief thrive. Not all claim to be scientific, and those that do are likely to be dismissed as pseudo-scientific by defenders of the orthodox scientific establishment. My discussion in 7.6 of the popular alternatives to science showed them to be a rich source of illustration of methodologically defective procedures. In particular, although each appeals to at least a section of society, none are able to produce consensus in favour of their knowledge claims.

Nevertheless, if orthodox science is to be regarded as fallible and incomplete as argued in this book, there is a chance that not merely is reality not as scientists say it is, but that one of the popular alternatives has come nearer to ultimate truth. For example, what if reality is very different from the cautious materialism that science is often used to legitimate? The mystical depths of the human mind are ignored by science and so are the higher planes of spiritual being that humans can access in a suitable meditational state.

I will briefly explore the possibility that the central insights of an alternative form of science with a different cultural basis might have captured reality and that, by developing those insights further, general agreement will eventually be produced. Such alternatives might be regarded as *protosciences*. I will end the section by discussing psychical research as an example.

Popular culture is full of speculative protoscientific ventures. In 7.6, I discussed some of the reasons why such ventures flourish in terms of the relationship between their social basis and the rationality of the participants. We were introduced to the viewpoint of the defender of a cherished traditional value, the 'seeker' who searches through religions, medical therapies or political movements looking for the answer to a problem, and the many people who enjoy the stimulus of the new and the imperfectly understood. People like these have hopes for alternative sciences, in case one may finally be vindicated. The incompleteness of science means that it has little to say on many of the practical questions in our lives. Protoscientific alternatives offer to help us to cope in areas outside present science.

Could there be viable alternative forms of science among such systems of belief? Certainly, in a scientific age, many of the alternatives claim scientific status. We have seen that they very often employ practices which are pathological as science. However, they do not *have* to be bad science.

To be science, they should not merely seek cautiously to conserve and apply a system of belief, but effectively be testing the limits of their claimed knowledge, extending the range of circumstances in which their beliefs apply to the world. To be orthodox science, they should do this in a way which is not limited to the congenial immediate reality of a particular interest group but which can be shared more generally, and in particular within institutionalised research practice.

Perhaps the way to take such systems seriously as alternative science is by appreciating that reality may well be quite different from the images currently constructed in science. There are more things in heaven and Earth than are currently dreamed of in scientific cosmologies. And perhaps, just perhaps, some of the popular alternatives to orthodox science contain truth, or at least half-truths. Perhaps the congenial immediate realities of some such belief systems are accessing a wider reality that orthodox science cannot.

Alternative sciences often flourish in restricted local realities. Many political, economic, and religious systems encourage people to act in ways which sustain prevailing beliefs, avoiding putting them under challenge. On the view offered in this book, immediate reality is sufficiently plastic to allow localised variants of belief. Few variants look so good outside the domain of action of those who sustain them. Such belief systems can expand by the enlargement of the community of believers, and, in this way, the corresponding immediate reality also expands. If the empirical understanding sustained in such immediate realities facilitates the construction of further immediate realities of the same kind, we may be persuaded that it corresponds to an important aspect of wider reality. Such alternative sciences would not be preoccupied with consensual certainty, but would claim to produce defensible results in the new domain of discourse opened up.

It is easy to see why orthodox science excluded this option, aiming instead for universal knowledge. Its emergence after the multiplication of variants of Christianity and of philosophical systems in sixteenth-century Europe had made it clear that non-consensual knowledge can quickly multiply into an indefinite number of factions. Orthodox science sought to limit itself to what we could all be sure of. If people cannot be brought into agreement on an issue, then that issue is redefined (for the moment) as non-scientific. If knowledge is to be something that we can all share (because it is true), then we cannot deliberately incorporate contentious supposition.

Although orthodox science works quite hard to limit itself to consensual knowledge, the growth of knowledge involves speculative and extrapolative steps beyond current orthodoxy. Conceptual understanding is applied to predictions of novel effects. Observational methods are pushed to their limits and beyond. Speculative theories are developed on tendentious bases. Sometimes, such extrapolations go wrong (as with N-rays*, or cold fusion*). At other times, observations turn out as predicted, observations based on uncertain methods are supported by further observations with different methods, speculative theories are reworked into a testable form. If science grows by the linking of steps none of which would be sufficiently rigorous on its own, some uncertain steps must normally be made *before* the interconnecting links are constructed.

The aim in speculative extrapolations of orthodox science is to consolidate them as quickly as possible into orthodoxy. New speculative possibilities that become fashionable talking points are given brief attention before scientific enthusiasms abandon them and move on. An extrapolation which is not soon consolidated comes to be regarded with increasing suspicion. Those who still remember it may find a way of incorporating it more satisfactorily into later developments. If that does not happen it is eventually lost from science.

Might the orthodox impatience with speculative possibilities make parts of reality inaccessible merely because the required level of consensual certainty is not available? Perhaps, for example, the mind really does access higher levels of being, but not in a reliable way. Must science exclude whole areas of ontological possibility?

Surely, if reality is actually as some alternative to science holds it to be, there would be practical advantage in presupposing such a reality. The goals we set ourselves would be more likely to be reached. In practice, if those who accept specific knowledge claims are thereby given some practical advantage in survival or in competition with non-believers, the belief system will increase its plausibility. Conversely, if we cannot point to any such practical differences, or if the practical differences do not turn out as the believers say, then there is little external point to holding that belief system. It becomes an isolated island of self-justified belief, and like many other self-justifying belief systems, no concern of science.

We can, then, consider the popular alternatives to science as speculative extrapolations away from scientific orthodoxy. They are external rafts drifting near the main craft of scientific knowledge. They are usually kept afloat because they tie in with non-scientific values. At some point in the extension of knowledge they may be incorporated into the main craft, but they may also float away, or sink. Just possibly, they may grow independently of scientific orthodoxy until we are able to say that we have two kinds of science.

Psychical research

The notion that there might be some spiritual aspect of an individual which survives death, to be reincarnated or to be moved on to another plane, is viewed with sceptical suspicion in science. A large number of religious and quasi-religious systems of belief, spanning many cultures, hold such beliefs dear. In the mid-nineteenth century, a high point in the age of modern science, spiritual phenomena became suddenly fashionable. Mediums made contact with the dead, who spoke through them and produced physical manifestations. Apparitions abounded. For those whose religious faith was under challenge from science, it suddenly seemed that science might find new evidence for the spiritual. All that was required was to subject the mediums to rigorous scientific scrutiny and so to produce a systematic empirical science of psychic phenomena. New institutions were set up to do just this. In 1869, the London Dialectical Society began a series of investigations into spiritualism, reporting in 1870. In 1882, the Society for Psychical Research was founded. All psychic experiences were investigated. Of central interest were the mental mediums and the physical mediums. The mental mediums called upon their spiritual contacts to say things to a subject at their seances which only the dead person (and the subject) could possibly have known. The physical mediums produced apparitions out of ectoplasm and spirit writing on sealed slates. Sometimes the physical medium floated in the air and other marvellous effects occurred. It was necessary to provide a congenial environment for the spirits, or they would not make their presence known. The room had to be dark, everyone held hands and suitable music and vocal summoning were central parts of the ritual. These conditions made scientific investigation difficult. Spirits did not usually stay obligingly still for long enough for photography, though photographs were taken. The darkness made it difficult for scientific investigators to work out what was going on.

Psychical researchers included a number of very senior scientists, some committed to the existence of the phenomena and others thoroughly sceptical. The topic was increasingly controversial. Physical mediums were frequently caught cheating (as revealed by suddenly illuminating the room), and others later confessed that they had been cheating. It was difficult to establish rules for distinguishing between a genuine psychic effect and an undetected conjuring feat. While physical mediums came into increasing disrepute, it seemed that effective tests might be possible for mental mediums. The best mental mediums often reported details of the past that could only have been known to the subject in the seance and to his or her deceased loved one. Could this be a trick, perhaps performed by assiduous research on the subjects before the seance? Could the medium simply be 'fishing' – saying plausible things and taking advantage of quite tiny visible reactions from the subject? However, suppose that a person, before death, put unguessable information in an effectively sealed container, telling no one what the message was. Then a medium who produced the information would surely have demonstrated psychic powers. Although this kind of feat was occasionally apparently accomplished, it was never done to the satisfaction of everyone, including the sceptics.

It became clear that there were simpler ways of producing the key psychic effects than by contacting the spirits of the dead. The mental mediums might be *telepathic*, reading other minds directly. The messages in sealed containers might also be read directly by *clairvoyance*. The physical manifestations might be produced directly by *psychokinesis*. As spiritualism became less fashionable, attention in psychical research turned to the direct investigation of these processes, as well as *precognition*, psychic knowledge of the future.

In the 1920s, a new pattern of investigation was established, especially by J. B. Rhine, at Duke University. The new task was to find the effect of psychic powers on guessing Zener cards. This was a pack of 25 cards, 5 each of 5 distinctive patterns (circle, square, cross, star, and wavy lines). Guessing the cards by chance would produce 5 correct guesses out of 25. Systematically to do better raised the possibility that more than guesswork was involved. One person would look at each card in turn, while another guessed (telepathy), or one person could guess a card that was not being looked at (clairvoyance). The use of one psychic power could disturb a result thought to be due to another, as when clairvoyance is simulated by reading the future state of mind of the experimenter when working out the results (precognition plus telepathy), or by influencing the order in which the cards are initially shuffled (psychokinesis). Ingenious methods were developed to try to test the existence of the powers independently.

The cards enabled studies to be done repeatedly, so that it could be shown that it was most unlikely that unusual results were being obtained by chance. Rhine obtained spectacular results from some of his early subjects, but the precautions against subjects (or assistant experimenters) cheating were weak in the early days, and the later more careful studies no longer produced such spectacular results. Rhine's contemporaries attempted to replicate his results. Some partially succeeded and others failed. The problems of making the investigations replicable for all experimenters was never solved.

The history of psychical research is the history of a form of investigation that never made it to scientific orthodoxy. A rich body of data was built up. It was simultaneously enough to convince believers and too heavily flawed to convince sceptics, who became ever more sceptical.

The directions of growth of science have been in directions which make it ever harder to assimilate such findings. Psychic effects are not recognised elsewhere in science. They would give yet another reason to say that the immediate reality of any experiment is excessively congenial. Even double-blind trials might be contaminated by nearby experimenters hoping that their pet ideas will succeed. (If such problems became generally accepted as pit-falls, doubtless scientific procedures could be devised to minimise their local effect, unless it turns out that everything in nature, even distant stars, are being influenced psychically.) Another difficulty is that the possibility of psychic powers of animals is even more contentious than of humans. When an animal is trying to detect a nearby predator, psychic powers would give an enormous survival advantage. Similarly, the predator could use *its* psychic powers to locate its prey. The normal evolutionary process quickly develops hereditary predispositions for such capacities once they emerge in competi-tive struggles for survival. They cannot remain marginal and unchanged. For reasons like these, it is hard to accept the attribution of subtle but functional psychic powers to animals. If our animal precursors do not have demon-strable versions of psychic powers, and we do, then something distinctive must have happened at some point in recent evolution to explain their appearance. Some emergent property of the growing capacities of mind must be involved. The most distinctive qualities of the human mind are dependent on our capacity to think and communicate symbolically. However, the development of language does not plausibly provide a mechan-ism for psychic communication. Indeed, it is sometimes speculated that it was the emergence of these language-based powers which reduced our need for psychic powers for survival.

Tenuous phenomena can often be made amenable to science by finding some special condition in which they can be produced and controlled unam-biguously. Another research area, the study of animal intelligence, which emerged at the same time and place as psychical research, successfully went through a process of being turned into a manageable scientific disci-pline. The initial context was one of looking for ways in which our animal precursors might display the evolutionary origins of our own level of intelli-gence. Researchers moved from collecting anecdotes (such as dogs finding their masters who had left them behind on moving house or cats opening windows to escape from locked rooms), to making personal observations of especially intelligent animals under controlled conditions (such as the famous 'Clever Hans, a horse appearing to be able to do arithmetic), to the experi-mental study of numbers of randomly chosen animals (such as the training of rats to run mazes). In the early stages of this process, it was very difficult to be sure if stories told by the doting owners of pet animals could be believed. The style of investigation became increasingly sceptical and limited to what was under the control of the investigator. Eventually, any experimenter could do any of the key experiments on any properly chosen sample of animals and be confident of getting comparable results. By the early twentieth century, the study was no longer about how intelligent animals could be, but of how stupid they were. In this properly objective context, the mechanisms of animal learning on which behavioural psychology was to be based were worked out.

Why was there no comparable process in psychical research? The same transitions can be found, from collecting anecdotes of hallucinations that involved contact with the spirits of the dead, to controlled studies of outstanding mediums, to statistically sophisticated experimental studies of psychic influences on the guessing of cards. The end result was that, as psychical research changed into parapsychology, the topic remained as controversial as ever. There are no uncontested facts in parapsychology. Sceptics worried about people reconstructing the evidence for their psychic anecdotes in terms of their belief that something paranormal had occurred. They worried that mediums had a commercial interest in cheating in seances, as a good living was to be had by performing for the public and for scientific investigators. They worried that the subjects and the investigators in card guessing games might have an incentive to cheat or at least to make mistakes in their own favour. The lack of replicability of non-random results of card guessing games was never sorted out. The same doubts remain at the end of the twentieth century.

Psychical research can, then, be regarded as an alternative science which has never made it to orthodoxy even after nearly one and a half centuries of trying. Because it remains controversial, the controversy-avoidance social mechanisms of science apply to it, and it becomes even harder to assimilate. Nevertheless, orthodox science cannot say that it has disproved the existence of psychic phenomena, only that if they exist they lie outside the domain of known science.

All this could change in a moment if an unambiguous form of psychic phenomena could be established, or if an independently testable mechanism could be supplied that worked out as predicted. (In the comic-book and film fictional versions, we meet a psychic alien or mutant, or invent a psionic amplifier. Then we have situations in which the content of telepathic messages unambiguously rises above the threshold of psychic noise.) But, unless some such thing was to happen, we have to admit that we simply do not know if there really are psychic phenomena.

Suppose that our minds really do have psychic powers and that these powers are unreliable. It might take orthodox science a very long time before its extrapolative growth could assimilate such matters. Would it not be worth developing an alternative science of parapsychology unconnected with orthodox science in order to speed this process up? Perhaps by studying the putative psychic phenomena in their own terms we might get a coherent set of results which would make their later assimilation by science easier?

Such a form of study lends itself to discussion in terms of the concept of immediate reality. Perhaps psychical researchers strive to surround themselves with a congenial immediate reality in which their ideas are seen to work well. The same ideas fail to work in less congenial circumstances.

It was suggested in chapter 5 that we tend to shape our immediate reality to suit ourselves. The immediate reality occupied by believers in psychic phenomena may be different from that of disbelievers, especially if the processes which link our actions to our perceptions are affected by our actions in ways we do not suspect.

Believers in psychic phenomena sometimes conjecture that sceptical scientists find the things they do because their minds have a negative psychic influence on the phenomena they study. One of the requirements of a successful psychic experiment appears to be that those producing the paranormal effects should be surrounded by a supportive atmosphere of believers. On this view, the sceptics do not appreciate that they are changing the immediate reality by their scepticism.

Sceptical scientists, too, may take advantage of the idea of an excessively congenial immediate reality. They may claim that 'psychic' phenomena are not produced paranormally, as the believers claim, but by the combination of sloppy thinking and self-fulfilling actions in an excessively congenial immediate reality.

Although it is clear that the psychic experiments of believers and sceptics occur in different immediate realities, we cannot yet conclude that the account of immediate reality offered by believers (namely that the sceptics interfere with the psychic process) is better or worse than the account of immediate reality offered by the sceptics (namely that believers set up their procedures so as to fool themselves rather too easily). We can, however, conclude that, as long as psychic phenomena only work in special local contexts, they are going to be difficult to build into a consensual science. In the mean-time, the alternative science will have to be a science of believers, with all the factional difficulties that that implies. The breakthrough would come if believers found a way to counter the negating effect of sceptical fields on psychic phenomena, or in some way amplified the effect so that it was beyond reasonable doubt.

10.5 Alien science

In popular fictional culture, the idea is often played with of extraterrestrial aliens who have their own science, appropriate to their quite different form and mode of living. It is a defensible speculative extrapolation of current science that there are very many intelligent extra-life forms in the galaxy, and so there are very likely to be many alternative forms of knowledge constructed by aliens. Although many forms of alien science would inevitably be too alien for us to make any sense of them, there might be other forms which are significantly different in their origins from our own, but which we could partially come to understand as our science becomes more general and more comprehensive. It is most implausible to think that ours is the only possible form of science. However, until we meet examples of such beings and such science, there are too few constraints on our speculations.

10.6 Summary

Science done by computers How much of science could be handed over to computers? Electronic instrumentation provides sensory and control devices, as well as calculating and modelling aids. Computers can substitute for humans in the repeated routines of science. A computer that did science would also have to contribute to the social levels of cognition rather as a slave could do in ancient societies. Perhaps, if people were prepared to attribute appropriate rights and responsibilities to them, future computers could be human companions, always dependent on support from the social system but contributing significantly to it. There is, then, nothing in principle to stop future computers doing as much science as people will allow.

Towards a feminist science When science was done almost entirely by men, the knowledge produced reflected the outlook of men in a manner sometimes contrary to the interests of women. This gender bias is beginning to change as women play a larger role in science. Could the *principles* of scientific knowledge construction also be modified to fit feminist principles? Throughout its history, science has absorbed the specific rationality of the many groups who practice it, and opening its institutions to feminist epistemology does not appear to pose special challenges.

Psychical research Perhaps reality is not well captured by the cautious materialism of present-day science? Could a popular protoscientific speculation such as parapsychology be nearer to the ultimate truth? In the absence of techniques to remove ambiguity from the underlying phenomena, parapsychology will always be controversial. The most that can be hoped for is a research community of believers operating independently of the rest of science. Perhaps scientific scepticism suppresses psychic fields in the immediate reality of such research, or perhaps the believers are combining sloppy thinking with the choice of excessively congenial local realities.

11

And in the long term?

11.1 Introduction

This book has extended a fallibilist realist conception of science in terms of the view that we change our immediate reality in terms of our ideas while simultaneously changing our ideas to match reality. It has conceded to scepticism that we do not reach certainty in scientific conclusions; the fragments of fully explicit scientific reasoning are set within frames of optimising judgements about wider considerations. The rigorous science that we explicitly consider depends on contextual assumptions which are subject to change. At least some values are present even in our most nearly objective conclusions. Inevitably, then, the way we evaluate science is affected by local and short term factors. I have argued that there are many cognitive frameworks in an open society and that we produce science by trying to reconcile them at several levels of cognitive activity. In this way we hope to minimise the dependence of science on very localised contexts. So far I have offered techniques for optimising scientific judgements in the present. Must we lose touch with the science of the remoter past? Must future science be unclear to us? Are we locked into our own historical period? It would be unfortunate if we were. Relativism across time is as problematic a philosophy as relativism across the spatially dispersed cultures of the present. On such a view, we could not hope to understand the past and we could not hope to anticipate the future.

A serious problem of the present day is that our judgements are too often made for the short term. Many institutions in modern society make themselves short-sighted in time. Democratic politics works for its worthwhile results for the reward of winning the next election. Economic thought can barely manage projections a year or so into the future. Humanitarian aid often seeks to relieve immediate suffering even if the methods used destabilise the social system of the people being helped. According to the short-term view, if we can only know what is happening now, there is no point in trying to provide for the future. It is quite defensible to cut down and not replace forests and to cause extinctions by the complete destruction of particular ecological niches. It is quite defensible to use up national assets. A short-term view might, however, be defended in the context of technology: the problems we produce for ourselves in the long term may not be worth worrying about too much now, for they may be easier to solve with the enriched technological resources of the future.

Scientific research suffers if there is a preponderance of short-term thinking. It has long had the reputation of occasionally yielding unexpected benefits which were impossible to plan for. To concentrate on immediately foreseeable benefits is to turn science into applied and mission-oriented science. Scientists worry that by concentrating too much on using the knowledge we have, we may reduce our capacity to generate unexpected new knowledge, and so dry up the source of future technologies.

The effects of science make short-term thinking even more problematic. For modern society undergoes fundamental and continuing change as a result of the continual birth and growth of science-based technologies. Societies have undergone radical transitions in the past, but, in present society, continuing rapid change is becoming the natural state of affairs. A consequence of this kind of change is that the period into the future about which we can be confident is growing ever shorter. The long-term view we too often ignore is becoming shorter than the lifetime of individuals.

If we think about the longer timescale of society, can we be sure that science will last long enough to be more than a transient factor? It is hard to be sure, but science should last at least as long as the present social system. Over longer timescales, the capitalist system, the technology it stimulates, the science which provides new technological possibilities, and the environment which is cumulatively exploited, all make the struggle between rival interest groups into a social game with more winners than losers. The benefits of science are not limited to its immediate patrons. Science and the technology it generates frequently escape the control of dominating power groups. Science does not, therefore, reliably stabilise the established social order, but usually facilitates social change. Any group may seek to advance its interest with the aid of science. In the long term, in a growing society, there should be a growth in interest groups indebted to science. Therefore, science will not easily be lost from society.

There will also, however, be groups which have been alienated from the interests of those who support science. Factions may also come to control

society who sustain power by suppressing science and other resources for
those who seek change. Science can never expect to become completely
uncontroversial. By making the future of society unpredictable, it makes its
own future uncertain. Nevertheless, the odds are stacked in favour of its
survival as long as society lasts.

The problem of short-termism was not so acute in previous centuries.
There were absolute frameworks available, which seemed perfectly adequate
for the longest-term judgements at a time when much of cultural change was
sufficiently gradual to be assimilated by successive generations with little
anxiety.

11.2 Absolutist frameworks and their problems

Ancient ideals of philosophical understanding were absolutes. The ancient
Greeks sought to see through the chaos of our immediate circumstances to
the eternal truths beyond them. This quest has remained at the heart of
philosophy, where it has become ever more problematic. As we have progressively discovered the extent of our own ignorance, it has become clear
that such a philosophical vision can function only as speculation, the knowledge we are actually gaining is at a much more provisional level. In particular, the empirical knowledge of science is always subject to revision. It is
plausible to argue today that we have no candidates anywhere in epistemology for the kind of transcendent philosophical certainty sought by Plato.

In the Christian society of Europe since the late Roman Empire, one
absolute framework came to be almost universally shared. The universe
and all that is in it was made by an omnipotent, omniscient, benevolent
deity, whose actions and intentions concerning mankind are reported in
the Bible. Within this framework, science is concerned with discovering the
laws of God's creation. Not every Christian was convinced that human
knowledge of the divinely created order should be sought or could be discovered. However, those whose stories are told in twentieth-century histories
of science built up arguments that we could come closer to God in our
understanding by studying His Creation. What we find out is not merely
more knowledge of His divine attributes, but, by knowing more, we put
ourselves in a more God-like state.

Christian frameworks also suggested a long-term pattern for human history. The world could be divided into periods, that leading up to the arrival
of Christ on Earth and that which occurred afterwards, in the context of His
revelation. There was also to be an end to history – the Christian view was

eschatological. This notion of a universal pattern to events became an important resource in Christian thought.

It was in this context that modern science appeared and thrived in the sixteenth and seventeenth centuries. Writers of the Enlightenment in the eighteenth century continued to build upon the Christian framework in their vision of progress, although they increasingly sought to distance themselves from the religious precursors of their ideas. From an Enlightenment point of view, we should escape from the religious framework into a more secular understanding. According to the idea of progress, we can move beyond our origins. Although societies rise and fall, the knowledge they produce is more easily regained by later societies. The art of writing text, for example, was never completely lost in the Dark Ages, so that recorded ancient knowledge could eventually be recovered. Slowly and uncertainly, we are cumulating the wisdom of the ages. The Enlightenment vision of knowledge and of society as progressive has been the dominant secular framework for science in the modern age.

In this book, I have not taken seriously the absolute frameworks of Christianity and of Enlightenment thought as a basis for the evaluation of scientific knowledge. The nineteenth and twentieth century collected arguments for their inadequacy for such a purpose. In the rest of this section, I will briefly present representative arguments for why neither is adequate for thinking about science in the long term.

One general form of criticism of both the theological and the progressive view of knowledge is that they suppose that the goals for science can be specified in advance. For Christian natural theology, science finds increasing evidence of God's power, wisdom, and benevolence in nature; for Enlightenment thinkers, science cumulates our power to act in the world for the benefit of society. However the effect of science on these goals makes them self-limiting. As we build up scientific understanding constructed in terms of God's plan in nature, we find that it is not simple and good, but complex and indifferent to the fate of humanity. We find that it is less like a human plan. There are, for example, natural processes at work which cause pain, destruction, and death. Nature at times appears evil. We find that the Christian presupposition of the central place of man in nature is ever harder to sustain. We find that some of the more orderly parts of nature are driven by natural mechanisms, so that if God is acting in the world to produce this order, He must be acting less directly than we first supposed. God disappears behind the machinery of His Creation. The conclusion, that we discover complexity rather than simplicity, was acceptable in religious terms. The complexity of nature suggests that the Creator, too, may be complex. And if we later find simple unifying principles that generate the apparent complexity, that too can be seen as due to the cleverness with which God acts.

Those who have been concerned to reconcile science and religion in the present century have found their task manageable. However, they have done it mainly by cutting back on the empirical claims of religion in order to protect its central moral claims, while leaving science less affected. Humanist critics of religion call this process of cutting back on its empirical claims, 'death by a thousand qualifications'.

Some of these points were already recognised by the Enlightenment philosophers. They sought to distance God from His Creation, or to leave Him out of the equations altogether. The more sceptical dwelled upon the problem of natural evil in the world – in God's Creation. However, their account of science still showed its religious ancestry. They took from the religious framework the idea that the cognitive essence of nature can be expressed in universal laws, laws from which every past and future state of the universe can be calculated, using our knowledge of the universe's present state. All we have to do is to discover those laws and to describe the present state of the universe. I argued earlier that the effect of centuries of search for such laws has been self-limiting. We have not discovered a cumulating pile of laws of nature, but have learned that nature does not, on the whole, obey patterns that we can *directly* observe. The regularities we thought we had found in the past have turned out to be over-idealised. The regularities we now look for are far behind appearances and can only be found by the less cumulative, more risky, strategies of theory construction.

The Enlightenment framework has developed other problems of its own. The ideal of fully explicit rationality, adopted and taken further by the positivists, has increasingly looked incapable of achieving the goals set for it, as I have argued earlier in this book. Science is uncertain, theory-laden and subject to rational revision. Even the central Enlightenment idea of progress has become problematic. A major weakness of the Enlightenment framework from the point of view of this book is that it idealises to contemplative knowledge rather than active knowledge. In a world in which we and other knowing agents are active, the situation we seek to know is always changing, its state being a function of the knowledge the agents have. It is indefinitely complex, and therefore not fully knowable, unless the agents collectively set about making it so simple that perfect, self-fulfilling knowledge is possible.

I suggest, then, that the Christian and the Enlightenment frameworks for science have passed their use-by dates. Whatever their value in other domains of human intellectual activity, they are inappropriate for the modern intellectual product of science. Where older ages looked for constancy now we need to understand science in a context of change.

11.3 The evolutionary framework

In place of Christian and Enlightenment frameworks, I defend a third wider view of science in the long term, the evolutionary framework. Versions of this framework have been around as long as the idea of evolution of species has been in science, although not all versions are acceptable for present purposes.[1]

The core idea of modern evolutionary theory is that each living organism develops from genetic material supplied by (one or) two parents, normally in the form of chromosomal packages of the chemical DNA. At a chemical level, DNA regulates our lives. Genes are sequences of DNA (or in some bacteria, RNA), each normally directing the manufacture of one of the proteins of life. Every individual organism has slight variations in genetic makeup, which can lead to differential success in life and in reproduction. Accidental changes to the DNA assembled into an individual from sperm and egg can occasionally be perpetuated by the advantage they give to the individual produced and to its descendents. The cumulation of changes in the individuals of a breeding population tends to lead to better adaptation to the environment. If the environment changes, the direction of selection will adjust correspondingly. The genetic material of life also influences the environment within which natural selection occurs, so that evolution is about the reciprocal interaction of genes and environment. The chemical drivers of life have evolved forms which survive best by encapsulating portions of their local environment so as to sustain vital functions. This happens at the level of the cell, the organism, the interbreeding population and ecosystems in local domains. As a result, life is a multi-level phenomenon.

On such an evolutionary view, life is not usefully to be understood as advancing according to a divine plan, nor is it progressing cumulatively. Because past evolution has made living systems into the kind of material they are, life is continually generating variants of itself, only some of which survive. The evolutionary classification of species was visualised by Darwin as being like a branching tree, each horizontal layer of which corresponds to a moment in time. As the tree grows upward through time, each new point

---

[1] For example, no attention is given here to the pre-Darwinian idea of evolution as the progressive cumulation of improvement. That view is now indefensible as an account of the evolution of species up to modern man. Evolution involves adaption of species to their curent environment and whatever cumulates may be lost as the environment itself changes.

can only grow out of an immediately lower point on the tree. Over time, branches and twigs are produced, most of them coming to an end in an extinction. The current highest horizontal section of the tree is to be seen as a product of the lower parts of the tree from which it emerged under the selective pressure of the present environment.

Analogies are often constructed between simple versions of this view of evolution and the growth of knowledge. It is plausible to suppose that just as humans evolved from animals, so human knowledge has evolved from animal knowledge. This view has been elaborated into the philosophical position known as evolutionary epistemology. Its key advocates are a group who acknowledge the pioneering work of J. T. Campbell (1974) and treat K. R. Popper as a godfather. The central metaphor is that, 'The highest creative thought, like animal adaptation, is the product of blind variation and selective retention' (Bartley, in Radnitzky & Bartley, 1987, p. 24).

Evolutionary epistemologists have argued persuasively for the inspirational qualities of this analogy. They have shown that thought about the details of evolutionary processes provides a rich set of interesting questions about the growth of knowledge. They have moved on to try to carve a niche in intellectual life for this new synthesis between science and philosophy (Callebaut & Pinxten, 1987).

The discussion of this book can often be linked to evolutionary analogies. For example, at every stage in science, from individual inspiration to social negotiation, more candidates come forward than survive. The selection processes are often competitive. The outcome of all this cognitive activity is knowledge which is normally better adapted to the prevailing circumstances than any of the competing variants. If the circumstances change, the selective pressures on knowledge will change. Furthermore, this book has sought to show that knowledge is comparable to evolution in that it is a multi-level process. Natural selection processes occur at the levels of the underlying chemistry, of individual organisms, of the survival of whole species, and of whole ecosystems. The current state of life on Earth is the result of all these processes.

However, analogies of the growth of knowledge with the evolution of species are not always helpful. For example, in evolution (before scientific man), selection occurs at many levels, while inheritable variations of organic forms are generated at the chemical level with scarcely any co-ordinated involvement of higher levels. That is why evolution has been Darwinian rather than Lamarckian up till now – changes which occur to whole organisms are not inherited by their offspring.[2] In contrast, I have argued that variation of knowledge can occur at every cognitive level, from perception to the processes of whole societies. The growth of knowledge can be Lamarckian as easily as it can be Darwinian.

Perhaps, however, analogies between the evolution of species and the growth of knowledge are sometimes conceived too narrowly. For example, the emphasis on selection mechanisms in evolutionary epistemology places great emphasis upon competition between individuals within a species. The discussion would benefit by supplementation with another metaphor, that of *the ecology of knowledge*. Perhaps knowledge is more analogous to interdependent species in an ecosystem. It may be illusory to seek a single unified form of knowledge that applies to all situations. Knowledge is constructed out of representations of particular situations. But situations result in part from the actions of other people who have constructed rather different forms of knowledge based on diverse values held within different perspectives. So there may be as many forms of knowledge as there are distinct cognitive situations. Science would then take the form of an ecology of knowledge, just as life takes many forms adapted to the diversity of ecological niches produced by the interaction of species.

Such an ecology of knowledge could even be seen as a prescriptive epistemology. A rich ecology, with great diversity of species, is generally argued to be better able to adapt than an impoverished ecology to extreme external pressures. Knowledge grows in quantity and in usefulness by feeding upon its own diversity. Since we would like science to provide the resources for action in all future contingencies, we should also desire a rich ecology of knowledge. If, in the twenty-first century, the human condition should become increasingly diverse, the metaphor of an ecology of knowledge may become increasingly appropriate.

Although narrowly conceived analogies between the growth of knowledge and the evolution of species are interesting but unreliable, a strong case can be made for using an evolutionary framework for the long-term understanding of science. First, however we must broaden our understanding of the evolutionary process. In essence, evolution does not merely involve natural selection of the packages of DNA which generate individual organisms. Evolution in the broad sense produced humans, for example. Humans developed means of shaping their environment into more congenial forms rather

---

[2] This is before the emergence of the genetic engineer, who is becoming able to modify genetic material as a result of the experience of whole organisms.

than merely being selected by environmental pressures. Among humans, evolution in the broad sense has now led to genetic engineers. Genetic engineers apply the techniques of knowledge construction (developed for the control of the environment) to the new challenge of re-engineering life. They are at once a fundamental product of the evolutionary process and a symbol of the current directions of scientific change. Evolution does not stop with the emergence of genetic engineers, it merely takes additional forms.

A more general understanding of evolution, such as this, must appreciate that *emergent structures* can appear which have their origin in earlier evolutionary forms but may not function in exactly the same way. As a result, the detailed pattern of evolution changes. Emergent structures have new properties which were not revealed by previous arrangements of their component parts.

I follow the tradition of evolutionary philosophy which sees the symbolic thought of human beings as such an emergent manifestation. I will shortly describe a fairly well understood emergent structure, the immune system, which can provide a model for discussion. But first, it will be helpful to introduce a brief discussion on evolution as adaptation to change.

Environmental change One of the challenges which evolving life faces is coping with change. The theory of evolution is simultaneously a theory of how organisms come to take a form well suited to the challenges of a particular environment and a theory of how organisms change as the environment changes. The evolutionary selective principles generated by adaptation to the present environment and to one that changes can be in tension. An optimally adapted individual may not be well adapted to new kinds of survival challenges. For example, in species in which one male defends a harem of females, greater size is of competitive advantage for males. But in some kinds of adversity, such as a sustained shortage of forage, it may be the smallest individuals which survive, because they need to find less food. The evolutionary trend to larger males can suddenly be upset by famine. If the only mechanism of evolutionary modification were the cumulation of presently favourable mutations, adaptation to rapidly changing circumstances would be slow, with the risk of extinction. In times of extreme change, those species which have suitable evolutionary mechanisms to allow survival in suddenly modified environments are less likely to become extinct. Any mechanism which maintains genetic diversity in a breeding population is likely to give such an advantage. This may be one reason for the widespread occurrence of sexuality in species. The gene pool is more effectively mixed up in sexual reproduction so that there are likely to be some individuals suited to changed circumstances, whatever the change. For example, new forms of parasites and diseases less often lead to extinction when there is sufficient genetic diversity in the target population for a proportion to survive. Bacteria, which do not have sexuality, are able to assimilate DNA as plasmids, which has similar advantages in rapid adaptation by drawing upon a wide gene pool.

In the history of evolution, changes in ecosystems occur so often that appropriate mechanisms to adapt to unexpected change help the species in which they emerge. In thinking about the manner in which species adapt to changing environments, thinking at the ecological level is very often helpful.

Analogous remarks apply to selective processes in the ecology of knowledge. Variant candidates for knowledge are selected with respect to a particular environment, but the selection process will change as the environment changes. Since the context of knowledge changes very fast, in part because modern knowledge itself generates change, modern forms of knowledge have to be able to cope with changing circumstances. For example, partly verbalised craft knowledge, even if well adapted within a stable tradition, adapts less well to changing circumstances than more abstract scientific understanding. As another example, forms of knowledge (such as more formalised knowledge) which can spread easily between local social contexts sometimes adjust well to historical changes in society. Perhaps, analogously, species which are adapted to the problems of migration to slightly different regions, sometimes have survival advantage when some widespread external pressure changes all those local environments.

So, like life, the forms of knowledge which are only adapted to the prevailing environment may become extinct more quickly than forms of knowledge which carry within themselves resources to cope with change.

It is worth looking in a little more detail at an example of an evolutionary mechanism for coping with change – the immune system.

The immune system as an emergent evolutionary structure One especially regular source of environmental challenge is the appearance of new forms of disease. Because pathogens are small organisms with short-lived generations, they can mutate rapidly on the generational timescale of their larger hosts. Host species might cope by maintaining enough genetic diversity for some individuals to survive every new form of disease. But, long ago, a more powerful defence against new forms of disease emerged by the cumulation of its component features under natural selection – the immune system. The immune system of modern species is a sophisticated and complex series of defences against disease. It can be regarded as having emergent properties because it can 'remember' new organisms to which the individual has been exposed in its lifetime, producing powerful and effective responses to them on subsequent exposure. Such 'learned' defences are added to the inherited defences against invasive organisms. One aspect of the immune system is that an immense diversity of antibodies are circulating through the body, ready to identify invading organisms, including ones to which that individual has never been exposed. Once identified, other parts of the immune system attack and destroy the invader. The diversity of antibodies is produced by mechanisms in which a limited number of molecular fragments, coded by DNA, are combined in pairs on the arms of the antibody molecule to produce a much larger number of antibody forms. To make this system effective, there are also mechanisms to increase the numbers of any antibody which actually finds an invader to which it is matched, as well as mechanisms which suppress immune reaction to the body's own tissues.

The immune system gradually increased its ability to cope with changes in invading organisms. However, once it had emerged, it changed the survival prospects for the species involved. Now larger species could more often hold their own against rapidly mutating disease organisms.

The idea that evolution can produce new structures which are sufficiently superior to the sum of their parts to open up new possibilities, is referred to as *emergence*. It is especially important in the process of evolution, for it changes the conditions under which natural selection operates. We claim a number of emergent characteristics for our own species. For example, the human capacity for language is an emergent property of evolution, changing the conditions under which natural selection affects our species.

Evolution is a function of environments as well as of individual organisms
Seeing human language in evolutionary terms brings another feature of the broadened conception of evolution to our attention. We tend to pay too much attention to the individual organisms on which natural selection works, and not enough to the environment. This has distorted the popular view of life itself. Natural selection is a holistic process in the sense that an individual organism survives in a particular environment. The environment is as important in the equation as the organism. Very often, organisms survive and reproduce by changing their immediate environment to their advantage. It must be conceded that in some species each individual organism is able to tolerate a wide range of environmental variation. Many primitive organisms can survive in an inert form as spores or seeds under highly adverse conditions, growing and replicating only when the situation is favourable. In such cases, it seems that we are correct to pay attention to the organism rather

than to the environment. But most organisms need a highly congenial environment for at least some phases of their life-cycles. If the environment supplies what they need, they do not have to be so self-sufficient. Human beings need vitamins among their food intake, because they cannot make these essential materials for themselves. Viruses are interesting as life-forms because so much of the necessities of their lives are provided by the cells which they invade and within which they reproduce. Perhaps the extreme of dependence on environment is the prion theory of the spongiform encephaly family of nervous diseases, which include Scrapie in sheep, BSE (Bovine Spongiform Encephalitis) in cattle, and Creutzfeldt-Jakob disease and related human disorders. According to the prion theory, the infectious agent in this unpleasant illness is an abnormal protein which cumulates in the brain as it triggers nervous tissue to make more copies of itself, perhaps from a more useful form of the same molecule. The disease comes from the way the cellular environment reacts to the chemical; the life-like qualities of replication of the infectious entity are in the environment and not in what is transferred in infection. The abnormal prion molecule can be acquired by the organism from mutation of hereditary material or from the environment.[3]

In general, evolutionary success is equivalent to interactions of organisms and their immediate environment which bring about the perpetuation of the hereditary material of the organism and of whatever is required to generate or sustain environments conducive to the survival of descendant organisms.

The emergent quality of human symbolic thought is linked to how it has allowed new and more powerful ways for the language-using community to modify their environment so as to enhance their prospects for survival. Indeed, human evolution in the last few thousand years has had more to do with the interaction between this emergent system and the environment than with changes in the genetic makeup of human beings. We are now at the threshold of a transition in which this ability to modify our environment will enable us to modify our own genetic material, through genetic engineering. And, with the emergence of genetic engineering, the basic rules of evolution are undergoing one of their biggest changes ever. For now a whole organism can influence its own genetic form rather than evolve new forms only through accidental changes to its hereditary material. Evolution can at last be Lamarckian if we so wish.

The emergent system of human symbolic thought can be entirely understood as a product of evolution and as part of the evolutionary process. There is no need to invoke religious or transcendental factors in order to

---

[3] The prion theory is discussed in Parry (1983) and Prusiner (1992).

account for it. We are the product of evolution, that is to say, the way a physical system has been cumulatively transformed by the survival of variants suited to self-perpetuation. Where our earliest precursors produced behaviour as chemically triggered responses, our animal ancestors combined instinctive and learned behaviour. With our capacity for language, we have learned the trick of modelling situations in collective discussion and in thought, and learning what to do before we have to face the actual physical challenge. We developed human language out of the social signalling systems by which our precursors kept one another aware of their presence, their needs, and their emotional states. Like all animals, our remote ancestors were keenly aware of their surroundings. The emerging repertoire of verbal representations facilitated an enhanced mutual awareness. At some point, words came to be used as keys to memory, so that language could go beyond immediate awareness. At some later stage, meanings based on ordered combinations of words according to grammatical rules emerged, and then new powers became possible, such as abstract reasoning. At some time we learned the trick of being able to signal to ourselves, of listening to our own speech.

The human organism has evolved over the last hundred thousand years (or so) by selection for its capacity to sustain this emergent process. Those individuals who were good at language and other productive collective practices were more successful socially and reproductively. As with the immune system, the process of symbolic thought has led to principles of selection which did not apply previously in natural selection. In the competitive struggles between genes to affect the DNA carried by individuals, emergent systems change the rules. Among humans, it is unlikely that there is group selection of competing communities, as was argued earlier in this century. However, natural selection favours human individuals capable of benefiting from shared cultural resources in facing survival challenges. Language empowers societies to develop diverse belief systems, some of which enhance survival. Many belief systems contain within themselves resources for adjusting to new circumstances. Whatever happens, some such systems, and the individuals who sustain them, are likely to survive.

Within each society, what matters most in evolutionary terms is that in the diversity of human thought and human responses to crisis there are answers to be found somewhere to the problems of survival. Bacteria can often pick up the answer to a survival problem by assimilating DNA fragments as plasmids; similarly, humans can sometimes pick up the answer to their problem from the culture to which they have access.

It is, then, to the advantage of human societies in changing times not to regiment human thought too precisely. Modern society leaves a great deal of free space for individual thought, which can develop in manifold ways. Individual freedom of thought is never complete, but at least some forms of freedom are encouraged. Modern social institutions benefit from the products of weakly constrained individual thought. Science is one such outcome of the emergent evolutionary system of human symbolic thought.

To persuade the reader to go along with such a view (at least while reading), it is helpful to try to show how some of the more distinctive features of human life can be represented. I will consider intentionality and values before applying the view to science.

Intentionality explained in evolutionary terms There has long been philosophical debate on whether intentionality can explain the difference between physical processes and human thought and action. We are said to understand physical processes in terms of material substances being forced to act according to physical laws. To represent a physical process there need be no other reference to anything outside it. In contrast, human thought processes are intentional; in perceiving, judging, loving, or hating, we direct our thought to an object, normally an independently existing object, but sometimes not (as in thinking of unicorns, which do not exist). Similarly, the words we use *mean*. That is, they refer to something. On an evolutionary view, this feature of our language is to be understood as a natural development of evolutionary processes, but one which has made new possibilities for evolution. In essence, the contrast of the intentional and the physical is between two kinds of discourse which can be applied to the same underlying processes. One form of discourse is holistic and the other capable of piecemeal reduction.

Consider again the abnormal prion chemical that is speculated to cause Scrapie and BSE. The chemical does nothing but trigger certain changes in the brain cells of its host. These changes happen to increase the amount of the chemical and to disperse it sufficiently for it occasionally to infect other individuals. No separate part of this process appears to be intentionally producing this effect, and yet the overall process is of the kind that we normally speak of in terms of an infectious agent that produces disease. If the agent of a disease had been a living entity instead of a mere chemical, we might well have referred to the organism's intention. When mammals produce highly goal-directed sexual behaviour, the temptation to describe what they do in terms of their intentions is especially great because the analogy with ourselves is so clear. The point I am seeking to bring out is that an intentional process can be mapped into a purely physical one if we look at what is going on as a whole. In purely physical terms, the system is so structured that it tends to bring about changes important for its replication.

Any external interference which reduces this tendency is either immediately corrected or the system eventually fails to replicate. In intentional language we refer this process to that part of the replicating system which is identifiable by analogy with ourselves as the agent responsible. We then concentrate our attention on the agent.

In human affairs, we have learned how to avoid getting into trouble with such usage. In a purely physical description of a non-human biological organism there seems to be no room for the attribution of intentionality. In a normal account of a human agent, our reference is continually to intentions. However, the tension we feel between the two forms of speech is the result of not matching like with like. The physical description leaves out the context, while the intentional account presupposes it.

Intelligence A similar but less abstract argument can be given in terms of concepts like 'intelligence'. We think of intelligence as a quality of an agent, which might, with some plausibility, be attributed to higher animals. And yet the criteria we devise to measure intelligence can also be applied to purely physical systems, such as computers. In the case of computers, what we see as intelligent is in part the result of the processes carried out by the computer and in part due to social inputs from the programmer who sets up the instructions and the user who runs the program in an appropriate context. Some insect communities also appear intelligent, even though the individuals involved are acting like automata. For example, decision-making among bees, such as in deciding when to swarm and where the swarm should go, appears intelligent, depending on the communication of the direction and desirability of prospective new sites by the famed bee language of dance. No one bee makes the decision. The hive's intelligence is collective.

In human life, too, the individual is partly dependent on the social community, but also assists in collective processes by learning how to be a language-using group of one. The deliberations of each individual can, in suitable circumstances, enhance the collective intelligence of the community.

I have argued then, that mental terms like intentions, meanings, and intelligence are holistic – they do not just refer to an inner state of the organism, but also to a functional relationship between organism and context. In many cases, the primary context is society.

Values Modern philosophy has constructed a sharp distinction between fact and value. Facts concern what *is* the case; values relate to what *ought* to be the case. Philosophers tell us that we should strive to keep the distinction between fact and value clear, in particular, it is fallacious to argue from 'is' to 'ought'. In the Enlightenment tradition, sustained by positivism, science is, or should strive to be, value-free. However, this book has argued that if we think of science as the outcome of active processes, it cannot be value-free. There are at least the procedural values of seeking knowledge in a particular way. In the context of its use, scientific knowledge is also incontestably linked to values. When we try to evaluate science, we inevitably impose our values upon it.

Let us briefly consider values in general and in terms of the processes of evolution in particular.

If the values linked to an individual in a particular situation could be calculated, they would be complex functions of the values of other people, other values held by the same person, and even of themselves – in terms of our anticipations of their consequences. They would be analogous to, but more complex than, the principles of behaviour that evolutionary theorists extract from the application of games theory (Smith, 1982). Our society does not agree that values can be calculated from some naturalistic basis, but assumes individuals have free will and exhorts us to follow socially sanctioned general principles in our choice of values. If not everyone agrees on a value, then its involvement in any task leads to discordant results. Philosophers look perpetually for some transcendent system of values, so that we may freely come into consensus in our judgements and actions.

In the evolutionary approach offered here, the fact that we may develop values unconducive to our own survival, harmful to others in our gene pool, or even likely to lead to the extinction of the species, demonstrates that mankind is *not*, in the main, locked into instinctive patterns of behaviour. Where instincts are strong and unmistakable, we do not normally talk of values. It is, for example, increasingly hard, second by second, to stop breathing as a voluntary action. (If anyone succeeded, they would fall unconscious, and resume breathing.) As a result, it is not appropriate to refer to the urge to breathe as a value. However, in conditions in which breathing becomes difficult, the need to breathe can take priority over what normally takes our attention as a value. The basic priorities in evolution function as values in the manner that breathing does. The survival of ourselves and our gene line naturally take priority when they become an issue. Most of the time, however, our values are concerned with *how* we live rather than *whether* we live.

An evolutionary approach can be relevant to arguments over values. It can be used to criticise some naturalistic approaches to values, such as utilitarianism. Extrapolations from limited sets of evolutionary ideas have been the basis of systems of value. For example, evolutionary ethics put a value on the behaviour which maximises an individual's prospects of survival, while ecological ethics attempts to put a value on those human actions which do least damage to existing ecosystems. The values of ecological environmentalism diverge radically from those of evolutionary ethics. The values extracted from evolutionary theory therefore appear to be a function of the particular analysis.

One well-known system of naturalistic values is *utilitarianism*, according to which we value happiness and the avoidance of suffering, and should regard as good those actions which maximise the sum total of human happiness and minimise total human suffering. Utilitarianism can be criticised in evolutionary terms, for pain and happiness may be no more than internal systems for the generation and control of adaptive behaviour. In environments similar to those of our evolutionary origins, acting in ways which make one happy is a natural thing to do. Suppose, however, we were in a more artificial future environment in which people may be permanently cared for by machines while they gained intense pleasure whenever they wanted through non-addictive chemicals which acted directly on the brain. Utilitarianism would seem to recommend everyone encouraging others to put themselves into such a state of drug-induced euphoria. Not everyone is attracted to such a prospect. Most people would prefer a life style which engages and overcomes the challenges of external reality. Happiness, in biological terms, is better regarded as a means to an end rather than an end in itself.

In the late nineteenth century, Herbert Spencer and others sought to establish an evolutionary ethics, another naturalistic system of values founded on what is conducive to survival in the struggle for existence. Their critics, among whom was T. H. Huxley, complained that it is logically unsatisfactory to argue from what is to what ought to be. The value placed in evolutionary ethics upon fitness for the competitive struggle worked against the value of co-operative interdependence. By seeking to make the struggle for existence less harsh, by giving its members a higher quality of life, society imposes a new set of moral values in place of the initial natural values.

From a modern perspective, Spencer was, indeed, mistaken. The values he extracted from his evolutionary philosophy were actually taken from the way he formulated it on the model of *laissez-faire* Victorian society. There is evolutionary advantage for competitive aggression only in specific situations, typically where competing individuals are unlikely both to be seriously damaged. The evolutionary costs of unrestricted aggression have, in many species, generated forms of behaviour which avoid aggression, or ritualise it into a less dangerous form. Competive struggle is not always demanded of animal species. Even instinctive self-sacrifice can be selected for, provided that it occurs in ways which give greater benefit to other carriers of the relevant genes than it costs the animal which is driven to altruism.

One aspect of the problem of extracting values from the theory of evolution is that the selection of hereditary material is always in a particular environment. This implies that any values humans extract from behavioural strategies which give hereditary advantage, are likely to be a function of that environment. However, humans are forever reshaping their environment and employing values to do so. Therefore the evolutionary values extracted are not universal but are actually a function of those pre-existing values.

Evolution does not, then, provide a naturalistic basis for values. The present formulation of evolution, which emphasises the way we survive by changing our environment through our interactions and which concedes that genetic engineers are a product of the evolutionary process, certainly cannot extract the values of evolutionary ethics.

If a group of people were to decide that the point of human life is to end it all immediately, and acted on that value, then they would no longer around to bother the rest of us. Their value has been selected against. Perhaps such a selective process leads to some kind of minimal evolutionary value, shared by all undestroyed societies.

Is this value the avoidance of death? Well, no. Death is (currently) inevitable for all of us. Perhaps, then, the minimal evolutionary value is the avoidance of extinction – the complete loss of the genetic material and associated culture which sustains life. Such a value would have to include the survival of knowledge as well as of our DNA. The knowledge we employ in adapting our environment to suit our needs is a part of what is at issue in survival. At some point, in the near future, that knowledge might include knowledge of the human genome. We would only become finally extinct if all such knowledge was irrecoverably lost.

We play our culture-bound games and they keep us busy. We can keep the proposed underlying evolutionary value sharp by modelling the big challenges of survival in our games. But we are not forced to. Even this value can be accepted or rejected.

There are, then, no *absolute* values in the evolutionary frame. Evolution is not going anywhere. We are not obliged to avoid complete extinction. All that evolutionism has to offer is a partial account of what it is for a human to have a value. Such values as the goals we seek and the means we employ to achieve them are an emergent feature of human life.

Science in an evolutionary framework In traditional ways of judging science, we would assess it in terms of our values. Within an evolutionist framework, however, this does not achieve final answers. In the short term, we judge it against currently held values. In the long term we do not know if those values will continue to hold and have no absolute values.

In evolutionary terms, science is an emergent characteristic of the emergent system of human symbolic thought. In the brief discussion which ends this section, I wish to emphasise two points. (a) Science changes our sense of what it is that is carried forward in evolution. (b) Science can reflexively change the reality we seek to understand. I will then go on to apply the evolutionary framework in the next section.

(a) Throughout most of the history of life on Earth, DNA (or sometimes, its complement, RNA) has been the primary substance that is carried forward to the next generation in evolution. But, with science, it would seem that knowledge can also function this way. Sequences of DNA coding can be stored electronically or in print as indefinitely long words formed out of just the four letters A, C, G, and T. The original DNA molecules can be reconstructed from the code. In this way, information is made interchangeable with DNA. New processes can go on in evolution in which key steps involve science.

(b) Because science can be reflexive, our self-understanding can now not merely change our environment but ourselves. Sciences which put the agent back into the picture instead of hiding him or her in the frame now find that reality has no limit to its complexity. We are coming to be able to use our partial comprehension of reality to change ourselves as well as our surroundings.

11.4 The evolutionary framework applied

I will now show how this view of the place of science within a generalised account of evolution may be applied to questions about science in the long term. I will consider first the question of 'progress vs adaptation', secondly 'long-term hindsight', and thirdly 'facing up to the future'.

Progress vs evolutionary adaptation as views of long-term scientific change

In the 1960s and 1970s, there was some discussion of whether the long-term process of scientific change was best regarded as progressive or evolutionary. The Enlightenment vision of progress supposed that there was cumulation of knowledge as judged by some absolute standard. The evolutionary view looked back over the past of science and noted that the goals we set science change, particularly with modification of the wider social situation. Early aims for science were often abandoned as unworkable, and new aims, which showed the present to be superior to the past were clearly unfair to apply retrospectively, as earlier scientists were not aware of the later agenda. In the evolutionary view, we cannot pronounce for the future, for we do not know what future goals will be, and our present goals may be rejected. Progress, on such a view is an illusion. Let us consider this issue briefly, applying the evolutionary framework developed here.

If we are to evaluate science in the long term, we find that the lack of a settled system of values makes our task extremely difficult. We cannot say that science progresses towards any specific goal if the goals which are set at different cognitive levels, in different geographical regions, and at different times are not all the same. In the more Christian contexts of the seventeenth and eighteenth centuries, natural philosophy was progressing by cumulative discovery of design, revealing the power, wisdom, and benevolence of the Creator. In the more professional context of the late nineteenth century, science was progressing by discovery of ever greater complexity, which could best be handled by increasing specialisation. In the more technologically oriented late twentieth century, science progresses by cumulating competitive commercial advantage to the economies that invest in it. Progress in each of these senses is to some extent at the expense of the others, for they involve values in partial tension.

One answer to this problem is to find some more minimal sense of progress, which transcends such variation of value. Collingwood (1961, p. 329) offered a criterion based on the idea that, although we cannot see into the future, we can compare the present with the past in terms of an analysis of problems and their solutions. If, for example, present science solves without loss the problems that were solved by past science and in addition solves new or previously unsolved problems, then present science has progressed beyond the past science. Problem solutions cumulate.

One problem with the application of this criterion is that the notion of occasional scientific revolutions, popularised by Kuhn, involves the view that some old problem solutions are now abandoned, and this loss must be set against the gain of new problem solutions. Those who commit themselves to a revolutionary paradigm judge that the new problem solutions are to be preferred to the abandoned problem solutions. In the transition from phlogiston theory to oxygen theory in late eighteenth-century chemistry, for example, the question of weight relations in chemical reactions was dealt with far more satisfactorily by the new theory, which did not have to attribute negative weight to anything. However, the fact that so many combustible substances, like metals, are dark, shiny, and conductors of heat, which had previously been explained by their all containing phlogiston, now became unexplained (Kuhn, 1970a).

In later discussion (Kuhn 1977, chapter 13), Kuhn expressed the view that, even in scientific revolutions, the theories preferred by scientists are superior to the old ones in terms of their accuracy, consistency, scope, simplicity, and fruitfulness. These criteria, according to Kuhn, are not wholly objective, for their interpretation is affected by the context of their application. Theory choice in the context of a conceptual revolution involves a holistic balancing of procedural values in the context of the values of wider society.

In post-Kuhnian science studies, then, the Collingwood answer to the problem of the progress of knowledge is no longer viable. For we see that not merely must we balance new problem solutions against abandoned old problem solutions, but that the balancing process involves co-ordinating a complex set of procedural values, themselves informed by the context in which they are applied.

For reasons such as these, the fashion in the long-term evaluation of science turned in the 1970s towards the established alternative of evolutionary adaptation. Science changes to fit the society (and the factions within society) in which it is embedded. As the society changes, so does the direction of scientific change. This view was often linked to social relativism. Every society, if it ever resolved its internal conflicts, would eventually get the science it deserves. There is no more to it than that.

And yet that answer is not universally satisfying. In this sense, every society has the analogue of Western science, and to the extent that its science has stabilised it has the science it wants. There is no fundamental difference between the witchcraft and magic of the Zande and Western science and technology, except that they are the cultural products of different societies. But surely, Western science is more distinctive than this?

The view offered here is that science should be understood within an evolutionary framework. It is a resource, one significance of which is that it can aid in avoiding extinction. Suppose we see it as our task to maximise this resource. We should then expand the horizons of conceptual possibility, using every trick we know. In the process, we will discover new possibilities of our own errors; that is not an indication of failure but a sign of the growth of our capacities. We should enrich our capacity for practical action and know enough about what we are doing to see its potential and its risks. We should try to produce reliable knowledge where we can, even though it is an uphill struggle. And we should try to find ways to manage the growing complexity of our conceptual universe. We should simplify, we should judge what is better and what is worse. We should strive to make knowledge manageable.

On such a view of science as a resource, could it develop by adapting to conditions set in each age and still be progressive?

There *is* a sense of cumulation in evolution. Traces of our entire evolutionary history are carried in our DNA. Although evolution is full of losses through extinctions, all living species have a continuous history of successful adaptations since the origin of life, each new development growing out of the resources provided by the existing DNA of its precursors. When we extend the concept of genetic material to incorporate scientific knowledge, as we must, the notion of cumulation is even clearer (on human timescales). For we are learning how to conserve genetic material which would otherwise have disappeared.

Although genetic material is partially preserved in its evolutionary descendants, the environments to which it is adapted are inevitably transient. It is difficult for us to reconstruct the past in the period of recorded history; our knowledge of the whole of prehistory is fragmentary. Our engagement with concerns of the present makes the past less accessible to us.

On an evolutionary view, then, only part of what is going on is cumulative – the genetic material – while another part, the past environments, is forever being lost.

We cannot say science is progressive in an evolutionary framework until we have filled in the rest of the equation – until we have recovered past environments. We need to model the past in the present, to reconstruct as fully as possible a past situation so that we can test our ideas against it as well as we can test our ideas in the present. In the ideal state, when we have complete knowledge of each past period, we can say that science is progressive, for then all the achievements of the past will still be available to us. We can use them as resources when appropriate and not use them when they are of no help. Under these ideal circumstances, there is progress within an evolutionary framework, because there is no significant loss of the past. Since we would clog our cognitions with the mass of detail in such an ideal of a completely reconstructed history, we may simplify our knowledge of the past in more schematic models.

In summary, science does not demonstrably progress by the cumulation of certainty, for, as we learn to make knowledge more reliable, we also extend the limits of conceptual possibility and so appreciate more possibilities of error. Science does not demonstrably progress by the cumulation of information, for as we add to what we know we also abandon some of what we thought we knew in the past. Science does cumulate the capacity for action, but not for the best action, because the resources it provides for our construction of answers are never complete, and, as the horizons of possibility expand, it becomes more complicated to sort our answers out. Furthermore, with an increased capacity for action, we tend to make our immediate environment more complex, so that the old answers become inadequate. Finally, as long as the world (including the social world) can change in ways which we did not fully anticipate and which are beyond our direct control, science, too, must change or the life-forms it sustains risk extinction.

Scientific progress, then, lies more in enhancing our capacities for thought and for action than in providing any foundation of certainty. A world without science might be cosier. But like geese being fattened for Christmas, we might be short-sighted in our complacency.

Long-term hindsight

In evolutionary terms, humans have the emergent capacity to understand and control environmental factors relevant to their continued existence rather than merely to react to them as most animals do. Science is a more powerful resource adding to this capacity. The power is the greater if we can take account of the experience of past generations as effectively as we do the present one. The culture of each past generation has directed human activity in different ways which historians seek to reconstruct and represent to their own generation.

If we are to be able to think about science over longer timescales, we must have adequate ways of representing the thought of other times within the thought of our own. We must allow for changes both in science and in the society within which it was embedded. We cannot learn easily from what happened in the past because the original circumstances no longer apply.

Consider what is required in the judgement that some aspect of modern science is superior to the corresponding science of an earlier period. If science is a growing resource, then a hypothetical minor modern physicist, Milly, who has specialised in optics, should know more about the subject than Isaac Newton did in his prime in the seventeenth century. No doubt Newton was an exceptionally clever person, and likely to be quicker on the uptake than our modern Milly. No doubt, also, there will be lots of subtle conceptual points about seventeenth-century optics, particularly in its metaphysical aspects, that Milly will be very weak on. But Milly has access to vastly more experimental effects, and has a richer repertoire of rigorous ways of explaining them than Newton. Milly's life in optical research could well be much richer than Newton's, for her culture is more specialised. If she wants to produce phenomena which it would be difficult for Newton to explain (for a while at least), then it is far easier for her than it would be for Newton to give her comparable challenges.

The impression that modern Milly knows more about optics than Isaac Newton will be even more plausible if Milly has made a hobby of the history of seventeenth-century optics. If she has learned about Newton's problems, methods, and solutions in optics, and can see what he actually managed to accomplish, then she will have a great advantage in constructing accounts of the weakness of Newton's optics in comparison with her own. She might even be able to see where her own optics looks weaker than Newton's in seventeenth-century terms, and set about giving an account of why the apparent seventeenth-century achievements were abandoned. Since Newton is long dead, he cannot study Milly's science of optics comparably.

Our judgements about long-term change in science require, then, an ability to recover the past in such a way that we can compare it with the present without distortion or loss in the historical process. It would not do, for example, simply to study records of the past with a modern eye and try to fit them into modern understanding. For that is to turn them into what they are not, attempts to produce knowledge suited to our own time, which was then the remote future. Present-day historiography of science attempts to reconstruct the past without appeal to any anachronism, that is, without explanatory reference to anything that happened later. An evolutionary historiography suitable for long-term hindsight will have to go further. Our task is to recover a representation of the past that is uncontaminated by anachronism, for use in the present. We need to be able to make comparative judgements between the present and representations of the past in the present. It requires the deployment of the skill of thinking in multiple frameworks which I have referred to several times as an example of informal rationality. In this way, we can hope to build up understanding of the past and of its relationship to the present. Our task is not merely holistic with respect to each period, but also reflexive with respect to our own period.

Within such an enterprise, long-term hindsight becomes achievable. If we collapse the past into the present, or lose the present in our search for objectivity about the past, then it is not.

One criticism that might be offered of this approach is that human knowledge is unlike earlier evolutionary forms in that it is done with some degree of self-understanding. We have learned to extend our animal awareness of our surroundings into self-awareness. The actions of past scientists, like other people, is not merely affected by their awareness of their social environment, but also by their image of how they relate to that environment. And, if they have an image of the future effects of their actions, that will affect what they do. Our historical understanding must also appreciate how people in the past related themselves to their future. Perhaps we, too, should think about our future.

Facing up to the future

The future of the domain of human action is not predictable even in the terms of deterministic science, for one causal factor is what we will do, which in its turn depends on what we think the future will be like. We cannot make firm predictions until after the consequences of our firm predictions have been worked out, which is only possible if we are sure that the prediction will make no difference. Prediction is further complicated by chance events whose origin lies beyond the horizons of our knowledge. The manner in which our own capacities grow (which we cannot know before it has happened), the way our own actions produce consequences in environments of our own making, and the way emergent new possibilities appear unexpectedly will perturb accounts of the future. Any knowledge we have of the future is uncertain knowledge.

Projections of the recent historical past suggest a future that will not be like the past, for we have constructed a society which contains an increasing number of multipliers of change which generate further unplanned changes. For example, our economic system hurries the development and introduction of new science-based technologies. Science and technology feed off themselves and off one another with no presently known limits. Our society will have to change drastically before it can remain the same.

In spite of these comments on the limitations of prediction, it is tempting to make tentative suggestions about the future. I anticipate that, in the near future, the problems of our understanding of our environment will not usually be because the impersonal world unexpectedly changes the conditions for human survival; but because we have to deal with crises of our own making. Unless we are suddenly threatened by unsuspected layers of complexity in wider reality, our problems will come from the ill-understood consequences of what we do. The knowledge we seek will be intimately bound to the social values which guided reality modification. We will have to cope with conflicts of value as well as lack of information.

The image I offer of the next century is of a world in which the problems of coping with change have become especially urgent. In earlier ages, we tried to understand the world in static and cyclic patterns. Now we are aware that the world is undergoing changes that may be sustained indefinitely and unpredictably, largely as a result of our own actions.

Evolutionary philosophy is a suitable framework for such a situation, for it recognises the difficulties in predicting future change, especially now that the original genetic mechanisms are supplemented by environmental knowledge and by genetic engineers.

Among the resources we have for facing the future are what we have gleaned from the past – not merely past knowledge and past genetic material, but the relationship of both to past environments. Our present competence is a function of this historical resource. The evolutionary framework suggests that, if we value survival, some strategies are riskier than others. We might develop strategies, making use of our knowledge of the past, which can reduce the more extreme consequences of taking risks.

What, then, of future science? Science is at present a powerful multiplier of social change. It helps destabilise society by producing changes with unforeseen and often undesired long-term consequences. It is also part of the solution to this problem, for scientific knowledge is one of the most valuable resources for adapting to changing circumstances. As I argued in the introductory chapter, it is too late to give up science, for the problems of present-day society will destroy us if we do not work out how to solve them.

Perhaps, the need to cope with future problems requires changing society even more than changing science. A widely expressed ideal of present-day capitalism is to concentrate funds for knowledge creation upon that which can be turned to immediate competitive advantage. This would appear to make science even more inextricably part of the problem which it is needed in solving. We should separate the growth of knowledge, regarded as resources for solutions to our problems, from knowledge viewed as the source of new problems. Can we construct a society in which dangerous ideas can be contemplated without irresponsibly being put into action, or must we suppress the dangerous ideas themselves?

A central symbol in this chapter has been the genetic engineer, who creates knowledge of how to re-engineer life. Today we are learning to read our genetic code as science. One day, we will be able to put information, including the then current science, into each individual's genetic code, together with the means to read it internally as a new form of innate personal knowledge. Then the gap between genetics and scientific information will have been closed completely. The genetic engineer is at once a fundamental part of the evolutionary process and an indicator of the current directions of scientific change. (S)he is also becoming a horrific image in present-day society, a modern-day Frankenstein, the realisation of science-fiction fears. The idea of genetic engineering raises a myriad of moral worries, fears of the unknown, of dangers we heedlessly bring upon ourselves.

I have just suggested that we should reverse the trend to integrate the construction and use of knowledge. This applies to genetics. Knowledge of genetic engineering, held responsibly, may give us the capacity to deal with problems that would otherwise overpower us.[4] We need ways to expand knowledge without putting it to immediate misuse. If we had such ways, there would be no great difficulty in seeing how to cut a pathway through the new moral jungle raised by genetic engineering. Those who have genes predisposing them or their children to dreadful disorders are bound to plead for the implementation of measures that might cure or prevent that disorder. Little by little, genetic knowledge will come to be used, building on what enough people want and others are not prepared to deny them. The future of the genetically engineered self may not be just round the corner, but it may not go away either. It is a suitable metaphor for the future of science.

11.5 Summary

In modern society, the readily foreseeable future does not extend far ahead, but some of our problems are best understood on a longer timescale. Traditional absolutist frameworks, such as the Christian and the Enlightenment views, are no longer so easy to apply to science, especially in the long term. The evolutionary framework, suitably generalised, is more suitable. We must appreciate, however, that, at all levels in evolution, the products of the genetic code tend to change the environment in which they are selected. Mankind, following a strategy of self-consciously modifying its surroundings, has produced a situation in which information about the environment is as important as the genetic code. The precise rules of natural selection change each time a new emergent system appears, making outcomes of evolution less predictable. Human language, science, and genetic engineering, are all emergent systems which rapidly transform the prevailing pattern of evolution. With the genetic engineer, scientific information about the environment and the genetic code about the organism are becoming interchangeable.

Many problems of the future may well be of our own making. Science can help solve such problems, especially if we can find ways to practice it so that it does not generate changes to society which produce even worse problems.

---

[4] Imagine, if you like, in an extrapolation of a present-day worry, a future in which persistent and universally distributed pollutants which act as weak oestrogens are found to have already affected the development of male children so that they will be unable to produce viable sperm in adulthood. Knowledge of the chemical basis of genetics could then make all the difference.

Appendix: Summary of cases of marginal and disputed science

Alchemy

An ancient art. It thrived in Alexandria in Hellenistic times, continued in Arab culture, and came to Europe in the fourteenth and fifteenth centuries. Its primary concern was with the correspondence between the macrocosm and the microcosm, so that knowledge of the transformations of material substances could uncover universal secrets. It became most famous for the search for the transmutations of base metals into gold and for the alkahest, or universal solvent. Its application to medicine (iatrochemistry) thrived in the sixteenth century. Its ideas were carried on after the seventeenth century especially in mystical cults, and its knowledge of chemical manipulation was a direct precursor of chemistry. Isaac Newton was among those in the seventeenth century who studied alchemical manuscripts and did experiments in the attempt to uncover secrets of the tradition (Dobbs, 1975, 1991). Its critics pointed out that its chief insights were expressed in ambiguous allegorical language, so that later alchemists were never quite sure what their precursors had actually achieved.[1]

Anti-fluoridation

In the 1940s, American public health authorities judged that naturally occurring fluorides in local water supplies may be reducing dental decay. A test was set up in which fluoride was artificially added to a town's water supply. It was so obviously successful that untreated towns being used as 'controls' soon adopted the same water treatment, bringing the test to a premature end. As the practice of fluoridation spread in the 1950s and 1960s, it was

---

[1] For historical introductions, see Burland (1967); Caron & Hutin (1961); Holmyard (1968); Raed (1966); Waite (1926).

found that those water authorities who held a local referendum on fluoridation often lost. Right-wing political movements were helping build up organised opposition. Those in favour saw it as a cost-effective public health
measure to supplement levels of this 'natural' ingredient of water to about
1 part per million. Those opposed saw it as involuntary 'mass medication', in
which a highly poisonous chemical with unproved benefits was being given to
people at concentrations quite close to those known to be dangerous. When
asked to vote, a majority of the general public regarded fluoridation as a
potentially dangerous step into the unknown.[2]

Astrology

In ancient times, astrology was important in many cultures. The core of the
Western tradition came from Babylon to Greece, and later to Rome. A
systematic exposition and defence of astrology was given by the
Alexandrian, Claudius Ptolemy in the second century AD (Ptolemy, 1940).
Ptolemy also wrote treatises on mathematics, astronomy, and geography. His
summary of arguments of how the planets and stars influence civilisations
and individuals shows forms of thinking (the sympathetic action of like on
like) that would now be dismissed as belonging to magic.

In the sixteenth and seventeenth centuries, astrology underwent a resurgence, stimulated by the recovery of ancient astrological texts and the ready
dissemination of popular topics by printing. Much attention was given to
astrological portents of Christian eschatology. Astrological medicine thrived.

In the modern era, astrology continues to thrive in popular culture, many
people finding entertainment value in horoscopes and zodiacal signs, a small
proportion running their lives according to it. Eastern astrology flourishes.
Astrology has been controversial since Roman times, when predictions of the
death of emperors was not appreciated by the authorities and Christian
authorities frowned on its fatalistic implications. Modern science has been
very sceptical of the feasibility of any mechanism by which planetary positions could have the required effects, and has been scathing about the weakness of the evidence in its support.[3]

---

[2] For further reading, see, Crain et al. (1969); Gotzsche (1975); Martin (1988); Mausner (1955); Sapolsky (1969).
[3] For histories, see Allen (1966); Curry (1989); Garin (1983); Lindsay (1971); Naylor (1967); Tester (1987).
For discussions of modern astrology's scientific basis see, Eysenck & Nias (1984); Gauquelin, (1983);
Jerome (1977).

The Bermuda Triangle

In the 1970s, many books and articles appeared describing how an exceptionally large number of mysterious disappearances had occurred at sea and in the air in the triangle between Bermuda, Miami in Florida, and Puerto Rico, perhaps stretched to include Cuba. Writers kept adding to the number of cases and elaborating the old ones. A wide range of explanations, some very far-fetched, was offered to explain what was going on (e.g. Berlitz, 1975; Jeffrey, 1975; Winer, 1974). One of the more influential criticisms, by L. Kusche (1975), argued, by tracing cases back to their sources, that virtually all the disappearances had a natural and logical explanation which had been ignored by those trying to make entertaining reading out of the Bermuda Triangle mystery.

CETI (Communication with Extraterrestrial Intelligence)

Since ancient times, the possibility of life on other worlds has been a controversial extrapolation of our understanding. Its plausibility has fluctuated with changes in prevailing belief systems. It rose when we established that other planets in the solar system were worlds like our own, it declined when closer observation showed that they would be inhospitable to terrestrial life. In the late nineteenth century, in the wake of H. G. Wells, *War of the Worlds* (1898), popular culture dwelled upon the fear of alien invasion. Mid-twentieth-century science-fiction movies picked up the theme. More recently, the view has been taken that a civilisation stable enough to reach the stars could not be so warlike. Current judgements can be organised around an equation produced by Frank Drake. With an estimated 400 billion stars in our own galaxy, and a guess (based on current theories of star formation) that a tenth of them may have planets, one per system of which may be suitable for life analogous to that on Earth, there is a very high potential for life elsewhere in the galaxy. If we speculate that one in 10,000 such planets might go on to develop intelligent life capable of communicating between the stars, and, if they last long enough for a reasonable proportion to be communicating now, then it is worth looking for interstellar communications by electromagnetic radiation. This calculation is controversial because it compounds a number of plausible speculations into an extreme extrapolation of current thinking. The CETI project has no firm information on which to build except the fact of our own existence. The search for signals at suitable wavelengths requires significant levels of funding. However, if it were true, and we made contact, the impact on our culture would be sure to be dramatic. This is a project which needs funding *before* it has an adequate scientific basis. If successful, it would become a science. If unsuccessful, it would be money wasted, for we might have set about it in an inappropriate way. Perhaps we are unaware of the most suitable medium for interstellar signalling. The biggest difference between it and projects like parapsychology is that it is judged more plausible by the scientific orthodoxy.[4]

Cold fusion

In March 1989, M. Fleischmann and S. Pons of the University of Utah announced at a press conference that they had fused deuterium ions (a heavy isotope of hydrogen) at room temperature in a simple electrolytic cell in which the cathode was made of palladium (or one of the other metals known to absorb hydrogen at very high concentrations). The evidence for fusion was the detection of neutrons, gamma radiation, and a great deal of heat. The theory was that the deuterium nuclei were being forced so close together that the same nuclear reaction as goes on in a hydrogen bomb was occurring at low temperatures. The announcement was part of a race for priority with S. Jones of nearby Brigham Young University, who reported similar results the next day. The details of these highly unexpected results seemed strange, but the possibility that a prolific source of exploitable energy had been discovered by an experiment needing only an unusual combination of readily available resources to check out, encouraged many scientists to try. Soon reports of positive (and negative) results were coming in from labs across the world. The story developed so rapidly that normal methods of scientific publication were initially bypassed in favour of news on computer networks and in the press. But, by the end of 1989, major laboratories in the USA and northern Europe concluded that there was no convincing evidence for cold fusion.[5]

Creation science

In a reaction against the teaching of evolutionary theory as orthodox science, a movement among Fundamentalist Christians in the USA has built up, especially since the 1970s, which seeks to establish a scientific basis for a rival Creationist account of the history of life on Earth and to show that evolutionary theory is mere speculation (e.g. Gish, 1979, Morris, 1974). The

---

[4] For a popular introduction, see Asimov (1981). For histories, see Crowe (1986); Dick (1982); For introductions to the scientific issue, see Christian (ed.) (1976); McDonough (1987); Sagan (1973).
[5] For fuller accounts of the story, see Close (1992); Huizenga (1993); Mallove (1991); Morrison (1990).

intellectual position constructed is analogous to the main stream of Christian natural history of the eighteenth century, buttressed by critiques of many later scientific developments which have made such a view more difficult to defend (Dolby, 1987). There is superficial plausibility in its arguments (a) for a shortened timescale for life on Earth, (b) for no species to have emerged from any other by 'descent with modification', (c) for the human species to have coexisted with the other species of the fossil record, and (d) for a global catastrophic flood. But it is difficult to find anyone who is persuaded by the case made who is not already committed to a traditional literal reading of the account of creation in Genesis 1 in the Bible (or its equivalent in other religions).[6]

Däniken on historical visitation by extraterrestrials

Chariots of the Gods? (Däniken, 1969) was the first of an immensely successful series of books by Erich von Däniken. In them, he claimed that alien beings had visited the Earth 10,000 years ago and had created man by altering the genes of apes in their own image. They did not leave immediately and were subsequently worshipped as gods. Archaeological study reveals traces of their technology. Däniken's ideas were backed up with copious evidence. Most of it is more plausibly given an orthodox interpretation, and some has been found to be completely spurious. Däniken admitted to using a great deal of journalistic license. Däniken's success is widely thought to have been due to his marketing a product well suited to popular yearnings of the age (*Der Spiegel* Editors, 1973).

Evolution before Darwin

The idea that a progressive force might occur in nature was a widespread popular scientific speculation since the late eighteenth century. Its application to the progressive evolution of species was developed most fully by J.-B. Lamarck around the beginning of the nineteenth century. Lamarck accepted that species are continually evolving from simple to complex within the (separate) hierarchies of plant and of animal species. The main mechanism of change he discussed was that the intensive use of an organ, especially if accompanied by effort, leads to that organ being inherited in a more developed form – as giraffes stretching to reach the leaves at the tops of trees became long-necked. In disputes in the early nineteenth century, the orthodoxy, especially as represented by Cuvier in France and Lyell in Britain,

---

[6] For further reading, see Godfrey (1983); Kitcher (1982); La Follette (1983); Nelkin (1982); Ruse (1982); Ruse (1988).

judged that the idea of evolution was contrary to the general run of evidence. Nevertheless, some writers continued to speculate on evolution; most infamously, the anonymous *Vestiges of the Natural History of Creation* (Chambers, 1994, originally 1844). This work was labelled 'pseudo-science' by those reviewers who defended the scientific orthodoxy. With his *Origin of Species* (1968, originally 1859), Darwin and his early supporters initially had to distance the new scientific orthodoxy of evolution by natural selection from the older view of progressive evolution. In the later nineteenth century the two views temporarily blurred into one another.[7]

Eugenics

With the widespread acceptance in the late nineteenth century of the theory of evolution by natural selection, the question naturally arose of how evolution is currently affecting mankind in modern society, and of whether we might direct that process by selective breeding as we do with domesticated plant and animal species. Negative eugenics is the discouragement of breeding by the least fit members of society, and positive eugenics is the encouragement of breeding by those individuals with the most desirable qualities. The key pioneer of both the science of human inheritance and its political application was F. Galton (e.g. Galton, 1985, originally 1909). Eugenics gained wide support in the early twentieth century, especially in England, Germany, and America. Critics asked, 'Who has the right to decide what human characteristics are most and least desirable?' Their complaint was given added force by general repugnance for the racist application of eugenics in Nazi Germany. A modern form of negative eugenics flourishes in genetic counselling, which assumes that potential parents who carry deleterious genes have a right to make informed decisions about the risks of conception and whether to abort a foetus affected by a deleterious gene.[8]

Freud's psychoanalytic theory

Sigmund Freud developed the theory and practice of psychoanalysis in the 1890s and the decades following. He turned it into a general theory of mind after the turn of the century. At the heart of his theory was the idea that the instinctive sexual drive goes through important stages in early childhood, and that many circumstances can perturb its natural development, to produce

---

[7] For an introduction to the history, see for example, Gillespie (1951); Millhauser (1959).
[8] For an introductory reading on the history of eugenics, see, Nlacker (1950; Chesterton (1922); Kevles (1986); MacKenzie (1976); Searle (1976); Farrell (1979).

neuroses. Psychoanalysis as a therapy seeks to bring disturbing past experiences to consciousness, so reintegrating the fragmenting personality of the patient. Freud made this into a general theory of mind. We are not fully conscious of the instinctive drives of the id. Evidence for their existence in all of us is given by the way dreams express in disguised form the fulfilment of the wishes the id generates (most famously the Oedipus complex), and in the hidden purposes revealed by the mistakes we make in everyday life. The ego is the conscious self, operating independently of others. The superego is the manner in which the ego internalises socially required self-control, especially as learned from parents in early infancy. Freud's theory was heavily criticised, (a) because it was not exposed to the kinds of empirical test normally required of a science, (b) because psychoanalysis was never shown to work better than non-psychoanalytic therapies or than natural recovery, and (c) because it was not adequately integrated into academic psychology, but flourished in the less demanding milieu of popular culture at the fringes of medicine. Freud's theory is discussed in chapter 8 in the context of Popper's claim of its unfalsifiability.[9]

Geller and psychic metal-bending

In the mid-1970s, a young Israeli, Uri Geller, became an international media phenomenon by apparently displaying a psychic ability to bend metal objects such as spoons and forks in front of witnesses and even while being filmed. His powers were examined scientifically, most famously his telepathy at the Stanford Research Institute in 1972 (Targ and Puthoff, 1974) and his metal bending in colleges of London University in 1974 (Taylor, 1975). Some members of his broadcast audiences would suddenly find that they had psychic metal-bending powers as well (e.g. Hastead 1981; Collins and Pinch, 1982). Geller provoked controversy, in part because he persuaded many people, including a few scientists, that his powers were genuine, while others, such as the conjuror, James Randi (Randi, 1978) insisted that conjurers could easily duplicate the tricks he claimed to be psychic powers. Randi pointed out his use of standard techniques of misdirection and his insistence on the presence of a close associate, Shipi Shtrang. Attempts to set up a test of metal-bending powers that was proof against trickery never gave results satisfactory to all. Geller was occasionally said to have been caught cheating but not blatantly enough for a conclusive exposé. Just as with the physical mediums

[9] For an introduction to the discussion of Freud's theory as science, see, for example, Cioffi (1970); Eysenck (1985); Eysenck & Wilson (1973); Grünbaum (1993); Lambert et al. (1986); Masson (1984); Rachman (1963); Sulloway (1991).

of the nineteenth century, a good time was had by all, but nothing much was achieved.

Goethe's theory of colour

The eminent German playwright and poet of the late eighteenth and early nineteenth century, J. W. Goethe, also made contributions to science. One of these was an attempt to overthrow the orthodox Newtonian account of how normal white light is actually compounded out of the coloured rays of the spectrum (Goethe, 1987, originally 1810). Goethe tried using a prism for himself, and, seeing rainbow hues only at the edges of objects, concluded that Newton's theory, as he vaguely remembered it, was wrong. Newton, he went on to argue, had been misled by the excessive abstraction of mathematical forms of thinking. His own theory assumed the primacy of natural human sensation, as in the doctrine that white is a simple sensation and cannot be compounded out of colours. He claimed that colours are produced by the interaction of light and dark, as dark mountains appear blue in the distance when seen through stray white light in the air, and the setting sun appears red when seen through the relatively dark sky. Goethe's powers of literary expression gained him many followers, but Newtonian scientists regarded it as a good example of the self-delusion of an individual made self-confident by the popularity of his literary works, who would have benefited from a judicious guide (T. Young, 1814).[10]

Intelligence testing

The general idea that it would be useful to measure the intelligence of individuals emerged with the rise of the meritocracy in modern society. The issue was sharply focussed by Francis Galton in Victorian times as an aspect of eugenics, but the first satisfactory test, by Alfred Binet (Binet and Simon, 1916, originally 1905), used a scale of normal child development in the capacity to answer simple graded problems as a standard against which to measure a child's ability. This was the basis of the IQ test. It soon stimulated a major industry of mental testing, especially in America. Critics of IQ questioned whether rapidly giving answers to simple problems of language, mathematics, and spatial judgement is a good measure of intelligence, whether measurements of IQ difference between social and racial groups

---

[10] For an historical introduction, see Helmholtz (1893); Magnus (1949); Nisbet (1972); Sepper (1988); Wells (1967).

shows anything about innate biological differences, and whether the effects of labelling a child as clever or stupid on the basis of a test result might not be self-fulfilling as the child learns to live up to the label.[11]

Kammerer and the mid-wife toad

Paul Kammerer was a Vienna-based biologist who claimed early in this century, controversially, to have produced experimental evidence of the Lamarckian (non-Darwinian) genetic mechanism of the inheritance of acquired characteristics. He worked on sea squirts, salamanders, and mid-wife toads. For example, the mid-wife toad is land-dwelling and the male lacks the dark and rough nuptial pads on the knees of similar species that breed in water. Kammerer claimed to have bred nuptial pads into his (male) mid-wife toads by forcing them to live and mate in water. Other biologists could not check his findings, for they were not as skilled as he in keeping such animals in abnormal conditions. Kammerer's work was disrupted by the World War I and by subsequent Austrian inflation. His claims were vigorously attacked by William Bateson in 1913. Debate with English geneticists continued after the war. In 1926, it was found that the knee pads on a preserved specimen had been artificially darkened with ink. Kammerer denied all knowledge of the apparent fraud but soon after committed suicide. People drew their own conclusions. The case was the subject of a book by Arthur Koestler (Koestler, 1971), which claimed Kammerer was a misjudged martyr for Lamarckianism (Gould, 1972).

Lysenkoism

In Stalinist Soviet Union in the mid-twentieth century, T. D. Lysenko proclaimed a series of agricultural projects linked to a version of genetics which accepted that the environment changes the hereditary nature of organisms. His first project was 'vernalisation', in which winter wheat seeds were moistened and chilled before being planted in the spring, so that the plants were not exposed to hard winters, but were able to begin growing faster than spring wheat. Lysenko's practical ideas were heavily criticised by orthodox geneticists in the USSR. In the West, Lysenko was judged an ignorant charlatan who exploited to the full the backing given him by Stalin (e.g. Huxley, 1949). But Lysenko and his supporters argued that the doctrines were com-

---

[11] For an introductory reading on the history and criticism of IQ, see example, Block & Dworkin (1977); Evans & Waites (1981); Gould (1981); Kamin (1974); Marks (1981); Montagu (1975); Taylor (1980).

patible with the state ideology of Marxist dialectical materialism and met the practical needs of the state. Lyskenkoism was officially accepted in the USSR in 1948. After the death of Stalin it took several years before it was discredited there. The case is discussed in chapter 4.[12]

Martian canals

During the opposition of Mars in 1877, G. Schiaparelli observed a fine network of lines on the planet's surface. In spite of being a committed believer in life on other worlds, he presented his description with scientific caution. However, the neutral Italian term, 'canali' was translated into English as 'canal', implying that the lines were artificially constructed. Committed believers in the plurality of inhabited worlds soon built up an elaborate picture of Martian life in the irrigated strips on each side of the canals, which carried water from the Martian polar regions to the drier equatorial areas. The next occasion for good telescopic observation (when Mars was suitably close to the Earth) was in the opposition of 1892. Martian astronomy boomed, some observers seeing canals and others being unable to. They were not seen in the same places as Schiaparelli's, encouraging the speculation that we were seeing the changing pattern of seasonal crops growing along water channels. Schiaparelli later commented that the Mars discussions had attracted many charlatans. Photography was still not sensitive enough for astronomy, so that there was no way to eliminate the problems of human observation. However, by the end of the century, the idea gained currency that the lines seen on Mars were optical illusions, an artefact of trying to draw outlines of vague patches which are very close together (as in such a telescopic image). Although orthodox astronomy moved towards describing Mars as a dry and dusty planet with many transient marks that were changed by widespread storms, popular culture retained until the 1970s the late nineteenth-century stereotype of a dying planet kept alive by global irrigation schemes.[13]

Marxism

Karl Marx had a profound political influence in the late nineteenth and twentieth centuries. He also had great intellectual influence in the social sciences. His critique of capitalism set it up as a stage in history.

[12] For an introduction to the history, see Gaissinovitch (1980); Joravsky (1970); Lecourt (1977); Medvedev (1969).
[13] For an historical introduction, see Crowe, 1986.

Capitalism followed feudalism, which followed ancient slave-based societies. In its turn capitalism would be followed by communism. The driving force of these historical changes lay in the economic base of society, that is, the means and relations of production. In the case of capitalism, the exploitation of wage-earning workers by the capitalists, who owned the means of production, would inevitably increase as competitive pressures squeezed the profit margins of capitalists. With the rise of working-class consciousness of exploitation, revolution was increasingly likely. It was Marx's friend and patron Friedrich Engels, who most vigorously proclaimed Marx's views as science, as in his speech at the graveside of Marx in 1883. Subsequent events did not turn out as Marx had anticipated, partly because of the influence of his ideas, especially in the political movements he supported. Capitalism grew from the wealth gained from geographical expansion and greater efficiency of machinery in which much capital had been invested. Wages rose (union bargaining helped this) and workers also became consumers, helping sustain demand. Although the system continued to exploit workers in the underdeveloped world, a majority of workers in the West had too much to lose in revolution. Only a minority of people in the West were obviously unfairly exploited. (They are now referred to as the underclass.) In the meantime, the revolutionary end to Tsarist Russia led in 1917 to a new regime that called itself Marxist and communist. Under Stalin in particular, the Soviet Union became a totalitarian state of the most repressive kind. Its political message was exported to many less developed parts of the world, most successfully to China. Although the Soviet Union created a powerful military and industrial state machine, it became mired in its own bureaucracy and no economic competitor to Western capitalism. With the collapse of the Soviet Union, especially after 1989, the image of Marxism suddenly declined throughout the world.

As a science (and also as a political programme), Marxism has never presented a single unified image. Thinking Marxists have internalised their own version of how Marx would see the problems they wish to deal with. Marxism has a tendency to fall into factional divisions because modern problems now take a quite different form from those of the mid-nineteenth century, because Marxism is often learned in derivative form from a diversity of present-day movements, and because it invokes combinations of value on which people do not readily agree.

In my own opinion, Marxism should not be thought of as a science, but as a set of values developed into a plethora of rival frameworks, some of which can act as frameworks for social science. I suggest that Marxism does not provide authority for answers to our social, political, and economic problems; rather it is a resource for critical thinking about these matters, an irritant to the complacent assumptions of orthodoxy. Some Marxist-inspired challenges to intellectual orthodoxy have taken hold in the wider culture.[14]

Mesmerism

The precursor of hypnotism, Mesmerism was a medical treatment exploiting a universal magnetic fluid that the Austrian, F. A. Mesmer, claimed to have discovered. The therapy became a craze in Paris for a few years in the early 1780s. Mesmer claimed to be able to produce convulsions in the afflicted parts of his patients by drawing magnetic fluid through them with several passes of his hands. The technique quickly developed further, including, in 1784, the production by the Marquis de Puységur of a state of artificial sleep, somnambulism. However, criticisms also developed. The main commission of inquiry in 1784 could find no trace of the magnetic fluid and concluded that the effects were produced by what the patient imagined had happened. Interest in the controversial phenomenon continued to spread. In the mid-nineteenth century, it was proposed by J. Braid that what is now called hypnotism is a subjective phenomenon involving a special form of suggestibility readily produced by fixation of the attention on a single object.[15]

N-rays

In 1903, in the context of the discovery of many new forms of radiation, R. Blondlot 'discovered' N-rays, naming them after his university town, Nancy. The rays were analogous to X-rays, with a spectrum of frequencies. They could be detected by the way they made an electrical spark or a phosphorescent patch brighten, or by the way they increased the sensitivity of the eyes in dim light. Many objects, including healthy living tissue, were found to emit N-rays, while lead and water absorb them. Many tens of publications on the new radiation appeared, virtually all from researchers at Nancy, or researchers who had been there to learn how to detect them. Other scientists claimed priority over Blondlot. The French Academy of Sciences prepared to give a very substantial prize to Blondlot. But scepticism also grew, especially among people outside France. R. W. Wood, on a visit to Blondlot in Nancy in 1904,

[14] For an introduction to the issue of Marxism as science, see, Cornforth (1968); Colletti (1972); Conze (1935); Popper (1960); Popper (1966).
[15] For an introductory reading on the history of Mesmerism, see, Binet and Féré (1905); Crabtree (1993); Darnton (1969); Ellenberger (1994, originally 1970); James (1950, originally 1890), vol. 2, chapter 27; Owen (1971); Walmsley (1967).

took advantage of the darkness of the laboratory temporarily to remove the prism producing the N-ray spectrum. This made no difference to Blondlot's success at making the measurements he expected. Wood's subsequent letter to Nature was widely regarded as legitimating the discrediting of the claims of Blondlot and his associates. Blondlot objected, and set about developing further his method of photographing the brightness of sparks to produce objective demonstrations of the rays. But belief in N-rays died away within a year. The case is discussed in chapter 9.[16]

Nazi racism

In its period of power in Germany between 1933 and 1945, the Nazi regime carried out an increasingly extreme racial policy, culminating in the most infamous act of genocide of recorded history, in which 6 million Jews and a similar number of Slavs were systematically slaughtered. Nazi racism was not, however, an isolated aberration. The Nazi ideologists produced a coherent and systematic body of thought (e.g. Rosenberg, 1970; Hitler, 1992, originally 1925–7), drawing upon a number of older racist themes (e.g. Chamberlain, 1899; Gobineau, 1970). Right-wing political theory had earlier emphasised the dangers of mixing the blood of different races. The Aryan race, which predominated in northern Europe, was to be the focus of German national pride. Other racial groups, and in particular, the Jews, were singled out as sources of Germany's recent problems. Popular feelings of anti-Semitism, widespread in central and eastern Europe, were turned into state policy.

The Nazi state was established after a century of German pride in the quality of its science. The Nazis accepted this, though they made a distinction between the scorned Jewish science, which tended to be abstract and theoretical, and the more respected Aryan science, which was more experimental and practical.

Were their ideas of race, which appeared to their foes to be an especially dangerous example of pseudo-science (e.g. Huxley, 1941), good science by the standards of Nazi culture? The theoretical underpinning of their main racial distinctions were not in accord with the physical anthropology or the biology of other European societies of the time, and have no basis in more recent genetics. Yet their views were the expression of a coherent cultural viewpoint. On the practical side, their racist 'science' systematically developed the potential of eugenical ideas, based on Darwin's theory of evolution, and current in

---

[16] For further reading, see, Nye (1980); Price (1961); Rostand (1960).

Germany since Haeckel. They established a programme of selective breeding of captive populations (especially from Poland) with valued racial characteristics (positive eugenics) and the systematic killing of those they regarded as racially inferior (negative eugenics). Their actions put eugenics in lasting disrepute, for it illustrated dramatically how difficult it is to establish objective values to which we may appeal in judging biological superiority and inferiority in human populations.

I suggest that, in spite of its systematic cultural basis, Nazi racism cannot stand as science, even as 'Nazi science'. Nazi racism operated in terms of unquestioning acceptance of the kind of simplistic stereotypes that are so often produced by the pressures of political polarisation of thought into black and white extremes. People in Germany just before 1945 did not dare publicly to question the assumptions underlying official Nazi racist views, fearing that they would then be subject to the even stronger pressures of thought control in a totalitarian society. Within the Nazi party, only Hitler was in a position to rethink and modify the racist ideas, and the final trend of his thought was ever more extreme.[17]

Perpetual motion

Seekers after perpetual motion did not merely want a system which would keep running forever, but a machine which, once set in motion, would go on

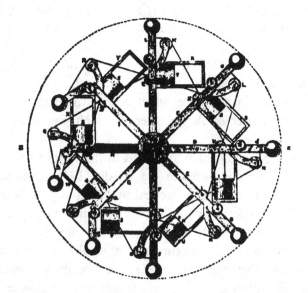

[17] For discussion of the scientific or pseudo-scientific basis of Nazi racism, see, for example, Gardner (1957); Gasman (1971); Zmarzlik (1976).

doing useful work without drawing on any external source of power. Although the idea was never universally regarded as ridiculous, leading figures of science since S. Stevin in the sixteenth century, have developed arguments on the assumption of its impossibility. Since the industrial revolution, in particular, less educated inventors have offered plausible ideas in seeking funds to develop actual perpetual motion machines. Working models have always been judged by sceptics to be fraudulent. However, occasional scientific speculations still appear which would make perpetual motion possible – as with F. Hoyle's steady-state model of the universe with its continual creation of matter-energy. The case is discussed in chapter 4.[18]

Phrenology

In the early nineteenth century, F. J. Gall proposed that specific aptitudes (such as, say ability with foreign languages) are localised in different organs within the brain, that the size of that organ is a measure of our capacity for that aptitude and that the bulges and dips in the skull reflect the underlying contours of the brain, so that study of the shape of the skull can inform us about individual aptitudes (see Figure). Based on a wide range of evidence,

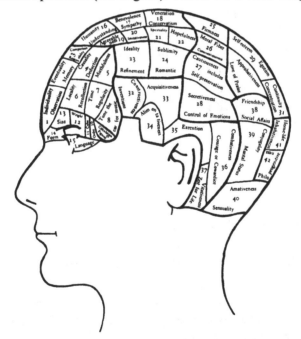

---

[18] For historical introductions, see, for example, Angrist (1968); Dircks (1870); Hering (1924); Ord-Hume (1977); Smedile (1962).

phrenology became very popular in the nineteenth century, but was never completely respectable. While its main supporters came to concentrate on popularising its practical applications (in aptitude testing, educational reform, vocational guidnce, etc), evidence against its assumptions cumulated. By the late nineteenth century, it was generally regarded as discredited. Curiously though, modern neuropsychology now has its own localisation of psychological functioning in the brain, with historical connections back to phrenology.[19]

Piltdown man

One of the most famous scientific frauds of the twentieth century, the fossil remains discovered at Piltdown between 1911 and 1915 were interpreted as proof of the ape-like origins of man, for the creature had a thickened, but clearly human skull, with an ape-like jaw (Woodward, 1948). The uniqueness of the fossil pieces and the protectiveness of its custodians at the British museum meant that not until a new generation of more sceptical custodians in the 1950s made crucial tests, was it shown that the jaw and teeth were artificially stained and from a modern orangutan and the skull was a medieval human bone fragment (Winer, 1955). By then, the skull had become anomalous in its relationship to fossils found in the rest of the world, so that, apart from the embarrassment to British science from the initial gullibility and the long delay in the fraud's exposure, the main subsequent interest seems to have been in whether the fraud was by the original discoverer, Charles Dawson alone, or with the aid of one of the more eminent scientists involved in the affair (e.g. Millar, 1972; Spencer, 1992).

Psychiatry

Generally accepted as an important professional practice traditionally concerned with the institutional care and treatment of the insane, psychiatry has always had critics. Its claims to be science-based come from its links to medicine, especally through the claim that many forms of madness have a demonstrable brain pathology and that even more psychiatric disorders benefit from drugs designed to act on brain function. Criticism was especially intense in the 1970s, and was focussed on arguments that many forms of psychiatric care and treatment were judged successful largely because it was

---

[19] For reading on the history of phrenology, see for example, Cantor & Shapin (1975); Cooter (1984); De Giustino (1975); Parssinen (1974); Shapin (1979a); Shapin (1979b); Young (1970).

in the interest of those in control of them to say that they worked. T. Szasz (e.g. Szasz, 1961, 1974) argued that we should not incarcerate the insane involuntarily in institutions, while saying that it was for the patient's own good. The courts would be preferable to psychiatrists in deciding on conflicts of interest between an individual and a society which wished to be rid of him. R. D. Laing argued (e.g. Laing, 1965) that schizophrenia could be seen as an inward escape that was a comprehensible response some people made to the impossible demands made of them by those around them. These criticisms were not generally accepted. More recent changes, such as the switch from institutional to community care, have led to new problems and to new criticisms of psychiatry. However, the practical need we continue to have for some way of coping with the insane means that, however much it is criticised, we will continue to need psychiatry.[20]

Psychical research/parapsychology

In the mid nineteenth century, there was an epidemic of paranormal phenomena. Spirits would knock when summoned and tables were being turned by psychic means. Seances became the rage, in which the spirits contacted by physical mediums were producing spirit writing and materialising, and those contacted by mental mediums were giving messages to members of the seance, often that no one living but the spirit's loved one could have known. In that age of scientific enthusiasm it seemed that it might be possible to establish an empirical scientific basis for phenomena that appeared to be closely linked to religion. In Britain, the Society for Psychical Research was founded in 1882. Cases were sought for which only a paranormal explanation seemed possible.

The focus of interest slowly moved from the most dramatic cases to those for which the evidence seemed strongest. For example, J. B. Rhine's series of tediously repetitious studies (e.g. Rhine, 1934) mainly using Zener cards (25 cards, 5 each of: star, cross, square, circle, and wavy lines) from the 1920s claimed to have built up a great weight of statistical evidence for telepathic, clairvoyant, psychokinetic, and precognitive phenoena. However, all the positive claims continued to be controversial with accusations of fraud abounding. Although plausible scientific mechanisms were conjectured, none was ever established. In the modern more secular age, people's attitudes

---

[20] For an introduction to the controversial nature of psychiatry, see Boyers and Orrill (1972); Radical Therapist Collective (1974); Clare (1980); Dain (1989); Foucault (1967).

to psychic phenomena can be linked to any doubts about the excessive materialism of science. The case is discussed in chapter 10.[21]

Pyramidology

The idea that the Great Pyramid of Giza was constructed according to a very precise quantitative plan, and that its measurements encode a message about future events for posterity, has been a popular point of discussion among pyramidologists for more than a century. Sceptics have said that the freedom to reinterpret the precise value of the ancient Egyptian units of measurement and the complete freedom in choosing which measurements should be seen as significant makes it possible to read any message at all into the pyramid's structure. At a more popular level, the idea that pyramid-shaped containers might have preservative power has been offered as something which each of us can check out personally.[22]

Scientology

In 1950, the science-fiction writer, L. Ron Hubbard, advocated a new form of psychotherapy, Dianetics (e.g. Hubbard, 1950). With publicity from J. W. Campbell, editor of *Astounding Science Fiction*, it became very popular, especially in the science-fiction network. In essence, the therapy was similar to Freud's early method of abreactive therapy. It involved identifying and bringing to consciousness traumatic experiences around which painful associations had cumulated. Hubbard's techniques for identifying such 'engrams' and clearing them began with pre-natal experiences and worked forward through the subject's life to make the patient into a 'clear'. Many self-proclaimed therapists took up the new cult of Dianetics. In order to regain control, in 1951, Hubbard rejected the Dianetics movement and founded Scientology. Scientology presented itself as a religion, at least in part to take advantage of religious toleration. In addition to the therapeutic techniques of Dianetics, now aided by a primitive lie detector, the 'E-meter', Scientology offered an elaborate theory of how we are also affected by the engrams cumulated in our previous lives. The personal aim of the Scientologist is to clear all his lives and so become an 'Operating Thetan',

---

[21] For an introduction to the history and criticism of psychic research and parapsychology, see, for example, Alcock (1981); Edmunds (1966); Grattan-Guiness (1982); Hansell (1966); Mauskopf & McVaugh (1980); Oppenheim (1985); Owen (1989); Rawcliffe (1959); Smythies (1967).

[22] The more scholarly literature of pyramidology includes Davidson & Aldersmith (1925); Edgar & Edgar (1923); Petrie (1881); Rutherford (1957-). Examples of the more popular literature include Lemesurier (1977), Toth & Nielson (1976).

comparable to the disembodied super-beings from whom we are all descended. The core theory was not offered as science for public scrutiny, but as a veiled religious mystery suitable only for higher initiates. Scientology was run as a tightly controlled business organisation, but frequently got into trouble, especially in the 1960s, for some of its more anti-social practices, comparable to those of other extreme religious sects.[23]

Sociobiology

Sociobiology was proclaimed in the mid 1970s as an interdisciplinary synthesis of the biological sciences applied to the social behaviour of animals (e.g. Dawkins, 1976; Wilson 1975). Its application to the human species was immediately controversial. The essential postulate is that if certain individuals display a form of behaviour which favours the preferential survival of their progeny, then any hereditary factor conducive to that behaviour will be preferentially selected. In its crudest form, this leads to such ideas as (a) that we are biologically predisposed to altruism if it favours enough close relatives, or (b) that, because a female can produce less offspring than a male, she is likely to favour strategies which maximise the chances of survival of each child, while a male also has the option of merely increasing the number of his children. Critics have pointed out that human sociobiological arguments are never watertight, and that the conclusions drawn may result from an interest in representing inequalities in present social life as biologically based, rather than being due to the weight of evidence (which would usually favour drawing no conclusion at all).[24]

Summerlin's patchwork mouse

A famous modern case of scientific fraud. William T. Summerlin had developed a technique for removing and culturing skin (and later also corneas) for several weeks, and then transplanting the tissue to a genetically incompatible individual without rejection. He claimed that he had repeated his successful early work on human subjects in studies with mice. But other people had been unable to reproduce his results. In 1974, when summoned to the office of his director (R. A. Good) to discuss lack of progress, he had taken along white mice with transplanted patches of dark skin, on which he had casually

[23] For a sociological study of scientology, see Wallis (1976). For examples of the judicial inquiries into Scientology, See Anderson (1965); Foster (1971); Powles (1969).

[24] For reading on the sociobiology controversy, see, for example, Caplan (1978); Clutton-Brock & Harvey (1978); Gregory et al. (1978); Montagu (1980); Rosenberg (1981); Ruse (1979); Sahlins (1976).

inked over the foreign skin to make it more obvious. When this action was later discovered by a discontented assistant, the scandal broke. Earlier examples of Summerlin's probable fraud were exposed and he was dismissed. Summerlin's claim that he had found a way of overcoming the rejection of genetically incompatible tissue was important and testable. The abuse of scientific patronage was a serious matter for the individuals and institutions involved, but fraudulent claims were never assimilated into science.[25]

Transcendental meditation (TM)

In a society that tends to regard stress as harmful, and that believes that traditional forms of meditation are effective at lowering stress, the TM movement marketed a very simple meditation technique. As part of its drive for acceptance, it encouraged scientific investigation of its claims for the meditation state as a physiologically and psychologically distinct form of consciousness (e.g. Orme-Johnson & Farrow, 1976). However, the movement could not afford to lose control of such investigations, as they were only of value if they yielded the desired result. Critics noted the high level of commitment to positive results by the key investigators. Later claims of TM include the reduction of crime and civil disorder in cities with a high enough proportion of meditators, and what is called 'flying' in which meditators make high bunny hops in a cross-legged pose.[26]

UFOlogy

The idea that strange objects seen in the sky might be a significant phenomenon for study arose out of the flying saucer craze that began in 1947. Most cases could be explained away as misperceptions of natural objects or as frauds, but those who took the phenomena seriously found that a small proportion of cases could not be dismissed so easily. The popular culture of UFOs has nurtured belief in an establishment conspiracy to suppress the evidence, together with a taste for stories of the bizarre actions of extraterrestrial visitors in UFOs. The serious study of UFOs raises questions about how best to use non-expert testimony about transient phenomena. In the present cultural setting, serious study cannot avoid looking disreputable.[27]

---

[25] For an introduction to the case, see, for example, Culliton (1974); Goodfield (1975); Hixson (1976).
[26] For an introductory reading on the scientific claims of TM, see, for example, Forem (1973); Russell (1976); Wallace & Benson (1972).
[27] For an introduction to the UFO issue, see, for example, Hynek (1972); Jacobs (1975); Sachs (1980); Sagan & Page (1972).

The Velikovsky affair

This arose from the works of I. Velikovsky, beginning with *Worlds in Collision* (Velikovsky, 1950). Velikovsky argued in long and scholarly works that the Earth was subject to violent catastrophes in historical times, and that evidence of these catastrophes can be found from the myths and legends of all cultures as well as from geological history. The key events took place between the fifteenth and eighth centuries BC. Initially, Venus had been a comet with a tail, it came from the vicinity of Jupiter, nearly colliding with the Earth. Later its contact with Mars led to that planet passing very close to the Earth. Velikovsky's ideas attracted a popular following and vigorous debunking by established scientists. The success of some of Velikovsky's 'predictions' led to a discussion among social scientists in 1966. His defenders argued that he had not been given a fair hearing. Heated debate continued another two decades. Part of the argument was over whether such radical ideas deserved a fair hearing, especially as Velikovsky's claims included the denial of assumptions on which critical arguments were based (such as the uniformitarian assumption that conditions in the present are a good guide to those in the past). The case is discussed in chapter 3.[28]

Vitamin C prevents colds

The idea that large doses of Vitamin C can reduce or eliminate the symptoms of colds and flu (e.g. Stone, 1972) was given great publicity by the eminent chemist L. Pauling in the 1970s and subsequently (e.g. Pauling, 1976). Pauling developed plausible circumstantial arguments in favour of the idea, and criticised the statistical basis of evidence claimed to discredit it. He was attacked vigorously by the medical establishment, who regard megavitamin therapies as dangerous fads. However, sales of large dose Vitamin C preparations rose. Many people seem to have judged that, since Vitamin C preparations are inexpensive and with virtually no risk, it is worth taking them, just in case the subjective benefits are not merely due to placebo effects.

Wegener's theory of continental drift

Before the modern revolution of plate tectonics of the 1950s and 1960s, the idea that the continents might move horizontally around the globe seemed

---

[28] For an introduction to the Velikovsky affair, see, for example, Bauer (1984); De Grazia (1966); Dolby (1975); Goldsmith (1977); *Pensée* Editors (1976); Polanyi (1969).

ludicrous. And yet prima-facie evidence has long been known, in the approximate match between the adjacent coasts of Africa and South America. A very full case for continental drift was put in the early twentieth century by the German glacial meteorologist A. Wegener (Wegener, 1978, originally 1915). Wegener assembled much of the modern evidence (but not that about the sea floor spreading which was so important in the later scientific revolution). His arguments were discussed most fully in the mid-1920s (e.g. Waterschoot van der Gracht et al., 1928). Most (but not all) of those who heard them rejected them as the work of a crank. He did not have a satisfactory theory of the mechanism of the movement of continents. He totally ignored counter-arguments and counter-evidence, and appeared too enthusiastic to interpret ambiguous evidence favourably. One curious feature is that, by the standards of half a century later, Wegener was almost entirely correct and his critics wrong.[29]

[29] For an introduction to Wegener and the subsequent revolution in earth sciences, see, for example, Frankel (1976); Georgi (1962); Hallam (1973).

References

Académie des Sciences (1904), 'Prix Le Conte', *Comptes Rendus de l'Académie des Sciences*, **139**, 1120–2.

Alcock, J. E. (1981), *Parapsychology: Science or Magic? A Psychological Perspective*. Oxford: Pergamon.

Allen, D. C. (1966), *The Star-Crossed Renaissance: The Quarrel about Astrology and its Influence in England*. London: Cass.

Anderson, K. V. (1965), *Report of the Board of Inquiry into Scientology*. Victoria, Australia: Government publication.

Angrist, J. (1968), 'Perpetual Motion Machines', *Scientific American*, **218** (June), 114–22.

Asimov, A. (1981), *Extraterrestrial Civilizations*. London: Pan.

Atkinson, R. L., R. C. Atkinson, E. E. Smith, & D. J. Bem (1990), *Introduction to Psychology*. 10th edn., San Diego: Harcourt, Brace, Jordanovich.

Ayer, A. J. (ed.) (1959), *Logical Positivism*. Glencoe, Ill: Free Press.

Bacon, F. (1620), *Novum Organon*. London.

Bacon, F. (1974, originally 1624), *The Advancement of Learning and the New Atlantis*. Oxford: Clarendon Press.

Bar-Tal, D. & A. W. Kruglanski (eds.) (1988), *The Social Psychology of Knowledge*. Cambridge University Press.

Barnes, B. (1977), *Interests and the Growth of Knowledge*. London: Routledge & Kegan Paul.

Baron, R. M. (1988), 'A Dual-Mode Theory of Social Knowing', in Bar-Tal & Kruglanski (1988).

Bartley, W. W. (1968), 'Theories of Demarcation between Science and Metaphysics', pp. 40–64 in I. Lakatos & A. Musgrave (eds.), *Problems in Philosophy of Science*. Amsterdam: North Holland Publishing.

Bauer, H. H. (1984), *Beyond Velikovsky: The History of a Public Controversy*. Urbana: University of Illinois Press.

Berger, P. & T. Luckman (1971), *The Social Construction of Reality*. Harmondsworth: Penguin.

Berlitz, C. (1975) *The Bermuda Triangle*. Frogmore: Panther.

Binet, A. & C. Féré, (1905), *Animal Magnetism*. London: Kegan Paul, Trench, Trübner.

Binet, A. & T. Simon, (1916, originally 1905), *The Development of Intelligence in Children*. Baltimore: Williams & Wilkins.

Blacker, C. P. (1950), *Eugenics in Retrospect and Prospect*. London: Eugenics Society and Cassell.

Block, N. & G. Dworkin (eds.) (1977), *The IQ Controversy: Critical Readings*. London: Quartet.

Blondlot, R. (1905), *'N-Rays': A Collection of Papers Communicated to the Academy of Sciences*, transl. J. Garcin. London: Longmans Green & Co.

Bloor, D. (1991), *Knowledge and Social Imagery*. 2nd edn. London: Routledge & Kegan Paul.

Blum, J. (1978), *Pseudo-Science and Mental Ability: The Origins and Fallacies of the IQ Controversy*. New York: Monthly Review Press.

Boulding, K. E. (1978), *Ecodynamics: A New Theory of Societal Evolution*. Beverley Hills: Sage.

Boyers, R. & R. Orrill (eds.) (1972), *Laing and Anti-Psychiatry*. Harmondsworth: Penguin.

Brannigan, A. (1981), *The Social Basis of Scientific Discoveries*. Cambridge University Press.

Brannigan, A. (1989), 'Artificial Intelligence and the Attributional Model of Scientific Discovery', *Social Studies of Science*, **19**, 601–12.

Bridgman, P. W. (1927), *The Logic of Modern Physics*. New York: Macmillan.

Bridgman, P. W. (1950), *Reflections of a Physicist*. New York: Philosophical Library.

Bridgman, P. W. (1959), *The Way Things Are*. Cambridge, Mass: Harvard University Press.

Brillouin, L. (1962), *Science and Information Theory*. 2nd edn., New York: Academic Press.

Broad, W. & N. Wade (1982), *Betrayers of the Truth: Fraud and Deceit in Science*. Oxford University Press.

Burke, J. B. (1904a), Letter to *Nature*, **69**, 365.

Burke, J. B. (1904b), Letter to *Nature*, **70**, 198.

Burland, C. A. (1967), *The Arts of the Alchemists*. London: Weidenfeld & Nicolson.

Burtt, E. A. (1932), *The Metaphysical Foundations of Modern Physical Science*. London: Routledge & Kegan Paul.

Butterfield, H. (1931), *The Whig Interpretation of History*. London: Bell.

Callebaut, W. & R. Pinxten (eds.) (1987), *Evolutionary Epistemology: A Multiparadigm Program, with a Complete Evolutionary Epistemology Bibliography*. Dordrecht: Reidel.

Cameron, I. & D. Edge (1979), *Scientific Images and their Social Uses: An Introduction to the Concept of Scientism*. London: Butterworth.

Campbell, D. T. (1974), 'Evolutionary Epistemology', pp. 413–63 in P. A. Schilpp (ed.), *The Philosophy of Karl Popper*. 2 vols. La Salle, Ill: Open Court.

Campbell, N. R. (1920), *Foundations of Science*. Cambridge.

Cantor, C. & S. Shapin, (1975), 'Phrenology in Early Nineteenth-Century Edinburgh: An Historiographic Discussion', *Annals of Science*, **32**, 195–225.

Caplan, A. L. (ed.) (1978), *The Sociobiological Debate: Readings on Ethical and Scientific Issues*. New York: Harper & Row.

Caron, M. & S. Hutin (1961), *The Alchemists*. New York: Grove Press.

Cartwright, N. (1983), *How the Laws of Physics Lie*. Oxford: Clarendon Press.

Chalmers, A. F. (1973), 'On Learning from our Mistakes', *British Journal for the Philosophy of Science*, **24**, 164–73.

Chalmers, A. F. (1982), *What is this Thing Called Science?*. Milton Keynes: Open University Press.

Chamberlain, H. S. (1899), *Foundations of the Nineteenth Century*. 2 vols. London: Lane.

Chambers, R. (1994, originally 1844), *Vestiges of the Natural History of Creation*. University of Chicago Press.

Charlesworth, M. (1982), *Science, Non-Science and Pseudo-Science*. Victoria: Deacon University Press.

Chesterton, G. K. (1922), *Eugenics and Other Evils*. London: Cassell.

Christian, J. L. (ed.) (1976), *Extraterrestrial Intelligence: The First Encounter*. Buffalo, N.Y: Prometheus.

Christie, J. R. R. & J. V. Golinski (1982), 'The Spreading of the Word: New Directions in the Historiography of Chemistry 1600–1800', *History of Science*, **20**, 235–66.

Cioffi, F. (1970), 'Freud and the Idea of a Pseudo-Science,' pp. 471–99 in R. Borger & F. Cioffi (eds.) *Explanation in the Behavioural Sciences*. Cambridge University Press.

Clare, A. (1980, originally 1976), *Psychiatry in Dissent*. London: Tavistock.

Close, F. E. (1992), *Too Hot to Handle: The Story of the Race for Cold Fusion*. London: Penguin.

Clutton-Brock, T. H. & P. H. Harvey (eds.) (1978), *Readings in Sociobiology*. Reading: Freeman.

Colletti, L. (1972), 'Marxism: Science or Revolution?' pp. 369–77, in R. Blackburn (ed.), *Ideology in Social Science*. London: Fontana. .

Collier, A. (1977), *R. D. Laing: The Philosophy and Politics of Psychotherapy*. Hassocks, Sussex: Harvester Press.

Collingwood, R. G. (1961, originally 1946), *The Idea of History*. Oxford University Press.

Collins, H. M. (1974), 'The T.E.A. Set: Tacit Knowledge and Scientific Networks', *Science Studies*, **4**, 165–86.

Collins, H. M. (1990), *Artificial Experts: Social Knowledge and Intelligent Machines*. Cambridge, Mass: MIT Press.

Collins, H. M. (1992, originally 1985), *Changing Order: Replication and Induction in Scientific Practice*. London: Sage.

Collins, H. M. (ed.) (1981), 'Knowledge and Controversy: Studies of Modern Natural Science', special issue of *Social Studies of Science*, **11**, 1–158.

Collins, H. M. & G. Cox (1976), 'Recovering Relativity: Did Prophecy Fail?', *Social Studies of Science*, **6**, 423–45.

Collins, H. M. & T. J. Pinch, (1982), *Frames of Meaning: The Social Construction of Extraordinary Science*. London: Routledge & Kegan Paul.

Commoner, B. (1970, originally 1963), *Science and Survival*. New York: Ballantyne Books.

Comte, A. (1969, originally 1830–1842), *Cours de Philosophie Positive*. Paris: Culture et Civilization.

Comte, A. (1974), *The Positive Philosophy of Auguste Comte*. New York: Chapman & Hall.

Conze, E. (1935), *The Scientific Method of Thinking: An Introduction to Dialectical Materialism*. London: Chapman & Hall.

Cooter, R. (1984), *The Cultural Meaning of Popular Science: Phrenology and the Organisation of Consent in Nineteenth-Century Britain.* Cambridge University Press.

Cornforth, M. (1968), *The Open Philosophy and the Open Society. A Reply to Dr. Popper's Refutations of Marxism.* London: Lawrence & Wishart.

Coulter, J. (1979), *The Social Construction of Mind,* London: Macmillan.

Crabtree, A. (1993), *From Mesmer to Freud: Magnetic Sleep and the Roots of Psychological Healing.* New Haven: Yale University Press.

Crain, R. L., E. Katz, & D. B. Rosenthal (1969), *The Politics of Community Conflict: The Fluoridation Decision.* Indianapolis: Bobbs-Merrill.

Crombie, A. C. (1952), *Augustine to Galileo: The History of Science, AD 400– 1650.* London: Heinemann.

Crowe, M. J. (1986), *The Extraterrestrial Life Debate 1750–1900: The Idea of a Plurality of Worlds from Kant to Lowell.* Cambridge University Press.

Culliton, B. J. (1974), 'The Sloan Kettering Affair', *Science,* **184**, 644–50 and 1154–7.

Curry, P. (1989), *Prophecy and Power: Astrology in Early Modern England.* Cambridge: Polity.

D'Arsonval, J. A. (1904), 'Rapport', *Comtes Rendus de l'Académie des Sciences,* **138**, 884–5.

Dain, N. (1989), 'Critics and Dissenters: Reflections on Anti-Psychiatry in the United States', *Journal of the History of the Behavioral Sciences,* **25**, 3–25.

Däniken, E. (1969), *Chariots of the Gods?* London: Souvenir.

Darnton, R. (1968), *Mesmerism and the End of the Enlightenment in France.* Cambridge, Mass: Harvard University Press.

Darwin, C. (1968, originally 1859), *The Origin of Species by Means of Natural Selection: Or, The Preservation of Favoured Races in the Struggle for Life.* Harmondsworth: Penguin.

Davidson, D. & Aldersmith (1925), *The Great Pyramid: Its Divine Message.* London: Williams & Norgate.

Dawkins, R. (1989, originally 1976), *The Selfish Gene.* Oxford University Press.

De Giustino, D. (1975), *Conquest of Mind: Phrenology and Victorian Social Thought.* London: Croom Helm.

De Grazia, A. (ed.) (1978, originally 1963 and 1966), *The Velikovsky Affair.* rev. edn. London: Sphere Books.

Dear, P. (1988), *Mersenne and the Learning of the Schools.* Ithaca: Cornell University Press.

Debus, A. G. (1965), *The English Paracelsians.* London: Oldbourne.

Debus, A. G. (ed.) (1972), *Science, Medicine and Society in the Renaissance. Essays in Honor of Walter Pagel.* 2 vols. London: Heinemann.

Dennett, D. C. (1991), *Consciousness Explained.* Harmondsworth: Penguin.

Der Spiegel Editors (1973), 'Anatomy of a World Best Seller', *Encounter,* **41**, 9–17.

Descartes, R. (1954), *Philosophical Writings: A Selection.* Transl. & ed. by E. Anscombe & P. T. Geach. London: Nelson.

Dick, S. J. (1982), *Plurality of Worlds: The Origins of the Extraterrestrial Life Debate from Democritus to Kant.* Cambridge University Press.

Dijksterhuis, E. J. (1961), *The Mechanization of the World Picture.* Oxford University Press.

Dircks, H. (1861), *Perpetuum Mobile: Or, Search for Self-Motive Power.* London: Spon.

Dircks, H. (1870), *Perpetuum Mobile*. Second series. London: Spon.

Dobbs, B. J. T. (1975), *The Foundations of Newton's Alchemy: Or "the Hunting of the Greene Lyon"*. Cambridge University Press.

Dobbs, B. J. T. (1991), *The Janus Faces of Genius: The Role of Alchemy in Newton's Thought*. Cambridge University Press.

Dolby, R. G. A. (1971), 'The Sociology of Knowledge in Natural Science', *Science Studies*, **1**, pp. 3–21.

Dolby, R. G. A. (1975), 'What Can We Usefully Learn from the Velikovsky Affair?', *Social Studies of Science*, **5**, 165–75.

Dolby, R. G. A. (1976), 'Debates over the Theory of Solution: A Study of Dissent in Physical Chemistry in the English-Speaking World in the Late Nineteenth and Early Twentieth Centuries', *Historical Studies in the Physical Sciences*, 7, 297–404.

Dolby, R. G. A. (1977), 'The Transmission of Science', *History of Science*, **15**, 1–43.

Dolby, R. G. A. (1982), 'On the Autonomy of Pure Science: The Construction and Maintenance of Barriers between Scientific Establishments and Popular Culture', pp. 267–92, in N. Elias, H. Martins, & R. Whitley (eds.), *Scientific Establishments and Hierarchies. Sociology of the Sciences*, VI, Dordrecht: Reidel.

Dolby, R. G. A. (1987), 'Science and Pseudo-Science: The Case of Creationism', *Zygon*, **22**, 195–212.

Dolby, R. G. A. (1989), 'The Authority of the Scientific Rejection of Pseudo-Science', *Bulletin of Science, Technology and Society*, **9**, 283–93.

Douglas, M. (1987), *How Institutions Think*. London: Routledge & Kegan Paul.

Dreyfus, H. L. (1992), *What Computers Still Can't Do: A Critique of Artificial Reason*. Cambridge, Mass: MIT Press.

Easlea, B. (1981), *Science and Sexual Oppression: Patriarchy's Confrontation with Woman and Nature*. London: Weidenfeld & Nicolson.

Edgar, J. & M. Edgar (1923), *The Great Pyramid Passages and Chambers*. Glasgow: Bone & Hulley

Edmunds, E. (1966), *Spiritualism: A Critical Survey*. London: Aquarian Press.

Ehrlich, P. R. (1971, originally 1968), *The Population Bomb*. London: Ballantyne/ Friends of the Earth.

Einbinder, H. (1964), *The Myth of the Britannica*. London: McGibbon & Kee.

Elkana, Y (1974), *The Discovery of the Conservation of Energy*. London: Hutchinson.

Ellegård, A. (1957) 'The Darwinian Revolution and Nineteenth-Century Philosophy of Science', *Journal of the History of Ideas*, **18**, 362–93.

Ellenberger, H. G. (1994, originally 1970), *The Discovery of the Unconscious*. London: Fontana.

Engelhardt, H. & A. L. Caplan (eds.) (1987), *Scientific Controversies: Case Studies in the Resolution and Closure of Disputes in Science and Technology*. Cambridge University Press.

Engels, F. (1946), *The Dialectics of Nature*. London: Lawrence & Wishart.

Evans, B. & B. Waites, (1981), *IQ and Mental Testing: An Unnatural Science and Its Social History*. London: Macmillan.

Evans-Pritchard, E. E. (1937), *Witchcraft, Oracles and Magic Among the Azande*. Oxford: The Clarendon Press.

Evans-Pritchard, E.E. (1976) *Witchcraft, Oracles and Magic Among the Azande.* abridged edn. Oxford: Clarendon Press.

Eysenck, H. (1985), *The Decline and Fall of the Freudian Empire.* Penguin: Harmondsworth.

Eysenck, H. & G. D. Wilson (1973), *The Experimental Study of Freudian Theories.* London: Methuen.

Eysenck, H. J. & D. K. B. Nias (1984), *Astrology: Science or Superstition?* Harmondsworth: Penguin.

Farrell, L. A. (1979), 'The History of Eugenics: A Bibliographical Review', *Annals of Science,* **36,** 111–23.

Farrington, B. (1953), *Greek Science.* London: Penguin.

Feibleman, F. K. (1959), 'Darwin and Scientific Method.' *Tulane Studies in Philosophy,* **8,** 5–14.

Festinger, L. (1957), *A Theory of Cognitive Dissonance.* Stanford University Press.

Feyerabend, P. K. (1966), 'On the Possibility of a Perpetuum Mobile of the Second Kind', pp. 409–12, in P. K. Feyerabend and G. Maxwell (eds.), *Mind, Matter and Method.* University of Minnesota Press.

Feyerabend, P. K. (1975), *Against Method: Outline of an Anarchistic Theory of Knowledge.* London: New Left Books.

Feyerabend, P. K. (1978), *Science in a Free Society.* London: New Left Books.

Firth, I. (1969), 'N-Rays – Ghost of Scandal Past', *New Scientist,* **44,** 642–3.

Fleck, L. (1979, originally 1935), *The Genesis and Development of a Scientific Fact.* University of Chicago Press.

Forem, J. (1973), *Transcendental Meditation: Maharishi Mahesh Yogi and the Science of Creative Intelligence.* New York: Dutton.

Forman, P. (1971), 'Weimar Culture, Causality and Quantum Theory, 1918–1927', *Historical Studies in the Physical Sciences,* **3,** 1–115.

Foster, J. G. (1971), *Enquiry into the Practice and Effects of Scientology.* London: HMSO.

Foucault, M. (1967), *Madness and Civilisation: A History of Insanity in the Age of Reason.* London: Tavistock.

Frankel, E. (1976), 'Alfred Wegener and the Specialists', *Centaurus,* **20,** 305–24.

Frankfort, H. & H. A. Frankfort, (1949), *Before Philosophy.* London: Penguin.

Frazer, J. G. (1911–15), *The Golden Bough: A Study in Magic and Religion.* 3rd edn., 12 vols. London: Macmillan.

Gaissinovitch, A. E. (1980), 'The Origins of Soviet Genetics and the Struggle with Lamarckism', *Journal of the History of Biology,* **13,** 1–52.

Galilei, Galileo (1967, originally 1632). *Dialogue Concerning the Two Chief World Systems,* transl. Stillman Drake. Berkeley: University of California Press.

Galton, F. (1985, originally 1909), *Essays in Eugenics.* New York: Garland.

Gardner, M. (1957, originally 1952), *Fads and Fallacies in the Name of Science.* New York: Dover.

Garin, E. (1983), *Astrology in the Renaissance: The Zodiac of Life.* London: Routledge & Kegan Paul.

Gasman, D. (1971), *The Scientific Origins of National Socialism.* London: Macdonald.

Gauquelin, M. (1983), *The Truth About Astrology.* Oxford: Blackwell.

Georgi, J. (1962), 'Memories of Alfred Wegener', chapter 12 of S. K. Runcorn (ed.), *Continental Drift.* New York: Academic Press.

Gibson, J. J. (1979) *The Ecological Approach to Visual Perception*. Boston: Houghton Mifflin.

Giere, R. N. (1988), *Explaining Science: A Cognitive Approach*. University of Chicago Press.

Giere, R. N. (ed.) (1992), *Cognitive Models of Science*, vol. XV of Minnesota Studies in the Philosophy of Science. Minneapolis: University of Minnesota Press.

Gilbert, G. N. & M. J. Mulkay (1984), *Opening Pandora's Box: A Sociological Analysis of Scientific Discourse*. Cambridge University Press.

Gilbert, W. (1958, originally 1600), *De Magnete*, transl. P. F. Motteley. New York: Dover.

Gillispie, C. C. (1951). *Genesis and Geology*. New York: Harper.

Gillispie, C. C. (1960), *The Edge of Objectivity: An Essay in the History of Scientific Ideas*. Princeton University Press.

Gish, D. T. (1979), *Evolution? The Fossils Say No!* San Diego, Calif.: Creation-Life Publications.

Gobineau, J. A. de (1970), *Selected Political Writings*. New York: Harper & Row.

Godfrey, L. R. (ed.) (1983), *Scientists Confront Creationism*. New York: Norton.

Goethe, J. W. von (1987, originally 1810) *Theory of Colours*, transl. C. L. Eastlake. Cambridge, Mass: MIT Press.

Goldsmith, D. (ed.) (1977), *Scientists Confront Velikovsky*. Ithaca, NY: Cornell University Press.

Gombrich, E. H. (1982), *The Image and the Eye: Further Studies in the Psychology of Pictorial Representation*. Oxford: Phaidon.

Goodfield, J. (1975), *Cancer under Seige*. London: Hutchinson.

Gooding, D., T. J. Pinch & S. Schaffer (eds.) (1989), *The Uses of Experiment: Studies in the Natural Sciences*. Cambridge University Press.

Goodman, N. (1965), *Fact, Fiction and Forecast*. Indianapolis: Bobbs-Merrill.

Goodman, N. (1976), *Languages of Art: An Approach to the Theory of Symbols*. Oxford University Press.

Gotzsche, A. L. (1975), *The Fluoride Question*. London: Davis-Poynter.

Gould, S. J. (1972), 'Zealous Advocates: Review of The Case of the Midwife Toad', *Science*, **176**, 623–5.

Gould, S. J. (1978), 'Women's Brains', *New Scientist*, **80**, 364–6.

Gould, S. J. (1981), *The Mismeasure of Man*. New York: Norton.

Grattan-Guinness, I. (ed.) (1982), *Psychical Research: A Guide to its History, Principles and Practices*. Wellingborough: Aquarian Press.

Greenberg, D., (1967), *The Politics of Pure Science*. New York: New American Library.

Gregory, M. S., A .S. Silvers, & D. Sutch (eds.), (1978). *Sociobiology and Human Nature: Interdisciplinary Critique and Defence*. San Francisco: Jossey-Bass.

Gregory, R. L. (1973), 'The Confounded Eye', in R. L. Gregory & E. H. Gombridge (eds.), *Illusion in Nature and Art*. London: Duckworth.

Grosz, E. (1990), 'Contemporary Theories of Power and Subjectivity', in S. Gunew (ed.), *Feminist Knowledge: Critique and Construct*. London: Routledge.

Grünbaum, A. (1984), *The Foundations of Psychoanalysis: A Philosophical Critique*. University of California Press.

Grünbaum, A. (1993), *Validation in the Clinical Theory of Psychoanalysis: A Study in the Philosophy of Psychoanalysis*. Madison, Conn: International Universities Press.

Hacking, I. (1983), *Representing and Intervening: Introductory Topics in the Philosophy of Natural Science*. Cambridge University Press.

Hadamard, J. (1954, originally 1945), *The Psychology of Invention in the Mathematical Field*. New York: Dover.

Hagstrom, W. O. (1965), *The Scientific Community*. Brighton: Basic Books.

Hall, A. R. (1952), *Ballistics in the Seventeenth Century: A Study in the Relations of Science and War with Reference Principally to England*. Cambridge University Press.

Hall, A. R. (1983), *The Revolution in Science, 1500–1750*. 3rd edn., London: Longman.

Hallam, A. (1973), *A Revolution in the Earth Sciences: From Continental Drift to Plate Tectonics*. Oxford: Clarendon Press.

Hann, C. M. (ed.) (1994), *When History Accelerates: Essays on Rapid Social Change, Complexity and Creativity*. London: Athlone.

Hannaway, O. (1975), *The Chemists and the Word: The Didactic Origins of Chemistry*. Baltimore: Johns Hopkins University Press.

Hansell, E. M. (1966), *ESP A Scientific Evaluation*. New York: Scribners.

Hanson, N. R. (1958), *Patterns of Discovery: An Enquiry into the Conceptual Foundations of Science*. Cambridge University Press.

Harding, S. G. (1986), *The Science Question in Feminism*. Milton Keynes: Open University Press.

Hastead, J. (1981), *The Metal-Benders*. London: Routledge & Kegan Paul.

Hekman, S. J. (1990), *Gender and Knowledge: Elements of a Postmodern Feminism*. Cambridge: Polity Press.

Helmholtz, H (1853), 'On the Conservation of Force', translated in J. Tyndall & W. Francis (eds.), *Scientific Memoirs Selected from the Transactions of Foreign Academies of Science and from Foreign Journals: Natural Philosophy*. London: Taylor & Francis.

Helmholtz, H. von (1893) 'On Goethe's Scientific Researches', in *Popular Lectures on Scientific Subjects*, London: Longmans. First series, lecture II.

Hering, D. W. (1924), *Foibles and Fallacies of Science: An Account of Celebrated Scientific Vagaries*. London: Routledge.

Hesse, M. (1961), *Forces and Fields: A Study of Action at a Distance in the History of Physics*. London.

Hesse, M. (1963), *Models and Analogies in Science*. London.

Hesse, M. (1980), *Revolutions and Reconstructions in the Philosophy of Science*. Brighton: The Harvester Press.

Hessen, B. (1971, originally 1931), 'The Social and Economic Roots of Newton's *Principia*', in *Science at the Cross Roads*. London: Cass.

Hewish, A., S. J. Bell, J. D. H. Pilkington, P. F. Scott & R. A. Collins, (1968), 'Observation of a Rapidly Pulsating Source', *Nature*, **217**, 709–13.

Hitler, A. (1992, originally 1925–7) *Mein Kampf*, transl. R. Manheim. London: Pimlico.

Hixson, J. (1976), *The Patchwork Mouse*. Garden City, NY: Anchor Press.

Holloway, M. (1993), 'Sound Science?', *Scientific American*, **269**, August, p. 11.

Holmyard, E. J. (1968), *Alchemy*. Harmondsworth: Penguin.

Horton, R. (1967), 'African Traditional Thought and Western Science', *Africa*, **37**, pp. 50–71, 155–87.

Hubbard, L. R. (1950), *Dianetics: The Modern Science of Mental Health*. Los Angeles, Calif.: The American Saint Hill Organization.

Huizenga, J. R. (1993), *Cold Fusion: The Scientific Fiasco of the Century*. Oxford University Press.

Huxley, J. (1941), *The Argument of Blood: The Advancement of Science*. London: Macmillan.

Huxley, J. (1949), *Soviet Genetics and World Science: Lysenko and the Meaning of Heredity*. London: Chatto & Windus.

Hynek, J. A. (1972), *The UFO Experience: A Scientific Enquiry*. London: Abelard-Schuman.

Jacob, M. C. (1976), *The Newtonians and the English Revolution*. Ithaca, NY: Cornell University Press.

Jacobs, D. M. (1975), *The UFO Controversy in America*. Bloomington: Indiana University Press.

James, W. (1950, originally 1890), *Principles of Psychology*. 2 vols. Vol. 2, chapter 27. New York: Dover.

Jardine, N. (1991), *The Scenes of Inquiry: On the Reality of Questions in the Sciences*. Oxford: Clarendon Press.

Jasonoff, S. (1987), 'Contested Boundaries in Policy Relevant Science', *Social Studies of Science*, **17**, 195–230.

Jeffrey, A. T. (1975) *The Bermuda Triangle*. London: W.H. Allen

Jerome, L. E. (1977), *Astrology Disproved*. Buffalo: Prometheus Books.

Joravsky, D. (1970), *The Lysenko Affair*. Cambridge, Mass: Harvard University Press.

Jordanova, L. (1989), *Sexual Visions: Images of Gender in Science and Medicine between the Eighteenth and Twentieth Centuries*. London: Harvester Wheatsheaf.

Kahney, H. (1986), *Problem Solving: A Cognitive Approach*. Milton Keynes: Open University Press.

Kamin, L. J. (1974), *The Science and Politics of IQ*. New York: Halsted Press.

Kearney, H. (1971), *Science and Change, 1500–1700*. London: Weidenfeld & Nicolson.

Keller, E. F. (1983), *A Feeling for the Organism*. San Francisco: Freeman.

Kevles, D. J. (1986), *In the Name of Eugenics: Genetics and the Uses of Human Heredity*. Harmondsworth: Penguin.

Kitcher, Patricia (1992), *Freud's Dream: A Complete Interdisciplinary Science of Mind*. Cambridge, Mass: MIT Press.

Kitcher, Philip (1982), *Abusing Science: The Case Against Creationism*. Cambridge Mass: MIT Press.

Kitcher, Philip (1993), *The Advancement of Science: Science without Legend, Objectivity without Illusions*. New York: Oxford University Press.

Klein, M. (1970), 'Maxwell, his Demon and the Second Law of Thermodynamics', *The American Scientist*, **58**, 84–97.

Kline, P. (1987), 'Philosophy, Psychology and Psychoanalysis', (Review of Grünbaum, 1984), *British Journal for the Philosophy of Science*, **38**, 106–16.

Knorr-Cetina, K. D. (1981), *The Manufacture of Knowledge: An Essay on the Constructivist and Contextual Nature of Science*. Oxford: Pergamon Press.

Koestler, A. (1964), *The Act of Creation*. London: Hutchinson.

Koestler, A. (1971), *The Case of the Mid-Wife Toad*. London: Hutchinson.

Koyré, A. (1957), *From the Closed World to the Infinite Universe*. Baltimore: Johns Hopkins Press.

Kuhn, T. S. (1968), 'The History of Science', pp. 74–83 in *International Encyclopedia of the Social Sciences*, vol. 14. Reprinted in Kuhn (1977).

Kuhn, T. S. (1970a, originally 1962), *The Structure of Scientific Revolutions*. 2nd edn. Chicago University Press.

Kuhn, T. S. (1970b), 'Logic of Discovery or Psychology of Research?', and 'Reflections on my Critics', both in Lakatos & Musgrave (1970).

Kuhn, T. S. (1977), *The Essential Tension: Selected Studies in Scientific Tradition and Change*. Chicago University Press.

Kusche, L. D. (1975), *The Bermuda Triangle Mystery - Solved*. New York: Harper & Row.

La Follette, M. C. (ed.) (1983), *Creationism, Science and the Law: The Arkansas Case*. Cambridge Mass: MIT Press.

Laing, R. D. (1965, originally 1960), *The Divided Self*. Harmondsworth: Penguin.

Laing, R. D. (1967), *The Politics of Experience* and *The Bird of Paradise*. Harmondsworth: Penguin.

Lakatos, I. (1970), 'Falsification and the Methodology of Scientific Research Programmes', pp. 91–195 in Lakatos & Musgrave (1970).

Lakatos, I. (1981), 'Science and Pseudo-Science', pp. 114–21 in S. Brown, J. Fauvel, and R. Finnegan (eds.), *Conceptions of Inquiry*. London: Methuen.

Lakatos, I. & A. Musgrave (1970), *Criticism and the Growth of Knowledge*. Cambridge University Press.

Lambert, M. J., D. A. Shapiro, & A. E. Bergin (1986), 'The Effectiveness of Psychotherapy', in S. L. Garfield & D. E. Bergin (eds.), *Handbook of Psychotherapy and Behaviour Change*. 3rd edn. Chichester: Wiley.

Langevin, (1904), report on the existence of N-rays, *La Revue Scientifique*, p. 591.

Langley, P., H. A. Simon, G. L. Bradshaw, & I. L. Zytkow (1987), *Scientific Discovery: Computational Explorations of the Creative Process*. Cambridge, Mass: MIT Press.

Langmuir, I. (1989), 'Pathological Science', *Physics Today*, **42** No. 10, 36–48, transcript of a talk given in 1953.

Latour, B (1987), *Science in Action: How to Follow Scientists and Engineers through Society*. Milton Keynes: Open University Press.

Latour, B. & S. Woolgar (1986, originally 1979), *Laboratory Life: The Construction of Scientific Facts*. Princeton University Press.

Laudan, L. (1982), 'The Demise of Demarcation', pp. 111–28, in R. S. Cohen & L. Laudan (eds.), *Physics, Philosophy and Psychiatry*. Dordrecht: Reidel.

Le Noble, R. (1971), *Mersenne; ou, la Naissance de Mécanisme*. 2nd edn., Paris: Vrin.

Lecourt, D. (1977), *Proletarian Science? The Case of Lysenko*. London: New Left Books.

Lemesurier, P. (1977), *The Great Pyramid Decoded*. Tisbury: Compton Press.

Lewis, D. (1968) *Convention: A Philosophical Study*. Cambridge, Mass: Harvard University Press.

Lindsay, J. (1971), *Origins of Astrology*. London: Muller.

Locke, J. (1979, originally 1689) *An Essay Concerning Human Understanding*. Oxford University Press.

Lycan, W. G. (ed.) (1990), *Mind and Cognition: A Reader*. Oxford: Blackwell.

MacKenzie, D. (1976), 'Eugenics in Britain', *Social Studies of Science*, **6**, 499–532.

MacKenzie, D. & B. Barnes (1979), 'Scientific Judgement: The Biometry-Mendelism Controversy', in B. Barnes & S. Shapin (eds.), *Natural Order: Historical Studies of Scientific Culture*. Beverley Hills: Sage.

Magee, B. (1973), *Popper*. London: Fontana.

Magnus, R. (1949), *Goethe as Scientist*. New York: Schuman.

Mallove, E. F. (1991), *Fire from Ice: Searching for the Truth Behind the Cold Fusion Furore*. New York: John Wiley.

Mannheim, K. (1936), *Ideology and Utopia: An Introduction to the Sociology of Knowledge*. London: Routledge & Kegan Paul.

Marks, R. (1981), *The Idea of IQ* (1981). Washington, DC: University Press of America.

Martin, B. (1988), 'Analyzing the Fluoridation Controversy: Resources and Structures', *Social Studies of Science*, **18**, 331–63.

Masson, J. M. (1984), *Freud, The Assault on Truth: Freud's Suppression of the Seduction Theory*. London: Faber.

Mauskopf, S. (1990), 'Marginal Science', pp. 869–85, in R. C. Olby, G. N. Cantor, J. R. R. Christie, & M. J. Hodge (eds.), *Companion to the History of Modern Science*. London: Routledge.

Mauskopf, S. H. & M. R. McVaugh, (1980), *The Elusive Science: Origins of Experimental Psychical Research*. Baltimore: Johns Hopkins University Press.

Mausner, B. (1955), 'A Study of the Anti-Scientific Attitude', *Scientific American* (February), 35–9.

Maxwell, J. C. (ed.) (1867), *The Electrical Researches of the Honourable Henry Cavendish*. London: Cass.

McDonough, T. R. (1987), *The Search for Extraterrestrial Intelligence: Listening for Life in the Cosmos*. New York: John Wiley & Sons.

McKendrick, J. G. & W. Colquhoun (1904), Letter to *Nature*, **69**, 534.

Medvedev, Z. A. (1969). *The Rise and Fall of T. D. Lysenko*. New York: Columbia University Press.

Meek, R. L. (1953), *Marx and Engels on Malthus*. London: Lawrence & Wishart.

Merton, R. K. (1973), *The Sociology of Science: Theoretical and Empirical Investigations*. Chicago University Press.

Merz, J. T. (1965, originally 1904–1912.), *A History of European Thought in the Nineteenth Century*. 4 vols. New: York: Dover.

Mill, J. S. (1875, originally 1843), *A System of Logic, Ratiocinative and Inductive*. 2 vols. 9th edn. London: Longman.

Millar, R. (1972), *The Piltdown Men*. St Albans: Paladin.

Millhauser, M. (1959), *Just before Darwin: Robert Chambers and Vestiges*. Middleton, Conn.: Wesleyan University Press.

Minto, W. (1899), *Logic: Inductive and Deductive*. London: Murray.

Montagu, A. (ed.) (1975), *Race and IQ*. New York: Oxford University Press.

Montagu, A. (ed.) (1980), *Sociobiology Examined*. Oxford University Press.

Morris, H. M. (ed.) (1974), *Scientific Creationism*. San Diego, Calif.: Creation-Life Publications.

Morrison, D. (1990), 'The Rise and Decline of Cold Fusion', *Physics World*, February. 1990, 35–8.

Mulkay, M. (1979), *Science and the Sociology of Knowledge*. London: George Allen & Unwin.

Naylor, P. I. H. (1967), *Astrology, An Historical Examination*. Oxford: Maxwell.

Nelkin, D. (1982), *The Creation Controversy: Science or Scripture in the Schools*. New York: Norton.

Nelkin, D. (1992), *Controversy: Politics of Technical Decision*. 3rd edn. Newbury Park, Calif.: Sage.

Newton, I. (1968, originally 1687), *The Mathematical Principles of Natural Philosophy*, transl. A. Motte. London: Dawsons.

Nisbet, R. (1972), *Goethe and the Scientific Tradition*. London: Institute of Germanic Studies.

Nye, M. J. (1980), 'N-Rays: An Episode in the History and Psychology of Science', *Historical Studies in the Physical Sciences*, **11**, 125–56.

O'Hear, A. (1982), *Karl Popper*. London: Routledge & Kegan Paul.

Oppenheim, J. (1985), *The Other World: Spiritualism and Psychical Research in England, 1850–1914*. Cambridge University Press.

Ord-Hume, A. W. J. G. (1977), *Perpetual Motion: The History of an Obsession*. London: Allen & Unwin.

Orme-Johnson, D. W. & J. T. Farrow (eds.) (1976), *Scientific Research on the Transcendental Meditation Program. Collected Papers*. Switzerland: Maharishi European Research University Press.

Orwell, G. (1949), *Nineteen Eighty-Four, a Novel*. London: Secker & Warburg.

Owen, A. (1989), *The Darkened Room: Women, Power, and Spiritualism in Late Nineteenth-Century England*. London: Virago.

Owen, A. R. G. (1971), *Hysteria, Hypnosis and Healing: The Work of J-M Charcot*. London: Dobson.

Pagel, W. (1958), *Paracelsus: An Introduction to Philosophical Medicine in the Age of the Renaissance*. Basel: S. Karger.

Parry, H. B. (1983), *Scrapie Disease in Sheep: Historical, Clinical and Practical Aspects of the Natural Disease*. London: Academic Press.

Parssinen, T. M. (1974), 'Popular Science and Society: The Phrenology Movement in Early Victorian Britain', *Journal of Social History*, **7**, 1–20.

Pauling, L. (1976, originally 1970), *Vitamin C the Common Cold and the Flu*. San Francisco: Freeman.

Pensée Editors, (1976), *Velikovsky Reconsidered*. London: Sidgwick & Jackson.

Petrie, W. M. F. (1881), *Pyramids and Temples of Gizeh*. London: Field & Tuer.

Piercy, M. (1979), *Woman on the Edge of Time*. London: The Women's Press.

Poincaré, H. (1904), comment on N-rays, *La Revue Scientifique*, p. 682.

Poincaré, H. (1914, originally 1908), *Science and Method*. London: Nelson.

Poincaré, H. (1952, originally 1902), *Science and Hypothesis*. New York: Dover.

Poincaré, H. (1958, originally 1905), *The Value of Science*. New York: Dover.

Polanyi, M. (1951), *The Logic of Liberty*. London: Routledge & Kegan Paul.

Polanyi, M. (1958), *Personal Knowledge*. London: Routledge & Kegan Paul.

Polanyi, M. (1969), 'The Growth of Science in Society', in *Knowing and Being*. London: Routledge & Kegan Paul.

Popkin, R. H. (1979), *The History of Scepticism from Erasmus to Spinoza*. Rev. edn. Berkeley: California University Press.

Popper, K. R. (1959, originally 1934), *The Logic of Scientific Discovery*. London: Hutchinson: London.

Popper, K. R. (1960), *The Poverty of Historicism*. London: Routledge & Kegan Paul.

Popper, K. R. (1966, originally 1945), *The Open Society and its Enemies*. 2 vols. London: Routledge & Kegan Paul.

Popper, K. R. (1974), 'Science: Conjectures and Refutations', in *Conjectures and Refutations*. 5th edn., London: Routledge & Kegan-Paul.

Popper, K. R. (1976), *Unended Quest*. London: Fontana, originally 'Autobiography of Karl Popper', in P. A. Schilpp (ed.), *The Philosophy of Karl Popper*. Illinois: Open Court, 1974.

Popper, K. R. (1979), *Objective Knowledge*. Rev. edn. Oxford: Clarendon Press.

Popper, K. R. (1986, originally 1957), *The Poverty of Historicism*. London: Ark Paperbacks.

Powles, G. R. (1969), *The Commission of Inquiry into the Hubbard Scientology Organisation in New Zealand*. Wellington, NZ: Government publication.

Price, D. J. de S. (1961), *Science since Babylon*. New Haven, Conn: Yale University Press.

Price, D. J. de S. (1963), *Little Science, Big Science*. New York: Columbia University Press.

Prusiner, S. et al. (1992), *Prion Diseases of Humans and Animals*. New York: Ellis Horwood.

Ptolemy, C. (1940), *Tetrabiblos*. Transl. and ed. by F. E. Robbins. London: Heinemann.

Quine, W. V. O. (1960), *Word and Object*. Cambridge, Mass: MIT Press.

Rachman, S. (ed.) (1963), *Critical Essays on Psychoanalysis*. Oxford: Pergamon Press.

Radical Therapist Collective (eds.) (1974), *The Radical Therapist*. Harmondsworth: Penguin.

Radnitzky, G & W. W. Bartley III (eds.) (1987), *Evolutionary Epistemology, Rationality and the Sociology of Knowledge*. La Salle, Ill.: Open Court.

Randi, J. (1978), *The Magic of Uri Geller*. New York: Ballantyne.

Ravetz, J. R. (1971), *Scientific Knowledge and its Social Problems*. Oxford: Clarendon Press.

Ravetz, J. R. (1990), 'Orthodoxies, Critiques and Alternatives', pp. 898–908 in R. C. Olby, G. N. Cantor, J. R. R. Christie & M. J. S. Hodge (eds.), *Companion to the History of Modern Science*. London: Routledge.

Rawcliffe, D. H. (1959, originally 1952), *Occult and Supernatural Phenomena*. New York: Dover.

Read, J. (1966), *Prelude to Chemistry: An Outline of Alchemy*. Cambridge, Mass: MIT Press.

Reichenbach, H. (1938), *Experience and Prediction*. University of Chicago Press.

Revue Scientifique, La, (1904), 'Les rayons N existent-ils', a series of unsigned articles on the N-ray affair, pp. 545–52, 590–1, 620–5, 656–60, 682–6, 705–10, 718–22, 752–4, 783–5.

Rhine, J. B. (1934), *Extra-Sensory Perception*. Boston, Mass: Boston Society for Psychical Research.

Righini Bonelli, M. L. & W. R. Shea (eds.) (1975), *Reason, Experiment and Mysticism in the Scientific Revolution*. New York: Science History Publications.

Rose, H. & S. Rose (1976), *The Political Economy of Science: Ideology of/in the Natural Sciences*. London: Macmillan.

Rosenberg, A. (1970), *Selected Political Writings*. London: Cape.

Rosenberg, A. (1981), *Sociobiology and the Preemption of Social Science*. Oxford: Blackwell.

Rossi, P. (1978), *Francis Bacon: From Magic to Science*. University of Chicago Press.

Rostand, J. (1960), *Error and Deception in Science*. London: Hutchinson.

Roszak, T. (1975), *Unfinished Animal: The Aquarian Frontier and the Evolution of Consciousness*. New York: Harper & Row.

Rudge, W. A. D. (1904), Letter to *Nature*, **69**, 437–8.

Ruse, M. (1979), *Sociobiology, Sense or Nonsense?* Dordrecht, Holland: Reidel.

Ruse, M. (1982), *Darwinism Defended: A Guide to Evolution Controversies*. Reading, Mass: Addison-Wesley.

Ruse, M. (ed.) (1988), *But is it Science?: The Philosophical Question in the Creation/Evolution Controversy*. Buffalo, NY: Prometheus Books.

Russell, P. (1976), *The TM Technique: An Introduction to Transcendental Meditation as the Teachings of Maharishi Mahesh Yogi*. London: Routledge & Kegan Paul.

Rutherford, A. (1957–), *Pyramidology*. Dunstable: Institute of Pyramidology.

Sachs, M (1980), *The UFO Encyclopedia*. London: Corgi.

Sagan, C. (ed.) (1973), *Communication with Extraterrestrial Intelligence*. Cambridge, Mass: MIT Press.

Sagan, C. & T. Page (eds.) (1972), *UFO's: A Scientific Debate*. Ithaca, NY: Cornell University Press.

Sahlins, M. (1976), *The Use and Abuse of Biology: An Anthropological Critique of Sociobiology*. London: Tavistock.

Salmon, W. C. (1961), 'The Vindication of Induction', in H. Feigl & G. Maxwell (eds.), *Current Issues in Philosophy of Science*. New York: Holt, Rinehart & Winston.

Salmon, W. C. (1964), 'On Vindicating Induction', in H. E. Kyburg Jr (ed.), *Induction: Some Current Issues*. Middletown, Conn: Wesleyan University Press.

Sapolsky, H. M. (1969), 'Science, Voters and the Fluoridation Controversy', *Science*, **162**, 427–33.

Sayers, J. (1982), *Biological Politics: Feminist and Anti-Feminist Perspectives*. London: Tavistock.

Schenk, C. C. (1904), Letter to *Nature*, **69**, 486–7.

Searle, G. R. (1976), *Eugenics and Politics in Britain, 1900–1914*. Noordhoff International.

Searle, J. R. (1980), 'Minds, Brains, and Programs'. *The Behavioral and Brain Sciences*, **3**, 417–24.

Sepper, D. L. *Goethe contra Newton: Polemics and the Project for a New Science of Color*. Cambridge University Press.

Shapin, S. (1979a), 'The Politics of Observation: Cerebral Anatomy and Social Interests in the Edinburgh Phrenology Disputes', in R. Wallis (ed.), *On the Margins of Science*, 139–78. *Sociological Review Monograph*, No. 27.

Shapin, S. (1979b), 'Homo Phrenologicus: Anthropological Perspectives on an Historical Problem', pp. 41–71 in B. Barnes & S. Shapin (eds.), *Natural Order: Historical Studies of Scientific Culture*. Beverley Hills, Calif.: Sage.

Shapin, S. (1992), 'Discipline and Bounding: The History and Sociology of Science as Seen through the Externalism-Internalism Debate', *History of Science* **30**, 333–69.

Shapin, S. & S. Schaffer (1985), *Leviathan and the Air Pump: Hobbes, Boyle and the Experimental Life*. Princeton University Press.

Shattuck, R. (1980), *The Forbidden Experiment: The Story of the Wild Boy of Aveyron*. London: Secker & Warburg.

Shields, S. A. (1978), 'Sex and the Biased Scientist', *New Scientist*, **80**, 752–54.

Shields, S. A. (1982), 'The Variability Hypothesis: The History of a Biological Model of Sex Differences in Intelligence'. *Signs: Journal of Women in Culture and Society*, **7**, 769–97. Reprinted in Harding (1986).

Skinner, B. F. (1945), 'The Operational Analysis of Psychological Terms', *Psychological Review*, **52**. Reprinted in H. Feigl & M. Brodbeck (eds.), *Readings in the Philosophy of Science*. New York: Appleton-Century-Croft, 1953, pp. 585–95.

Smedile, S. R. (1962), *Perpetual Motion and Modern Research for Cheap Power*. Boston, Mass: Science Publications of Boston.

Smith, J. M. (1982), *Evolution and the Theory of Games*. Cambridge University Press.

Smythies, J. R. (ed.) (1967), *Science and ESP*. London: Routledge & Kegan Paul.

Spencer, F. (1992), *Piltdown: A Scientific Forgery*. London: Natural History Museum Publications.

Stark, W. (1958), *The Sociology of Knowledge*. London: Routledge & Kegan Paul.

Starr, S. L. & Griesemer, J. (1989), 'Institutional Ecology: Translations and Boundary Objects: Amateurs and Professionals in Berkeley's Museum of Vertebrate Zoology', *Social Studies of Science*, **19**, 387–420.

Stevens, S. S. (1935), 'The Operational Basis of Psychology', *American Journal of Psychology*, **47**, 323–30.

Stone, I. (1972), *The Healing Factor: 'Vitamin C' Against Disease*. New York: Grossett & Dunlap.

Sulloway, F. J. (1991), 'Reassessing Freud's Case Histories: The Social Construction of Psychoanalysis', *Isis*, **82**, 245–75.

Swinton, A. A. C. (1904a), Letter to *Nature*, **69**, 272.

Swinton, A. A. C. (1904b), Letter to *Nature*, **69**, 412.

Szasz, T. (1961), *The Myth of Mental Illness*. New York: Harper & Row.

Szasz, T. (1974), *Ideology and Insanity: Essays on the Psychiatric Dehumanization of Man*. Harmondsworth: Penguin.

Targ, R. & H. Puthoff, (1974), 'Information Transmissions under Conditions of Sensory Shielding', *Nature*, **251**, 602–7.

Taylor, H. F. (1980), *The IQ Game: A Methodological Inquiry into the Heredity-Environment Game*. Brighton: Harvester.

Taylor, J. (1975), *Superminds*. London: Picador.

Tester, S. J. (1987) *A History of Western Astrology*. Woodbridge: Boydell.

Tiryakian, E. A. (1972–3). 'Towards the Sociology of Esoteric Culture', *American Journal of Sociology*, **78**, 491–512.

Toth, M. & G. Nielson, *Pyramid Power: The Secret Energy of the Ancients Revealed*. New York: Destiny Books.

Toulmin, S. E. (1953) *The Philosophy of Science: An Introduction*. London: Hutchinson.

Truzzi, M. (1972), 'The Occult Revival as Popular Culture: Some Random Observations on the Old and Nouveau Witch', *The Sociological Quarterly*, **13**, 16–36.

Velikovsky, I. (1978, originally 1950), *Worlds in Collision*. London: Abacus.

Waite, A. E. (1926), *The Secret Tradition in Alchemy: Its Development and Records*. London: Kegan Paul, Trench, Trubner.

Wallace, R. K. & H. Benson, (1972), 'The Physiology of Meditation', *Scientific American*, **226** (February), 84–90.

Wallis, R. (1976). *The Road to Total Freedom: A Sociological Analysis of Scientology*. London: Heinemann.

Walmsley, D. M. (1967), *Anton Mesmer*. London: Hale.

Waterschoot van der Gracht, W. A. J. M., B. Willis, R. T. Chamberlin, & others, (1928), *Symposium on the Theory of Continental Drift*. Tulsa, Ok: American Association of Petroleum Geologists.

Watson, J. B. (1913), 'Psychology as the Behaviorist Views It', *Psychological Review*, **20**, 158–77.

Wegener, A. (1978, originally 1915), *The Origin of Continents and Oceans*. London: Methuen.

Wells, G. A. (1967), 'Goethe's Scientific Method and Aims in the Light of his Studies in Physical Optics', *Publications of the English Goethe Society*, **38**, 69–113.

Wells, H. G. (1993, originally 1898), *War of the Worlds*. Bloomington: Indiana University Press.

Westfall, R. S. (1980), *Never at Rest: A Biography of Isaac Newton*. Cambridge University Press.

Westrum, R. (1978), 'Science and Social Intelligence about Anomalies: The Case of Meteorites', *Social Studies of Science*, **8**, 461–93.

Wilson, E. O. (1975), *Sociobiology: The New Synthesis*. Boston: Belknap.

Winch, P. (1964), 'Understanding a Primitive Society', *American Philosophical Quarterly*, **1**, 307–24.

Winer, J. S. (1955), *The Piltdown Forgery*. Oxford University Press.

Winer, R. (1974) *The Devil's Triangle*. New York: Bantam.

Winograd, T. (1973), 'A Procedural Model of Language Understanding', pp. 152–86 in R. C.Schank & K. M. Colby (eds.), *Computer Models of Thought and Language*. San Francisco: Freeman.

Wittgenstein, L. (1953), *Philosophical Investigations*. New York: Macmillan.

Wittgenstein, L. (1971, originally 1921), *Tractatus Logico-Philosophicus*. 2nd edn. London: Routledge & Kegan-Paul.

Wollstonecraft, M. (1988, originally 1792), *A Vindication of the Rights of Woman*. C. Poston (ed.). New York: Norton.

Wood, R. W. (1904), Letter to *Nature*, **70**, 530–1.

Woodward, A. S. (1948), *The Earliest Englishman*. London: Watts.

Woolgar, S. (1976), 'Writing an Intellectual History of Intellectual Developments: The Use of Discovery Accounts', *Social Studies of Science*, **6**, 395–422.

Wynne, B. (1992), 'Misunderstood Misunderstanding: Social Identities and Public Uptake of Science,' *Public Understanding of Science*, **1**, 281–304.

Yates, F. A. (1964), *Giordano Bruno and the Hermetic Tradition*. London: Routledge & Kegan Paul.

Young, R. M. (1970), *Mind, Brain and Adaptation in the Nineteenth Century*. Oxford: Clarendon Press.

Young, T. (1814), 'Zur Farbenlehre, On the Doctrine of Colours', *Quarterly Review*, **10**, 427–41.

Zemansky, M. W. (1964), *Heat and Thermodynamics*. 4th edn., New York: McGraw-Hill.

Ziman, J. (1978), *Reliable Knowledge: An Exploration of the Grounds for Belief in Science*. Cambridge University Press.

Ziman, J. (1984), *An Introduction to Science Studies*. Cambridge University Press.

Zmarzlik, H-G. (1976), 'Social Darwinism in Germany: An Example of the Sociopolitical Abuse of Scientific Knowledge,' in G. Altner (ed.), *The Nature of Human Behaviour*. London: Allen & Unwin.

Index